T0320264

Aseptic Pharmaceutical Manufacturing II

Applications for the 1990s

Edited by
Michael J. Groves and Ram Murty

CRC Press
Taylor & Francis Group
Boca Raton London New York

CRC Press is an imprint of the
Taylor & Francis Group, an **informa** business

CRC Press
Taylor & Francis Group
6000 Broken Sound Parkway NW, Suite 300
Boca Raton, FL 33487-2742

First issued in paperback 2019

© 1995 by Taylor & Francis Group, LLC
CRC Press is an imprint of Taylor & Francis Group, an Informa business

No claim to original U.S. Government works

ISBN-13: 978-0-935184-77-8 (hbk)
ISBN-13: 978-0-367-40173-3 (pbk)

Library of Congress Cataloging-in-Publication Data

Catalog record is available from the Library of Congress

**Visit the Taylor & Francis Web site at
http://www.taylorandfrancis.com**

**and the CRC Press Web site at
http://www.crcpress.com**

Contents

Preface

The subject of aseptic pharmaceutical manufacturing has suddenly come into focus again following the epic decision by regulators in the United States and Europe that all sterile products should be terminally sterilized. This decision was easy to make but difficult to justify in many cases, and this is where the topic has become of interest and a matter for debate. Many of the newer biotechnologically derived drugs for injection are proteins or polypeptides that are destroyed by the application of heat. The technology still does not allow the terminal sterilization of perhaps as many as 80 percent of injectable products available today.

We take the pragmatic view that the arguments for and against this decision need to be widely discussed and debated. If the product cannot be heat-sterilized, then the aseptic processing itself needs to be improved to the point where the patient is not placed at any significant risk and the regulators can relax their present posture with confidence. As editors of this book, we feel that we are making a significant contribution to this debate, demonstrating how the technology has advanced and where it will advance to in the immediate future. In 1987 Wayne Olson and Mike Groves published the first volume of this series, with the subtitle "Technology for the 1990s."

The present editors feel strongly that they have compiled a book that will now take the technology into the next Millennium. In 1987 it was considered that form-fill-seal and isolator technologies would receive more attention in the future and lessen anxieties associated with aseptic operations. This present volume, eight years later, is the fulfillment of that dream, and there is no doubt that these areas will continue to expand and blossom. Technology cannot flourish in isolation, and we predict that products made by this process will continue to improve in "quality" and become demonstrably safer for the consumer. The improved safety may remain a statistical concept but, at the end of the day, producers, regulators, and users will know that the products they produce, regulate, or use will be safer because of the closer scrutiny resulting from today's discussions.

ACKNOWLEDGEMENTS

Collectively we would like to thank those authors whose efforts are evident between these covers. We know only too well just how difficult it is to find the time for not only writing a chapter but getting it ready for publication, reviewing proofs, and generally being involved—and all against the clock.

Individuals who have earned our undying admiration at Interpharm are Amy Davis and Jane Steinmann. Ms. Davis has been a driving force behind our efforts, and we are eternally grateful. Ms. Steinmann has edited with great care and we are all thankful. This book will constitute a monument to their efforts.

Michael J. Groves *Ram Murty*
Illinois *Kentucky*

May 1995

1

Introduction

Michael J. Groves
Ram Murty

The exact origins of parenteral therapy are lost in antiquity, but we do know that devices recognizable today as syringes, made from pewter or brass, were found recently in the wreck of the *Mary Rose*, a ship of the line that sank under the eyes of an embarrassed King Henry VIII during a battle with a French fleet in 1545. Intravenous administration of opium dissolved in wine was attempted by Sir Christopher Wren— better known as the architect of St. Paul's Cathedral in London, but, coincidentally, a mathematician and professor of surgery—in 1656. Over the next 200 years or so there were sporadic investigations of this route of administration, culminating in the intravenous administration by Latto in Edinburgh of salt solutions to patients during an outbreak of cholera. Alexander Wood, also at Edinburgh, is credited with the invention of a graduated glass syringe fitted with a hollow needle "like the sting of a wasp" in 1853. Solutions of morphine are believed to have been injected by surgeons during and after the American Civil War; the first compendial monograph for morphine injection first appeared in the *Addendum to the British Pharmacopoeia* (BP) of 1874. Although these early monographs suggested that injections be prepared carefully, there was no sterilization procedure and development of the technology did not proceed rapidly until the discovery of the anti-syphilitic "Salvarsan" at the turn of the present century. Had "Salvarsan" been active orally, it is doubtful if progress would have been as rapid; by the 1920s BP monographs for injections also included sterilization procedures. In the 1930s there was considerable discussion of the effectiveness of these sterilization procedures and the associated methods of testing for sterility. Over the next 20 years technology developed for the production and preparation of blood products and penicillin during World War II. This was coupled with the realization that postproduction sterility testing was unlikely to detect anything except the most badly contaminated product, a point emphasized by reports of deaths due to microbiologically contaminated products during the mid-1960s. Public inquiries on both sides of the Atlantic resulted in the promulgation of the first Good Manufacturing Practices (GMPs) by the Food and Drug Administration (FDA) in 1975. Over the past 20 years the subject has been extensively studied and developed. Today, the main issues are the requirements by the FDA that all sterile products be terminally heat sterilized unless the product is unable to withstand this treatment. Not really an issue with salt solutions, it is rather less so with sugar solutions and almost certainly not feasible with the vast majority of modern biotechnology-derived protein or peptide drugs. The need at this immediate point in time is to thoroughly discuss the problems and issues that arise from this requirement. This present book is designed to provide some insight into a number of aspects of this current situation.

HISTORICAL DEVELOPMENT OF ASEPTIC PROCESSING

Although terminal sterilization of injectable solutions was an established technical process by the 1920s, this was the period in which, arguably, the first injectable protein, insulin, became available. With very little to guide them, manufacturers very quickly discovered that it was necessary to prepare a sterile solution of the drug by filtration through porous ceramic filters, followed later by Seitz or compacted asbestos fiber filters. The solution required handling in a controlled environment with aseptic precautions and packaging in presterilized container components, such as glass vials fitted with rubber closures. Glove boxes and the use of ultraviolet germicidal lamps were explored in an effort to obtain a suitably "clean" environment; however, as testing methods developed, the limitations due to the techniques were revealed. The use of special clothing to protect the product from the workers themselves was slowly developed. Concomitantly, advances in heat sterilization technology were also being made. The Tyndalization process in which spore-forming organisms, resistant to heat below 95°C, were allowed to develop to the more sensitive vegetative forms by alternately exposing the product to heat (for killing) and incubation (for allowing the growth of survivors) over two or more cycles was shown to be less than efficient or effective. Tyndalization was, therefore, abandoned, although the process has been explored again more recently. Organisms were found that were resistant to heat exposure to temperatures below 120°C; this temperature became the norm for the autoclaving process. Then it was realized that the operation of a steam autoclave was not without its own problems and methods were devised for recording temperatures inside the autoclave and, eventually, inside samples of the product. The effect of air mixed with steam on the temperature of what should have been saturated steam had been known from work during the previous century on the construction and operation of railway engine boilers. The realization that a poorly operated autoclave, from which not all the air had been removed, could allow the product to be incorrectly considered to be "sterile" was a painful and, in some cases fatal, process. This must be considered unfortunate since a generation earlier, inadequately heat-treated dressings had resulted from the application of an insufficient heat treatment due to a failure to remove all the air prior to steaming. Application of the same temperature-measuring techniques to air ovens produced similar findings about the inadequacy of operation procedures and, again, resulted in improved and reproducible production techniques. The resultant effect of all of these observations was to move away from mere "cookery book" procedures to more intelligent applications of knowledge—essentially the first realization that validation of procedures was a necessary and appropriate way to go.

The sterilization process eventually became more controlled with the application of scientific measurements based on the response of selected, heat-resistant, microorganisms. The microbial death process was considered to be a kinetic process, requiring time to go to completion. Calculations were made of D, the decimal reduction time, or the thermal destruction value, Z, and the F_0 or number of minutes at a specified temperature above, say, 100°C, equivalent to holding the product at 121°C (250°F) for 15 minutes. At the same time it became evident that the success of a sterilization process, by definition killing *all* microorganisms in a product inside a sealed container, would critically depend on the bioburden or the number of organisms originally present inside the container before sterilization. This led to an exploration of the effectiveness of filtration procedures for the removal of microorganisms, dead or alive. In the 1970s a membrane filter with an average pore diameter of 0.45 μm was considered to be suitable for the purpose of removing bacteria; later this limit was reduced to a pore diameter of

0.2 μm. Membrane manufacturers also had problems in characterizing their products. For example, ideally the nominal diameter of 0.2 μm should be that of the largest pore, not the average pore size. Many of the arguments and discussions at that time were only resolved by refined characterization and measurement techniques. Nevertheless, although particles larger than 0.2 μm diameter and, as it happens, significant quantities of particles smaller than this diameter, will be removed on passage through a membrane of this nominal limiting pore size, particles corresponding to viruses and viable cell fragments could, theoretically, pass through into the filtrate. This consideration alone represents the biggest limitation to the production of any product by an aseptic manufacturing process. Nevertheless, filter manufacturers now routinely challenge their products with suitable suspensions of small microorganisms in order to support their claims of effective sterilization. These claims, perhaps wisely, have not been extended to suggesting that viruses are removed from the product. The unfortunate realization that viruses, such as HIV or hepatitis, were left in blood products and could subsequently infect patients has stimulated research in this area; thus far an effective filtration technology has not been developed.

As noted earlier, contamination of a product with viable microorganisms, either as a result of faulty processing or, in the case of aseptic filling, by the intrinsic nature of the process, is unlikely to be detected by postproduction sampling and sterility testing. The concept of sterility assurance levels (SALs) has arisen over the past decade and numbers provided. For example, although there is really no method to accurately and precisely measure SALs, a terminal heat sterilization process is considered to provide a SAL of 10^{-6}, indicating that there is a 1 in 1,000,000 chance of finding a contaminated container in a batch of product. Batch sizes of 1,000,000 units are not common but, with modern production methods, need not necessarily be impossible. A SAL of 10^{-6} would, therefore, suggest that there would be one contaminated container in that batch. This is, of course, unacceptable to the patient who receives that one container. What the concept should mean is that there is a 1:1,000,000 chance of finding a contaminated container *each time the batch is sampled,* irrespective of the number of containers waiting to be sampled. By the same convention aseptic processes are considered, without, one might add, a lot of evidence, to provide a SAL of 10^{-3}. Thus, in a 3000 lot assembled aseptically, 3 of the containers would be contaminated. This would not be true if, in fact, the chance of contamination at each sampling point was considered to be 1:1000. The FDA considered this to be unacceptable and this is the main reason why terminal heat sterilization of the product was suggested at one point as being mandatory.

The issue is not insignificant because approximately two-thirds of sterile pharmaceutical products are filled aseptically, with up to 87 percent of small-volume parenterals (SVPs) (<100 ml) being filled in the same way in the United States [data from a Parenteral Drug Association (PDA) survey by Agalloco and Akers in 1993]. Indeed, it was suggested that, as more biotechnology-derived therapeutic drugs enter the marketplace, more products will need to be processed aseptically. The FDA requirement is, therefore, inopportune; it will undoubtedly stimulate deeper research into, and understanding of, the topic.

THIS PRESENT BOOK

It should be noted that *Aseptic Pharmaceutical Processing*, edited by W. P. Olson and M. J. Groves and published in 1987, contained a number of useful chapters that could, with benefit, be read before turning to this second edition, which is effectively an updated and expanded review of the subject.

The older glove boxes, noted earlier, have made a reappearance under the rubric of "barrier technology." Discussed at length in the Olson and Groves book, the subject has been developed here by J. P. Lysfjord and his colleagues in the chapter on isolators and filling lines. Lyophilization and barrier systems are also discussed by J. W. Snowman. Both of these chapters are important because, even if a sterile solution could be placed into a sterilized container, for a few moments in time, the product is exposed to the environment during filling and subsequent drying. The use of barrier technology today is intimately connected with the success or otherwise of a sterilizing gas used to sterilize all working surfaces inside the barrier. The most promising sterilizing gas in use currently is hydrogen peroxide vapor, a subject discussed by L. M. Edwards and R. W. Childers.

The ultimate method for minimizing this environmental exposure is the filling procedure described by Leo in Olson and Groves known as form-fill-and-seal. In this machine a container of plastic is actually blown and formed in a clean environment immediately prior to filling it with a sterile solution. The solution also serves to cool the extruded plastic and the final container is sealed immediately after filling. The actual environmental exposure is, therefore, reduced to milliseconds. More to the point, the whole apparatus is automated and enclosed in a controlled environment that also has the effect of minimizing contaminant exposure. The chance of contaminated product coming from a form-fill-and-seal machine is, therefore, very small; indeed, the problem has been the perennial difficulty of how to validate any claims made for the process. Some progress toward providing a solution to this exceptionally difficult issue has been made by filling many thousands of containers with sterile bacterial growth medium and determining how many of these sealed containers demonstrated growth on subsequent incubation. Issues associated with this process are described by C. S. Sinclair and A. Tallentire in their chapter. These authors have deliberately contaminated the controlled environment surrounding the machine in order to simulate a worst-case situation. Even in this case their data demonstrate a high degree of protection of the filled product and it is likely that, under favorable conditions of operation, a form-fill-and-seal machine could produce a product aseptically, while at the same time providing a SAL of approximately 10^{-6}. This subject is obviously ongoing, but future developments suggest that automation of part or all of the production process in a controlled environment is what is needed to produce an aseptically filled product in accordance with modern regulatory requirements.

Validation is required, however, for each and every aspect of the process. From a scientific perspective, this offers the biggest challenge because it is often very difficult to prove, with appropriate scientific and logical conviction, that a specific claim is valid. This topic is discussd by V. Kumar and R. Murty; sight should not be lost of the fact that adequate validation of any process is difficult, requiring a sound knowledge of the basic sciences and an intimate knowledge of the purpose and procedure of the process under review. Unfortunately, this subject has often been obscured and blurred in recent years so that the actual purpose is sometimes lost in misplaced rhetoric. This is another regulatory issue that will only become acceptable from the broader perspective of time.

It will be evident that the FDA requirement to convert aseptic processes to terminal sterilization, discussed at a FDA conference in October 1993, is not a local issue; R. Dabbah discusses the international perspective of this topic. Dabbah, a staff member of the United States Pharmacopeial Convention, is well positioned to interpret the international scene; it should be noted that the current USP 23 contains a general chapter on sterile drug products for home use, <1206>, and a chapter on sterilization and sterility assurance (of compendial articles), <1211>. It is, therefore, an invaluable resource used by the FDA with the backing of law. This latter chapter, incidentally, defines the SAL as a "microbial survivor probability" so that a 10^{-6} probability would suggest that there is an assurance of less than one chance in one million that viable microorganisms are present in the sterilized article. "Survival" and "removal" have quite separate meanings and the use of expressions such as SAL or microbial survivor probability ignore the likelihood that a single surviving organism would proliferate after exposure to a heat sterilization process or, in the case of a filtration process, would be left in the product to proliferate. In the case of heat sterilization, SALs of 10^{-12} are often sought in what is, obviously, an "overkill." However, many products do not survive these excessive exposures to heat. Filtration processes are now being judged by the number of logarithmic reductions of exposure that can be accomplished. For example, if a filter fails to allow any organisms to pass through when exposed to a test solution containing 10^7 organisms/cm^2, then the log reduction value (LRV) is 7. This emphasizes the point made earlier that the cleaner the solution (i.e., the lower the bioburden) before the final sterilization process, the greater is the likelihood that the process will be successful. In the case of a terminal filtration process through, say, a 0.2 μm nominal pore size membrane, it is extremely unlikely that only one filter-collecting surface would be used; assemblies of filters are more likely to be used in current practice. Because of the thinness of membrane filters, the supporting surfaces required careful and accurate machining. Since stainless steels must be used, the high cost of the supporting filtration equipment has tended to inhibit the use of multiple or redundant filtration units. However, realization that the cost of modern biotechnology-produced products is also high has tended to change the perspective on what has become an important capital item in the production plant.

Crucial issues are those of personnel and, especially, personnel training. There is now a realization that the product should have minimal exposure to "people" since people represent the major source of microbiological contamination. However, even in a highly automated production line, people are needed at key points in the process. M. J. Akers deals with the training of personnel, emphasizing a vital need for workers in an aseptic process to understand the purpose of their work and the need to develop the appropriate skills necessary for aseptic product assembly. This subject leads naturally to the laboratory techniques associated with aseptic processing, discussed by R. Mehta and R. Murty. A rather more specialized involvement on a laboratory scale, because of some unique issues, is reviewed in the chapter on radiopharmaceuticals by P. O. Bremer. Here the issues are scale and the fact that the shelf-life of a radioactive diagnostic agent is often too brief to allow testing of the product prior to administration to the patient. Procedures, therefore, have to be in place that guarantee safety and efficacy, allowing validation "after the fact."

The problems of handling modern biopharmaceutical drugs have been alluded to throughout this discussion. In a chapter by N. M. Lugo, the issues are reviewed and placed into perspective since, after all, insulin, a complex protein drug, has been on the market for nearly 70 years.

This volume contains rather less information on the technical issues that are involved in aseptic production of the sterile product. Nevertheless, there have been some significant advances in lyophilization over the past five or so years and the subject is updated in the review by E. Trappler. Clearly this has come about through a deeper understanding of the subject brought about by academic studies allied with the involvement of many equipment manufacturers. This has been to the net benefit of the producer and, ultimately, to the product consumer.

This general need to improve "quality" from the consumer's perspective has not been confined to aseptically produced sterile products in just one industry; it is a worldwide movement seen in many, if not most, industries at this time. For the most part, this is associated with the ISO 9000 initiative; this movement is reviewed by K. Stephens from a GMPs point of view. This chapter is also valuable for the overall perspective it provides of the movement toward ISO 9000 certification. The author has introduced an element of demystification about the process and summarizes the overall benefit it provides for the producer and, ultimately, the consumer. However, the ISO process itself is, by the very nature of international collaboration, a very complex and involved business. Some feeling for this is conveyed in the chapter by M. Korczynski and his colleagues, who have been involved in the certification process for aseptic manufacturers in the United States. Finally, we have included a review of packaging and labels by R. Murty, an area of much development in recent years that will continue to provide a challenge into the future.

CONCLUSIONS

The general theme of this introduction has been the way stepwise technical progress has occurred since the very earliest attempts to inject solutions into the human body. Generally, attention has been focused on one or two limited aspects of the overall subject. These have been resolved and understood first before turning attention to other issues or aspects of the technology. We feel this book will make a valuable contribution to understanding the issues involved in aseptic manufacture. Although the regulatory bodies, on both sides of the Atlantic one might add, are suggesting that assurance of "sterility" be enhanced by terminal heat sterilization, dealing with product stability in the real world is not always technically feasible. There will have to be some understanding reached between producers and regulators based on a mutual understanding of the technical issues involved; here we might venture the hope that this book will play a part in generating and increasing this understanding. There is little doubt that many of the modern proteinaceous drugs will not withstand terminal heat treatment, even when carefully controlled to provide a minimum SAL of 10^{-6}. This may also be true for some of the smaller new chemical entities. The need for aseptic assembly of the product will continue to exist; all that remains is for the manufacturer to optimize conditions. For example, automatic or semiautomatic production processes combined with barrier technology will limit exposure of the product to an environment inevitably contaminated by human operators. The environmental controls, such as air filtration methods, are better understood, as are gowning and manipulative procedures. The ultimate form-fill-and-seal methods are not suitable for all products. Where they can be applied, however, these machines are probably optimal although their capital cost is undoubtably substantial. The industry is still characterized by small or relatively small production runs of a wide variety of different products. Automated filling and inspection lines are available, but it seems very likely that the cost of this equipment will rarely be justified. Manually intensive processes will, therefore, continue to be used; but it seems at this point that some rethinking and redesign of the basic processes will be required in the immediate future.

2

Controlled Environments in the Pharmaceutical and Medical Products Industry: A Global Review from Regulatory, Compendial, and Industrial Perspectives

Roger Dabbah

The issue of a controlled environment in the pharmaceutical industry in the processing of drugs or medical devices that would be ultimately terminally sterilized or not terminally sterilized has taken center stage in increasing the anxieties of regulators and manufacturers alike. The rationale is without any doubt an attempt to increase the protection of users to potential contamination by either microorganisms or particulates.

The increase in the "safety" of patients that is sought by the regulators is balanced by the cost to manufacturers of increasing the safety margin for their products, resulting ultimately, according to the manufacturers, in an increase in healthcare costs to government and patients alike. Although it is true that advances in technologies of controlled environments have considerably reduced the risk to patients, regulators will err toward a more conservative approach in their attempts to reduce these risks, while manufacturers will respond through a cost/benefit analysis, where the costs are very well known and can be readily estimated, while the benefits are more elusive, albeit predictable.

The ethical issues raised above will not be discussed, but must be taken into consideration as regulators and manufacturers make decisions today that will impact on the future of the industry and the costs of healthcare.

There is an axiom in healthcare to which most rational healthcare practitioners and providers subscribe: If the prescribed treatment puts a patient at a risk higher than the risk presented by the disease, then the treatment is not appropriate. The application of this axiom to controlled environments is that if a drug or a medical device is processed in noncontrolled environments, the risk to patients would increase. The issue here is not that regulators or manufacturers do not agree on this issue, but in the extent of governmental intervention in matters that are best left to the discretion of manufacturers.

In the adversary system between regulators and manufacturers, the intent of controlled environments in the pharmaceutical and medical product industry seems to be lost in exquisite debate on the merits of regulations and guidelines that often attain biblical proportions.

The issue of controlled environments, complex as it is already, becomes even more complicated in the current climate of global harmonization of regulatory and compendial requirements and/or guidelines and of the new age requirement that markets and manufacturers have to think globally while also satisfying national or local requirements.

The need for controlled environments for aseptically processed medical products is not in dispute in any rational discourse. However, the global discussion revolves around the level of control that must be excercised from a particulate and a microbiological perspective. This is especially important when aseptically processed products are not terminally sterilized. This issue is critical, especially for biotechnology-derived products, since these are often proteins that are heat labile and cannot be terminally sterilized.

The role of pharmacopeias in the development of requirements and/or guidelines for controlled environment will also be examined and discussed.

DEFINITIONS

The terminology used, especially on a global level, is sometimes as important as, if not more than, the contents of regulations, guidelines, and other assorted technical information vehicles that have been published or proposed. It is through a common terminology that misunderstandings can be avoided. It is not by accident that the International Organization for Standardization (ISO) insists in including a glossary within each of their voluntary standards. In this section we will not develop a glossary, but actually discuss some of the critical terms that have been batted around in the evolving debates on controlled environments.

Controlled Environments

Controlled environments are environments where the processing of drugs and medical products are conducted, where a specified number and type of measurable indicators are continuously or intermittently measured and compared against standards or targets to provide a historical status of critical conditions that are related to the quality of final products. That relationship between these indicators and the quality of final products need not be a direct or a measurable relationship, but can also be inferred based on a logical and scientific examination of these parameters.

Clean Rooms

Clean rooms are controlled environments where parameters such as airflow, microbiological and particulate quality of air, equipment surfaces and other room surfaces, and personnel equipment (including gowns, boots, and masks) are monitored and compared to standards or targets. The flow of personnel and material as well as access to clean rooms are defined and strictly adhered via standard operating procedures.

Aseptic Processing

Aseptic processing is a process that combines a presterilized product with a presterilized container that is then closed with a presterilized closure in a clean room. Aseptic-processed products can be terminally sterilized or might not be processed further.

Terminal Sterilization

Terminal sterilization is a process that subjects the combined product/container/closure system to a sterilization process that results in a specified assurance of sterility. Sterilization processes include steam, ethylene oxide, radiation, other sterilizing gases or chemicals. The sterility assurance level (SAL) of terminally sterilized products is 10^{-6}.

Sterility Assurance Level

The SAL is a term that has been conventionally used to describe the capability of a sterilization process, that is measurable through validation of the sterilization cycle, to assure that the probability of assurance of a sterilization process will allow not more than one unit in 1,000,000 units produced to be nonsterile. This concept has been applied, unfortunately by many, to aseptic processing, where it is said that aseptic processing is capable of delivering an SAL of 1 in 1000 units produced. The basis of that SAL description of a processing system is theoretically and operationally erroneous. The SAL for a terminal sterilization process is based on the theoretical assumption that has been verified, that microorganisms, especially resistant spores, are affected by a sterilizing agent in a manner that can be mathematically described and predicted and that can be extrapolated to a probability of 10^{-6}.

In aseptic processing the so-called SAL is based on media fills of 3,000 units that yield not more than three nonsterile units. It is also erroneously indicated by many that if you cumulate for a given process a large number of media fills results, that you could approach or even exceed an SAL of 10^{-6}, thus making aseptic processing a process with the same SAL as terminal sterilization. This "leap of faith" does not take into consideration that a media fill only represents a direct evaluation of the microbiological status of the aseptic processing environment, which cannot be extrapolated to a SAL.

Controlled Environment-at-Rest

The critical parameters in a controlled environment-at-rest are measured when no personnel are present in the controlled environment and no processing is taking place. These critical parameters or indicators are made in order to "commission" a controlled environment. Contractors who build controlled environments use at-rest standards or targets to design and build the controlled environments.

Controlled Environment-at-Work

In a controlled environment-at-work critical parameters or indicators are measured when the full processing system is operating with the full complement of personnel in the controlled environment.

Media Fills

A media fill consists of replacing the product with a sterile growth medium (usally agar broth) and aseptically processing the liquid agar medium in the same fashion as the real product. The product/container/closure system units are then incubated at a specified temperature for a specified time, then examined visually for microbial growth. It is said, in error, that if not more than one unit in 1000 produced is nonsterile, the aseptic process gives an SAL of 10^{-3}. In general 3,000 units are tested and the target is not more than 3 units showing microbial growth.

Samplers

Samplers are devices that sample the controlled environment air to measure directly or indirectly the microbiological or particulate status of that environment's air. Samplers can be passive, such as agar settling plates, or active. Active samplers obtain a measured amount of the environment's air for a specified period of time and collect the particles in the air (viable and nonviable) in a liquid or solid medium that permits the determination of a count.

The microbial status of surfaces is assessed using agar contact plates or equivalent devices or swabs that sample a specified surface. Determination of viable counts estimates the microbial status of surfaces.

Regulatory Requirements

Regulatory requirements are enforced by regulatory agencies to protect patients and public health. These requirements are general as in the Good Manufacturing Practices (GMP), or very specific when included in New Drug Applications (NDA), 510(K) applications, product marketing approvals (PMA), licenses, or any other appropriate documents allowing for the marketing of products.

Compendial Requirements

In the U.S., pharmacopeial standards for quality, identity, purity, and potency are enforceable by regulatory agencies under the adulteration or mislabeling provisions of the Food, Drugs, and Cosmetic Act. In the USP, these standard requirements are either in the General Notices, the General Chapters, or the individual monographs and they apply to all units in the market during its entire shelflife.

In other countries, pharmacopeial requirements are also enforced by appropriate regulatory bodies.

Regulatory Guidelines/Guidances

Guidelines issued by regulatory agencies help manufacturers evaluate their processes and products in line with what the regulatory agencies believe are appropriate. These are not regulatory requirements; however, they are sometimes used by regulatory field inspectors as targets against which processes and products can be evaluated from a regulatory point of view.

When technology advances too rapidly, as in the biotechnology-derived product industry, regulatory agencies will issue "Points to

Consider" that are specific, but can be changed as the technology advances.

Compendial Guidelines/Guidances

Compendial information included in the General Information chapters in pharmacopeias are providing regulatory agencies and manufacturers with the background necessary to produce quality products without being a requirement that is enforceable by regulatory agencies.

Current Good Manufacturing Practices

Current Good Manufacturing Practices (cGMPs) are used by regulatory field inspectors to evaluate processing and products. The cGMP regulations are controversial since they leave wide open the application of the evolution of practices in one organization that might be more technologically advanced, to the application to all organizations in the industry. This gray area in the application of GMPs has created controversial inspection reports from regulatory agencies and has increased the anxieties of manufacturers who feel that they are always shooting at a moving target.

Voluntary Standards

Voluntary standards are generally developed under the aegis of industrial and professional societies within the industry. These general standards might or might not be accepted or acceptable to regulatory agencies, but are an honest attempt to bring the industry perspective in the development of standard requirements. They can be national or international and can provide a blueprint for the harmonization of regulatory and compendial requirements and guidelines.

REGULATORY REQUIREMENTS FOR CONTROLLED ENVIRONMENTS IN THE UNITED STATES

The Food, Drugs, and Cosmetic Act provides the Food and Drug Administration (FDA) with the authority to set legal requirements in manufacturing, processing, packing, or holding of drugs. These general requirements, which are also applicable to controlled environments, are in Part 210: Current Good Manufacturing Practice in Manufacturing, Processing, Packing, or Holding Drugs, and in Part 211: Current Manufacturing Practice for Finished Pharmaceuticals.

The status of cGMP regulations is clearly indicated in Part 210.1(a):

regulations in Part 211 through 226 contain the minimum cGMP for methods to be used in, and the facilities or controls to be used for, the manufacturing, processing, to assure that such a drug meets the requirements of the Act as to safety, and has the identity and strength and meets the quality and characteristics it purports or is represented to possess.

The authority to enforce cGMP for drug products is indicated in section 501 (a) (2) B of the Act:

Methods used in or the facilities or controls use for manufacturing, processing, conforms to GMP.

A specific application to controlled environments is in section 211.42, as follows:

C—BUILDINGS AND FACILITIES
211.42 Design and construction features
. . .

    (c) . . .
        (10) Aseptic processing, which includes as appropriate:
            (i)   Floors, walls and ceilings of smooth, hard surfaces that are easily cleanable;
            (ii) Temperature and humidity controls;
            (iii) An air supply filtered through high-efficiency particulate air filter under positive pressure, regardless of whether flow is laminar or non-laminar.
            (iv) A system for monitoring environmental conditions;
            (v) A system for cleaning and disinfecting the room and equipment to produce aseptic conditions;
            (vi) A system for maintaining any equipment used to control the aseptic conditions.

Although controlled environments are not specifically highlighted in the GMP, there are a number of other sections that are applicable. For example, in section B (Organization and Personnel) under 211.22(c), the "quality control unit shall have the responsibility for approving or rejecting all procedures or specifications impacting on the identity, strength, quality, and purity of the drug product." Under 211.25 (Personnel qualifications) the specified qualification for personnel involved in processing of drug products are indicated and they apply to aseptic processing and processing under controlled environments.

The GMP requirements under section 211.46 (Ventilation, Air Filtration, Air Heating and Cooling) indicate that adequate systems for contaminant controls must be designed in the system. The requirements under 211.56 (Sanitation) apply directly to controlled environments. More detailed requirements are also in sections under F (Production and Process Controls) in 211.100 (Written procedures; Deviations) and 211.113 (Control of Microbiological Contamination).

Personnel involved in processing under controlled environments must be very familiar with all the requirements of cGMP that apply to these environments.

REGULATORY GUIDELINES FOR CONTROLLED ENVIRONMENTS IN THE UNITED STATES

The FDA has issued a number of guidelines that are designed to supplement and give information on the subject of aseptic processing. A guideline dated June 1987 has been functioning as a de facto requirement, especially because it has been used as background by FDA field inspectors as a checklist during inspections.

The 1987 FDA guidelines on Sterile Products Produced by Aseptic Processing include the application of the various cGMP requirements to aseptic processing, but also expand these to include the type of facility to be used, the issues of critical and noncritical areas, the validation of processing, and environmental testing of controlled facilities.

The FDA issued in 1987 a guideline on the Principles of Process Validation, including features that are applicable to the processing of drugs in controlled environments.

However, a proposed rule on the Use of Aseptic Processing and Terminal Sterilization in the Preparation of Sterile Pharmaceuticals for Human and Veterinary Use was published by the FDA in the *Federal Register* in October 1991 (Vol. 56, No. 198, 51354). The proposed policy statement indicates that sterile products must achieve an SAL of 10^{-6} via terminal sterilization of aseptically processed products. It also indicates that the aseptic processing of drugs will not be permitted unless the manufacturer shows that the product cannot be terminally sterilized; data to establish this must be provided to the FDA. This is a clear reversal of the 1987 guidelines that accepted aseptic processing without terminal sterilization for all products regardless of their ability or inability to be terminally sterilized. This policy has not yet been finalized and has been the subject of a plethora of comments, mainly from manufacturers, and mostly unfavorable. These comments revolve around the economical impacts of such a policy, especially in increasing the cost of drugs to the healthcare system, thus to consumers. Other comments revolve around the issue of "SALs" for aseptic processing as compared to terminal sterilization.

The issue of SAL for aseptic processing is a complicated issue. The meaning of an SAL has been misused in its application to aseptic processing, since it is derived from the results of media fills where not more than 3 units in 3,000 processed can be nonsterile is translated into an SAL of 10^{-3}. This SAL is generally compared to an SAL of 10^{-6} for terminal sterilization, thus making aseptic processing a process less desirable than terminal sterilization. There are no theoretical or even operational reasons to compare these SALs or to accept the premise that aseptic processing presents a higher risk to patients than terminal sterilization. If the regulatory approach is flawed, and I believe it is, the response of the industry is even more flawed. That response is to propose to give to aseptically processed products a mild heat treatment, that when combined with the so-called SAL of 10^{-3} for aseptic processing, would somehow achieve an SAL of 10^{-6}. This response is not mathematically, operationally, or even conceptually valid.

The fate of the proposal of October 1991 is not known at this time, but is confounded by the FDA issue of another guideline—in this case a draft guideline—in June 1993 on Recommendations for Submitting Documentation for Sterilization Process Validation in Applications for Human and Veterinary Drug Products. The status of a draft guideline as compared to a guidelines presents some interesting issues, but revolves around the use of this draft guideline by field inspectors and/or by inspection checklists issued by various regional FDA compliance groups. Assuming that this draft guideline becomes a guideline in due time, it is not very clear how the FDA would use it in their enforcement policies, even if the proposed policy of October 1991 in the *Federal Register* becomes official. This new draft guideline was presented at various "dogs and ponies shows" put together by the FDA. It is also not very clear if the draft guidelines will apply to new New Drug Applications, Abbreviated New Drug Applications, Animal New Drug Applications, or Animal Abbreviated New Drug Appli-cations only, or to all sterile products on the market, retroactively. It might even be applied to Investigational New Drug and Investiga-tional New Animal Drug applications. The recent draft guidelines discuss a proposed "Microbiological Monitoring of the Environment in Controlled Environments" for terminally sterilized products:

> *D. A microbiological monitoring program for production areas along with a bioburden monitoring program for product components and process water is considered essential. Frequency, methods used, action levels, and data summaries should be included. A description of*

the actions taken when specifications are exceeded should be provided.

A specific section in these draft guidelines is devoted exclusively to Information for Aseptic Fill Manufacturing Processes Which Should Be Included in Drug Applications. Nowhere in this draft guideline is mentioned the proposed policy of allowing only terminal sterilization unless the sterilization process is shown to be deleterious to the product. The multiplicity of signals that are received by industry and their apparent contradiction to each other can easily explain the anxieties of manufacturers of sterile products.

The information requested in this draft guideline is essentially a clarification of GMPs as they apply to aseptic processing. Since the document has been distributed and is available through FDA–CDER, I will only show an outline of the draft while highlighting some of the most relevant features in terms of monitoring controlled environments.

IV. Information for Aseptic Fill Manufacturing Processes Which Should Be Included in Drug Applications

A. Buildings and Facilities

1. Floor plan

    . . . The air cleanliness class of each area should be identified (e.g., Class 100, Class 10,000, Class 100,000).

2. Location of equipment

B. Overall manufacturing operation

    . . . material flow, filling, capping and aseptic assembly. The normal flow of product and components from formulation to dosage form should be identified.

1. Drug product solution filtration

2. Specifications concerning holding periods

3. Critical operations

C. Sterilization/depyrogenation of containers, closures, equipment and components.

    . . . Validation information for sterilization processes included.

D. Procedures and specifications for media fills

    . . . included with the data summary.

11. Microbiological monitoring

> Provide microbiological monitoring data obtained during the
> media fill run. See (F) below for further details.

E. *Actions concerning product when media fills fail*

F. *Microbiological monitoring of the environment*

    . . .

    1. Microbiological methods

    . . .

       a. Airborne microorganisms

       b. Microorganisms on inanimate surfaces

       c. Microorganisms on personnel

       d. Water systems

       e. Product component bioburden

    2. Yeasts, molds, and anaerobic microorganisms

    3. Exceeded limits

G. *Container/closure and package integrity*

H. *Sterility testing methods and release criteria*

I. *Bacterial endotoxins test and method*

J. *Evidence of formal written procedures*

COMPENDIAL GUIDELINES FOR CONTROLLED ENVIRONMENTS IN THE UNITED STATES

The role of the United States Pharmacopeia (USP) in the development of standard requirements and guidelines in the pharmaceutical and medical products industry is unique. First, the USP is a nongovernment organization while all the other pharmacopeias of the world are parts and parcels of their regulatory bodies. Standard requirements developed by the USP are enforceable by the FDA in the areas of quality, purity, identity, and potency of drugs and medical products. Compliance to the USP standard requirements are enforced by the FDA.

These USP standards are public standards developed by a Committee of Revision that is elected by a Convention that convenes every five years. The standards developed by the USP are mandatory standards and are expert standards. Volunteers that are elected to the Committee of Revision represent themselves, not the organizations to which they formally belong.

In addition to enforceable standard requirements, the USP develops and publishes information of interest to regulatory agencies and manufacturers alike in the General Information Chapters of the USP. The development of public standards as well as of Information Chapters follows an open process, where a Subcommittee develops documents that are then published for public comments in *Pharmacopeial Forum*. Comments from regulatory agencies, industrial organizations, trade and professional associations, and academia as well as from all other interested parties are received by the USP, sent to the relevant Subcommittee, reviewed, and considered. If changes are accepted by the Subcommittee and these changes are more than editorial changes, the Subcommittee republishes the proposals in *Pharmacopeial Forum* for additional comments. Depending on the nature and scope of the proposals, republication in *Pharmacopeial Forum* can occur several times, until the Subcommittee is satisfied that all comments have been responded to, favorably or not.

The background on the USP and its process in the development of standards and guidelines is necessary to understand the process of the development of the proposed USP guidelines on controlled environments. The historical development of this guideline, or in USP terminology, an Information Chapter, is worth recounting as an illustration of the openness of the USP process.

Following discussions with a national organization in the pharmaceutical industry, a Subcommittee of the USP Committee of Revision developed a proposal on "Microbiological Evaluation and Classification of Cleanrooms and Clean Zones that was published in the Sept–Oct 1991 issue *Pharmacopeial Forum* (Vol. 17, No. 5, p. 2399). The rationale for the publication of the proposal was that the aseptic processing of drugs and other medical products had come under scrutiny by regulatory agencies in terms of the assurance of sterility as compared to terminal sterilization. The Subcommittee felt that the controlled environments under which aseptic processing was done needed to be evaluated from the microbiological point of view to assure that contamination of product during filling operations was minimized. The introduction of a microbiological evaluation of clean rooms and clean zones to complement the classification of these areas using particulate levels as indicated under Federal Standard 209E was thought to be necessary since it would be directed specifically to the pharmaceutical industry. It was felt by the Subcommittee as well as by other organizations that Federal Standard 209E was designed for the electronic industry and was unduly restrictive for pharmaceutical aseptic processing.

Considerable discussions and public comments ranging from objecting to USP intrusion in so-called processing standards away from its traditional standard requirements for finished products to objections to proposed microbial limits for the different critical areas, resulted in modification of the proposal that was then republished in the Sept–Oct 1992 issue of *Pharmacopeial Forum* (Vol. 18, No. 5, p. 4042). Heated discussions at USP Open Conferences or following numerous presentations by USP staff and Subcommittee members at national and international meetings of a variety of associations in the industry and participation in a multitude of panel discussions highlighted the interest in this information chapter. One of the major objections from manufacturers was that the FDA field inspectors would use the information in this chapter and especially the proposed levels for microbiological quality of the environment as standard requirements and would inspect facilities against the proposed microbial levels in the Information Chapter.

The essential features of this Information Chapter for controlled environments were as follows:

1. Classification of clean rooms and clean zones on the basis of microbiological quality of air, surfaces, and personnel attire surfaces.

2. Three classes, namely MCB-1, MCB-2, and MCB-3, were defined on the basis of the criticality of the potential for microbial contamination of product.

3. Harmonization with guidelines used for controlled environments in Europe was attempted, especially in the proposed microbial level for air in MCB-1 areas. For example, a level of less than 1 CFU/m^3 corresponding to a count of not less than 0.03 CFU/ft^3 was proposed similar to the level recommended in Europe for these critical areas.

4. Microbial levels for air to be based using a slit-to-agar sampling system, since it was reported that different samplers of air provided different estimates of counts of air.

5. Periodic identification of flora in these areas was recommended.

6. Verification and reestablishment of microbiological status of each area were recommended at regular intervals or when significant changes were introduced in aseptic processing systems.

The publication of the modified proposal generated considerable debates and comments that led to the appointment and formation of a Microbiological Control Expert Advisory Panel to the USP Subcommittee to provide a scientific forum between industry, government, and USP that would result in a useful document on controlled environments in the pharmaceutical industry.

The proposed information chapter is being recast and modified following several meetings between the Advisory Panel and the Subcommittee and will take the following format:

Title: Microbiological Evaluation of Cleanrooms and Other Controlled Environments

A. Definition of controlled environments

B. Establishment of controlled environment via U.S. Federal Standard 209E

C. Importance of adjunct microbiological environmental program for routine monitoring

D. Training requirements of employees using and monitoring the controlled environments

E. Establishment of sampling sites; design of sampling plan; establishment of frequency of testing

F. Protocol for the establishment of microbiological Alert levels.

G. Considerations of room classification under U.S. Federal Standard 209E and microbiological status of the environment

H. Equipment characteristics for microbiological evaluation of controlled environments

I. Identification of microflora in controlled environments

J. Future of controlled environments in the pharmaceutical and medical products industry

The reconstructed Information Chapter has been published in the March–April 1995 issue of *Pharmacopeial Forum* (Vol. 21, No. 2) for additional public comments before it is proposed for addition to USP 23 and its supplements. (See the Appendix to this book.)

FEDERAL STANDARD 209E

The status of Federal Standard 209E "Airborne Particulate Cleanliness Classes in Clean Rooms and Clean Zones," published by the Institute of Environmental Sciences (IES) for application to controlled environments in the pharmaceutical industry is not very clear. This federal standard is approved by the Commissioner, Federal Supply Services, General Services Administration for the use of all federal agencies.

This standard for controlled environments is a very comprehensive document that appears to have been developed for the electronic industry, where cleanliness levels in terms of particulates are very critical, but also far exceed the needs or requirements of aseptic processing in the pharmaceutical industry. The wholesale application of Standard 209E to the pharmaceutical industry is problematic, but it does provide a clear classification of clean rooms that would help construction of these facilities to be used for aseptic processing of pharmaceuticals. The classification Class 100, Class 1,000, Class 10,000, and Class 100,000 refers to the number of particles of size 0.5 μm that must be achieved for the assignment of classification of the area.

Federal Standard 209E classes refer to the total particle count in these areas, without differentiating between nonviable and viable particles. The adoption of Federal Standard 209E by the pharmaceutical industry for aseptic processing allows the construction of facilities based on a universally accepted criteria and the "commissioning" of these facilities based on easily measured parameters. In the pharmaceutical industry, although particulates are important in terms of limits in injections, the contamination of products by microorganisms, bacteria, yeasts, or molds are more critical. Federal Standard 209E does not give guidance for the establishment of microbial levels in a controlled environment, nor to the monitoring of the environment for viable particles.

NASA STANDARDS FOR CLEAN ROOMS AND WORKSTATIONS FOR THE MICROBIALLY CONTROLLED ENVIRONMENT

The 1967 edition of these NASA standards (NHB 5340.2) is the grandfather document for all guidelines that refer to microbiological levels in controlled environments. It is referenced by the FDA documents on aseptic processing especially in terms of the recommended microbiological levels for these areas. The 1967 NASA document was developed because the original Federal Standard 209A, published in August 10, 1966, did not include recommended levels and procedures for microbiological evaluation of controlled environments.

This document is often quoted since it shows cleanliness classes based on particulates and microbial contents of the air. I have adapted a table of air cleanliness classes as follows:

Class (metric)	Max # of Particles per ft³ 0.5 μm and larger (per liter)	Max # of Viable Particles per ft³ (per liter)
100 (3.5)	100 (3.5)	0.1 (0.0035)
10,000 (350)	10,000 (350)	0.5 (0.0176)
100,000 (3500)	100,000 (3500)	2.5 (0.0384)

It is not very clear how the relationship, if any, between particulate counts and microbial counts was established.

PARENTERAL DRUG ASSOCIATION GUIDELINES

There are two major guidelines published by the Parenteral Drug Association (PDA). One of these guidelines was published in 1980 as PDA Technical Report No. 2 on "Validation of Aseptic Filling for Solution Drug Product," which reviews the validation methodologies, especially the media fill approach for aseptic processing. A second publication in 1990 on "Fundamentals of a Microbiological Environmental Monitoring Program" was published by PDA as Technical Report No. 13. This second technical report includes the development of microbiological surveillance procedures to determine microbiological alert and action levels for controlled environments. It also indicates that these alert and action levels are to be facility specific, but does not give specific targets for these levels. This approach appears to be highly practical, but somewhat lacking internal consistency and logic. It can lead, for example, to a given drug being aseptically processed in two different facilities under environmental microbiological alert and action levels that are different, thus presenting the patient with potential risks that are different, albeit very low.

The second document also develops the concept of corrective action following numerous excursions of the microbiological level that exceed the alert and/or action levels for the particular aseptic processing facility. The document also provides for a reevaluation of these alert and action levels as data are accumulated.

The interest of PDA members in environmental control for aseptic processing is well served by the publication of these guidelines and by holding at regular intervals public forums on the subject where new developments, either regulatory or compendial, are highlighted and discussed.

INTERNATIONAL ORGANIZATION FOR STANDARDIZATION VOLUNTARY STANDARDS

The International Organization for Standardization (ISO), through its numerous technical committees (TCs) develops voluntary standards, or guidelines, that are designed to be harmonized worldwide. The role that ISO can perform, and does perform in the arena of harmonization of standards, especially in the areas of controlled environments for aseptic processing, is now taking shape via a number of TCs.

The ISO does its work through a number of national standard setting organizations in various counties. The American National Standard Institute (ANSI) is the U.S. member to the ISO. Countries are assigned secretariat duties for specific TCs through the ISO national members. For example, two TCs that are vital in the areas of controlled environments have been assigned to ANSI. In turn, ANSI can assign these secretariats to U.S. standard setting organizations. The ISO-TC 198 on Sterilization was assigned to the American Association for Medical Instrumentation (AAMI). To advise the U.S. delegation on ISO-TC 198, AAMI implemented a United States Technical Advisory Group (USTAG).

Within the ISO-TC 198 work is performed by a number of working groups (WG). The WG dealing with controlled environments is WG-9 on Aseptic Processing of Healthcare Products. Without detailing the various levels of documents, it is suffice to say that WG-9 has developed a first committee draft (ISO-CD 13405) that includes guidelines for facility designs; critical processing zone definitions; product and process specifications; personnel; aseptic processing areas (APA); environmental air systems and controls; secondary contamination control relative to gowning, cleaning, and disinfection of aseptic areas; environmental monitoring programs; alert and action levels; media fill programs; and many more areas relevant to controlled environments in aseptic processing. It is too early in the development of these ISO voluntary standards to give specific details on controlled environments issues, since a final document for the ISO will have to proceed through a lengthy and complex system of voting and comments. This draft is very comprehensive and is essentially directed toward the healthcare industry and its aseptic processing practices.

A second TC of ISO, ISO-TC 209 has recently been established and the secretariat has been assigned to the U.S. and the Institute for Environmental Sciences (IES). Contrary to the WG-9 of ISO-TC 198 that is specifically directed toward healthcare products, ISO-TC 209 has a much broader mission that covers the development of standards for clean rooms and associated controlled environments that include standardization of equipment, facilities, and operational methods and the definition of procedural limits, operational limits, and testing procedures to achieve desired attributes to minimize microcontamination. These standards will apply to all controlled environments, thus its application to healthcare products is not its prime concern.

The regulatory implication of these standards when they become available, especially because of their nonmandatory status, is not very clear, especially in the U.S. Its implication in Europe is closely related to the European Committee for Standardization (CEN) acceptance of these ISO standards as European standards. If and when an ISO standard becomes a CEN standard, all members of CEN will have to use the ISO standard as a national standard without any alteration.

Since the establishment of the ISO-TC 209, WGs have been designated and formed and some have already advanced early drafts of proposals for standards. There are seven WGs in operations: WG-1, Airborne Particulate Cleanliness Classes (UK, convenor); WG-2, Biocontamination (France, convenor); WG-3, Metrology and Test Methods (Japan, convenor); WG-4, Design and Construction (Germany, convenor); WG-5, Clean Room Operation (USA, convenor); WG-6, Terms and Definitions (Switzerland, convenor); and WG-7, Clean Air Devices (U.S., convenor).

In the WG-2 on Biocontamination the concept of a Hazard Analysis Critical Control Point (HACCP) system is being discussed for possible application to biocontamination control in controlled environments. This approach, which has been pioneered in the food industry aseptic processing, might or might not be directly applicable to aseptic processing in the healthcare industry, although its general principles might serve as a reference point in its application to the pharmaceutical and medical products industry.

In essence, the HACCP concept is part of a risk assessment of an aseptic process in terms of hazards to the product and ultimately, the patient. It also calls for the identification of critical points in the process that can be controlled to eliminate or minimize these risks and the development of alert and action levels for each critical point that would indicate if the process is in or out of control. It also calls for the development of a monitoring system and the development of corrective actions when specified levels are exceeded (based on trend analysis), and for documentation of all procedures and records obtained during the application of the monitoring program.

MICROBIAL CONTROL IN THE MANUFACTURING ENVIRONMENT IN THE U.S. MEDICAL DEVICE INDUSTRY

Guidelines of the Health Industry Manufacturers Association

The Health Industry Manufacturers Association (HIMA) published in 1978 in the Medical Device Sterilization Monographs series, Report No. 78-4.3, which discusses programs designed to assess the microbiological environment of a device manufacturer for the control of the bioburden of products that will be sterilized or the bioburden of those that will not be subsequently sterilized. This HIMA report is a very valuable reference because it evaluates the various methods for monitoring sources of contamination, be it from air, surfaces, or employee equipment and gears. An updated version of this report is sorely needed since technology advances have occurred in this field since 1978.

Guidelines for the Compounding of Sterile Drug Products for Home Use

The emergence of home healthcare modalities of treatment and the increase of the percentage of the population in need of home care has resulted in the development of organizations that compound drug products on a customized basis. These compounding centers can be in hospitals or in special facilities that service, in general, in-state areas rather than interstate areas. These facilities are primarily under state regulations that specify requirements needed for the compounding of sterile drug products. The role of State Boards of Pharmacy in the development of these standard requirements is facilitated by the development of a number of guidelines such as Practice Standards for Home Sterile Preparation (published in 1992 by the American Society of Hospital Pharmacists [ASHP]) or Sterile Drug Products for Home Use <1206>, an Information Chapter in USP 23 (1995). This USP information Chapter and the ASHP guidelines are not mandatory, but often, for marketing as well as scientific reasons, organizations producing these sterile products for home use, choose to follow one or the other of these guidelines.

The USP information chapter <1206> discusses the responsibility of the pharmacist dispensing these products and defines two types of compounding: a low risk and a high risk in terms of the microbiological conditions under which the potential for introduction of microorganism is low or high. A section on validation sterilization or aseptic processing, including media fills validation, is discussed and indicated. Various operational quality control (QC) and monitoring programs are described and the maintenance of product quality and control after the product leaves the compounding center is also discussed. Among the methods discussed for assessing the microbial status of the controlled environments are settling plates, contact plates, and slit-to-agar or other impaction samplers.

REGULATORY REQUIREMENTS FOR CONTROLLED ENVIRONMENTS IN EUROPE

The European Union (EU)—formerly called the European Economic Community—bases its regulatory requirements on the Guide to Good

Manufacturing Practices for Medicinal Products for products that are terminally sterilized or aseptically processed. In addition, most individual countries in the EU have their own GMPs that introduce variances for the requirements for controlled environments. Following discussion of the EU GMP requirements for controlled environments, I will review the GMPs in the areas of aseptic processing in a number of European countries. One should not be surprised that the requirements for aseptic processing are in the European GMPs, in as much as the regulatory requirements for these areas are also in the U.S. GMPs.

European Union Requirements for Controlled Environments

The EU requirements for controlled environments attempt to harmonize these requirements among the Union members. In contrast to the U.S. GMP, the EU GMPs apply to large scale manufacturers and to hospitals and for the preparation of products for clinical trials.

In section 1 (Quality Management) under 1.2 (Quality Assurance) statements in (i) to (vii) show that these requirements apply to controlled environments, without controlled environments being specifically mentioned. For instance, in 1.2 (v) it is indicated that "All necessary controls on intermediate products, and any other in-process controls and validations are carried out." It is interesting to note that a "procedure for Self-Inspection and/or quality audit appraises the effectiveness and applicability of the Quality Assurance System" is cited in 1.2 (ix).

The duties of the QC department under 2.7 in section 2 (Personnel), include the "monitoring and control of the manufacturing environment, process validation, and the inspection, investigation, and taking of samples, in order to monitor factors which may affect product quality" which can be directly interpreted as applying to controlled environments.

The requirements in 2.10 (Training) directly apply to controlled environments and are as follows: "Personnel working in areas where contamination is a hazard, e.g. clean areas or areas where highly active, toxic, infectious, or sensitizing materials are handled, should be given specific training."

The requirements from Premises and Equipment, to Documentation, to Production, to Quality Control should be consulted since their general provisions apply to controlled environments.

Requirements for the manufacture of sterile medicinal products include specific requirements in the areas of controlled environments for terminally sterilized products as well as for aseptic preparations. These supplementary requirements stress specific points not covered in the preceeding sections.

The general requirements stress that "1. should be carried out in clean areas where entry should be through airlocks for personnel or for goods." Under 3 it is indicated that "Clean areas for production of sterile products are classified according to the required characteristics of the air, in grades A, B, C and D. The air characteristics are given in the table below." (I have adapted the table of air classification system for the manufacture of sterile products to show that in addition to particulate counts the air classification is also done on the basis of microbiological levels of the air.)

Grade	Max. Permitted # of Particles per m^3 equal to or above		Max Permitted # of Viable Microorganisms per m^3
	0.5 μm	5 μm	
A (laminar)	3,500	none	less than 1
B	3,500	none	5
C	350,000	2,000	100
D	3,500,000	20,000	500

Under the specific requirements for products that will be terminally sterilized, it is indicated that the preparation of solutions should be done in a grade C environment to assure low microbial and particulate counts. A grade D environment could be used if steps are taken to minimize contamination. Large-volume parenterals (LVPs) and small-volume parenterals (SVPs) should be filled under Grade A conditions (laminar workstation) in a Grade C environment.

For aseptically processed products, the handling of starting material is done in a Grade C environment if sterile-filtered later in the process. Otherwise a Grade A zone is required with a grade B background, with the same conditions being acceptable for the preparation of solutions. Handling and filling of aseptically prepared products for both LVP and SVP are done in grade A environments in a grade B background.

Personnel should wear clothing appropriate to the air grade where they will be working; the sanitation of these clean areas is to be monitored at planned intervals during operations using microbial counts. Specifically for aseptic processing products, monitoring should be frequent and results must be taken into consideration in the determination of batch approval.

It is interesting to note that for products that cannot be sterilized in the final container, first filtration through a sterilizing filter (0.22 μm or less) must be used prior to filling the containers. Since filtration does not remove all mycoplasma and viruses, it is also suggested that some degree of heat treatment be considered. For sterile finished products, especially those that have been aseptically processed, it is also suggested that samples tested for sterility should include containers filled at the beginning and end of the batch and after any significant interruption of work.

The Pharmaceutical Inspection Convention (PIC) is harmonizing its inspection scheme with the European GMPs; a review of the scheme indicates that the requirements for controlled environments are essentially those of the European GMPs.

GOOD MANUFACTURING PRACTICES REQUIREMENTS FOR CONTROLLED ENVIRONMENTS IN NATIONAL GMPs IN SELECTED EUROPEAN COUNTRIES

United Kingdom

The requirements for controlled environments are included in the Guide to Good Pharmaceutical Manufacturing Practices, from the Department of Health and Social Security. Other than the general GMP requirements that apply to all environments as well as to controlled environments, the specific requirements are under the section on Manufacture and Control of Sterile Medicinal Products.

A number of definitions are indicated relating to controlled environments. For example, an aseptic area is defined as "A room, suite of rooms or special area within a Clean Area designed, constructed, serviced and used with the intention of preventing microbial contamination of the product." A clean area is also defined as "A room or suite of rooms with defined environmental control of particles and microbial contamination, constructed and used in such a way as to reduce the introduction, generation and retention of contaminants within the area." The UK GMPs does, however, detail the Basic Environmental Standards required for a variety of functions. Other than these standards for aseptic areas, the GMPs cover Environmental Standards for Clean Areas for Solution Preparation (Grade 2 or Grade 3 with additional measures taken to minimize microbial contamination—these grades correspond to the EU GMPs Grade C and Grade D); for Clean Rooms for Filling Solutions (Grade 2 = EU Grade C) that will be terminally sterilized; for Clean Area for Components Preparation, there are no special requirements other than to assure that microbial and particle contamination are minimized.

In the Basic Environmental Standards for Aseptic Areas, it is indicated that "rooms with conventional filtered air-flows and with contained work stations in the form of filtered air hoods or laminar air flow protection at working points are usually more appropriate than laminar air flow rooms." It is also interesting to note that these GMPs also call for a Grade 1/B environment (similar to EU GMPs Grade B) when the aseptic area is at rest, while a Grade 1/A is required (similar to EU GMPs Grade A) under the contained areas when the aseptic area is at work (fully operational).

Microbiological Considerations for Basic Environmental Standards are also highlighted: "It is vital that microbiological contamination of clean and aseptic areas should not exceed acceptable limits." Monitoring of the microbial contamination is done using a variety of methods, such as settle plates, air samplers and surface swabs, and so on. Intensive microbiological monitoring is indicated when a new unit is put in operation, a new process is implemented, or new operators are included. Scaling down of the monitoring is indicated once suitable conditions have been established.

Under the Cleanliness and Hygiene section, specific requirements for aseptic processing areas are included. These areas "should be cleaned frequently and thoroughly in accordance with a written program. Where disinfectants are used, different types should be employed in rotation to discourage the development of resistant strains of microorganisms." The notion that rotation of disinfectants is needed for the stated reason has been generally discredited and should possibly be removed from GMPs document.

The importance of environmental monitoring according to specified levels for controlled environments is highlighted by the statement on Batch Release:

The decision to release a batch of sterile product for use should take account not only of the specific production records and results of tests performed on that batch, but also of information gathered before and during its manufacture from the monitoring of the environment, personnel, intermediate products, equipment and processes.

France

The various requirements of Quality and Good Pharmaceutical Manufacturing Practices apply to controlled environments as well as for other environments. However, in an Appendix to Chapter 15 on

Sterile preparations, there are specific requirements under 15.A.1 (Sterile Preparations not Sterilized in the Final Container) as well as under 15.A.3 (Controlled Environment Zones).

In 15.A.1.6 (Manufacturing Process) two items are interesting. Under 15.A.1.6.2 it shows that "A preliminary qualitative and quantitative study of the risks of microbial (and particulate) contamination carried out at each stage of the manufacture allows to determine the characteristics of the premises, equipment and constituents as well as optimal production conditions." This type of analysis is similar to the HACCP system that is under discussion in Biocontamination Control under ISO-TC 209 WG-2 discussed earlier in this chapter.

In the same section of the French GMPs, under 15.A.1.6.3, it is noted that "The manufacture and filling into containers must be carried out in a controlled and sterile environment (cf. 15.A.3) and with sterile equipment." The section on Controlled Environment Zones (15.A.3) classifies the various controlled environment zones. The characteristics of each environment, called here as alpha-1, alpha-2, beta, gamma, and delta, are indicated in terms of particulate levels and microbial levels. The multiplicity of the naming of different environmental zones depending on the country that issues the GMPs is obvious and creates some confusion that one hopes will be dispelled by the European-wide application of terminology as indicated in EU GMPs. In terms of specific values for each environmental zone, these GMPs clearly indicate that these values are only applicable to the critical points of these various zones. The French GMPs clearly state that "these values, given for guidance, constitute an average level of contamination and must be adapted to each case, particularly in relation to the methods used to verify them. In fact, the most important point is to determine trends and detect possible deviations from the fixed mean level."

The general characteristics of these controlled zones include (under 15.A.3.1.2) additional characteristics not related to contamination. For instance, it indicates the generally accepted values for relative humidity (40 to 60 percent); temperature (21°C ±2°C); relative pressure (15 Pa or 1.5 mm Hg) with respect to an adjoining room with higher contamination; and air change rate (at least 20 times for high critical areas) and depending on the nature of the product, it can be adjusted (hygroscopic, hazardous, heat sensitive, photosensitive, etc.).

The general aspects of control protocols included under 15.A.3.5.8 (Microbial Contamination of the Environment) indicate that the number and type of microorganisms present in the environment must be included in the control protocol. The methods of sampling are listed as impingement on a solid medium, filtration through a gelatine membrane, and collection in a liquid medium. The results of settle plates cannot be expressed in terms of the volume of air; the method is useful in the monitoring of critical points, but the culture medium should be as unselective as possible and the incubation time and temperature should be related to the usual type of contamination expected.

REGULATORY REQUIREMENTS FOR CONTROLLED ENVIRONMENTS FOR SOUTH-EAST ASIAN NATIONS

The Good Manufacturing Practice General Guidelines for the Association of South-East Asian Nations (ASEAN) defines clean rooms or clean areas as ". . . with defined environmental control of particulate and microbial contamination, constructed, equipped and used in such a way as to reduce the introduction, generation and retention of contaminants within the area."

Sterile products are to be processed in a specific area that is "specially equipped and maintained to ensure cleanliness and sterility of the area." Products that will be terminally sterilized may be processed in a nonsterile but clean area.

South Korea

The Good Pharmaceutical Manufacturing Practices in Korea (KGMP) are issued by the Ministry of Health and Social Affairs (MOSHA). Controlled environments are listed in the chapter on Standards under Special Premises refering to areas used for the manufacturing of sterile products. These areas have to be separated from other working areas, must have doors and windows that can be closed tightly, and must be in aseptic conditions by utilizing U.V. light, filters, and other systems. It is also indicated that an aseptic-conditioned box can be used in lieu of an aseptic area.

In the section on Manufacturing Process Control for aseptic operation, "special precautions should be taken to prevent microbial contamination, and microbial contents in the air of the areas should be regularly measured."

Malaysia

The Guidelines on Good Manufacturing Practice for Pharmaceuticals in Malaysia addresses the issue of controlled environments in a manner very similar to the ASEAN GMPs.

Under the Sterile Products section it is indicated that all sterile products must be produced under carefully controlled and monitored conditions. Special facilities are indicated for products that are aseptically processed without being terminally sterilized. From a microbial control point of view, it is indicated that "Routine microbial counts of the air in the areas should be carried out before and during processing operations. The results of such counts should be checked against established standards."

The Malaysian GMPs defines three specific areas, Gowning Room, Clean Room, and Sterile Room. It also indicates that "It is important that microbial contamination of sterile areas should not exceed acceptable limits. Such areas should be monitored for microbial contamination on a routine basis."

In the Processing section (under 6.9.11.14) the concept of media fill is introduced to "test the incidence of contamination." Regular revalidation or revalidation via media fills is also indicated when new processes are introduced or when significant changes in process are also introduced.

India

The requirements for Controlled Environments under India's Good Manufacturing Practices are in section 1.2 (Requirements for Sterile Products Manufacturing). They essentially follow the ASEAN GMPs requirements, although the specific language differs. "For all areas where aseptic manufacturing has to be carried out air supply shall be filtered through bacteria retaining filters (HEPA filters) and shall be a pressure higher than the adjacent areas." In addition, routine microbiological tests are required during manufacturing operations for these aseptic areas, and the results of these counts must be checked against established standards.

Documentation and standard operating procedures for Parenteral Preparations include "details of environmental controls like temperature, humidity, microbial count in the sterile working areas" as well as the "particulars regarding the precautions taken during the manufacture to ensure that aseptic conditions are maintained."

Australia

The regulatory requirements for Controlled Environments in Australia are under the Code of Good Manufacturing Practice for Therapeutic Goods (National Biological Standards Laboratory). These requirements are specified in Part 2—Sterile Drugs. First, under section 14

(Premises), under 14.1 (Aseptically Prepared Products), under 14.1.1 that ". . . filling, sealing and any part of the process during which the product may be exposed to contamination shall be carried out in a Class 3.5 Laminar Flow Work Station (AS 1386) which is located in a Cleanroom with an air cleanliness level of Class 350 (AS 1386)." The classes 3.5 and 350 seem to refer to particulate counts and are similar to the Class 100 and Class 10,000 in the U.S. Federal Standard 209E.

The microbial monitoring requirements are listed under 14.1.1.9. "systematically monitored at regular intervals for microbial contamination during actual or simulated processing operations, using air samplers or other methods of sampling acceptable to the inspecting authority. The air sampling shall be carried out at the site of critical operations."

In the Manufacturing Control section (under 18.9), aseptic processing and filling equipment procedures and environments are ". . . checked for efficiency at the time of initial commissioning and at regular intervals" using media fill runs. The interesting part in that requirement is that the value of not more than 6 units nonsterile per 3000 units tested is obtained after incubation at 32°C for 14 days. This limit of not more than 0.2 percent contamination rate is an unusual occurrence in a national GMP that constitutes a regulatory requirement.

New Zealand

The Code of Good Practice for Manufacturing and Distribution of Medicines in New Zealand addresses the requirements for Controlled Environments under section 4 (Sterile Products: Additional Requirements). In the requirements of Premises, under 4.1.1.1 (Aseptically Prepared Products) the characteristics of the aseptic areas are listed, with the microbial monitoring of these areas being listed under 4.1.1.9. "microbial content of the air in it is regularly counted and during typical manufacturing operations, using air samplers or other acceptable methods."

The New Zealand Code does include a section 7 on Supplementary Notes for Hospital Pharmacies, and especially for controlled environments under 7.6 (Sterile Products Manufactured in Hospitals). This is unusual to find such requirements under GMPs. These requirements include the training of personnel involved in proper aseptic techniques; attention to adequate control of starting materials and the processing environment; the concept of double barrier system, and other requiurements all designed to reduce, minimize and eliminate microbial contamination.

REGULATORY REQUIREMENTS FOR CONTROLLED ENVIRONMENTS IN JAPAN

The GMP regulations that apply to Controlled Environments in Japan are in the GMP Regulations of Japan prepared by the Inspection and Guidance Division, Pharmaceutical Affairs Bureau, Ministry of Health and Welfare.

Under section 3 (Regulations for Building and Facilities for Pharmacies, Etc.) in the subsection on Buildings and Facilities for Drug Manufacturing Plant (5) and especially under Buildings and Facilities in Manufacturing Plant for Sterile Preparations (6), the general requirements for controlled environments are listed. In addition to the general provisions, the manufacturing area for injectable drugs manufactured under aseptic conditions must meet a number of additional requirements. One is "The room for weighing of sterile starting materials and preparation, filling and sealing shall be aseptic room." It is also interesting to note that this does not apply if in the room is an "aseptic box." Special requirements for the processing of blood products, of radiopharmaceuticals, and for medical devices are also indicated in detail.

REGULATORY REQUIREMENTS FOR CONTROLLED ENVIRONMENTS BY THE WORLD HEALTH ORGANIZATION

The World Health Organization (WHO) in its Good Practices for the Manufacture and Quality Control of Drugs under section 4 (Premises), and specifically under 4.3 (Special) indicates the requirement for separate enclosed areas for the aseptic processing of drugs that are intended to be sterile but cannot be sterilized in their final containers. "These areas should be entered through an air-lock and should be essentially dust free and ventilated with an air supply through bacteria-retaining filters giving a pressure higher than in adjacent areas." There is a direct mention of routine microbiological testing of the air to be done in these areas before and during manufacturing operations. The results of these tests "should be checked against established standards."

DISCUSSION AND RECOMMENDATIONS

The similarities among the various national and international GMPs, perhaps not in all details but at least for the general principles of controlled environments in the manufacture of drugs are rather striking. These similarities will, in the long run, facilitate the harmonization of regulatory requirements in a worldwide fashion. The need for controlled environments is not in dispute between manufacturers and regulatory agencies. What is in dispute is the level of control for these environments.

If a clean room is designed properly according to specifications that are technologically feasible and possible, and are maintained in proper operations, the quality assurance of products manufactured in these environments will increase.

Although in the pharmaceutical industry nonviable and viable particulate are important aspects of quality for sterile injectable products, the degree of "cleanliness" necessary is not of the level necessary for the electronic industry. In the current climate of cost containment for healthcare, it would be irresponsible to require that aseptic processing of sterile injections be done in controlled environments similar to those used for the fabrication of computer chips. However, the cost factor should not be the major factor in the decision to use aseptic processing or terminal sterilization in the manufacture of drugs. For those products that cannot be sterilized in the final containers, aseptic processing should be acceptable from a regulatory point of view. This situation exists in the areas of biotechnology-derived products that are often heat labile. To improve the SAL for these products, controls on all the stages of the process are absolutely necessary, and perhaps a closed system, such as isolator systems, should be recommended. One, however, has to suspend judgement on the advantages and disadvantages of these isolators systems as more of them are becoming operational.

A middle of the road approach suggesting that a mild heat treatment be given to aseptically processed product which cannot be terminally sterilized might be an optimal solution. This is provided that attempts to quantify the additional increase in sterility assurance are abandoned because they are faulty in their reasoning and confuse the issues.

The microbiological assessment of controlled environments must retain its primary role in the aseptic processing of drugs. It is very easy to confine assessment of controlled environments using continuous or intermittent monitoring of nonviable particulates, because the technology of microbiological counting is very variable. Sterility of injections is a function of the presence or absence of microorganisms, albeit using a probability function. To decrease the probability that sterile injections are contaminated, it is imperative to validate the aseptic process, using media fills, without having to resort to increasing the number of media filled containers to satisfy statistical significance. Once the system is validated, then monitoring of the air, surfaces, and personnel must be operational, with specified alert and action levels and specified procedures when these levels are exceeded or trending upward.

The reluctance of regulatory agencies to specify in their regulations the microbiological standards for the different areas of "cleanliness," but the multitude of guidelines that indeed specify these levels, allows for some flexibility from the regulatory and manufacturers viewpoints. However, depending on the country of origin, these microbiological levels are different, presenting some problems for the future in the harmonization of operational requirements for controlled environments.

3

Quality Systems and Total Quality[1]

Kenneth S. Stephens

THE FOUNDATIONS OF TOTAL QUALITY

The cornerstone of quality control (QC), as it has evolved, was laid in the 1920s with the development of the control chart by Walter Shewhart and of sampling/inspection principles and techniques by Harold Dodge. But for more than just the techniques, for which these pioneers are well known, it is the principles of quality and QC and the basic foundations of what we now call "Total Quality" that were laid by these and other pioneers in the era. From the definition of quality to the study and understanding of variation to the scientific basis for control, from specifications to measurement, from sampling to decisions to feedback, the foundations of quality and QC were laid. An early paper by Shewhart (1927) was entitled *Quality Control*. His monumental, classic work, *Economic Control of Quality of Manufactured Product*, was published in 1931 (Shewhart 1931). An early paper by Dodge (1928) was entitled *A Method of Rating Manufactured Product*.

Many references trace elements of quality and QC to ancient times with the construction of major world structures—still standing today—and these among others bring the development up to the early

[1]Portions of the following are from Stephens, K. S., ISO 9000 and Total Quality, *Quality Management Journal*, Volume 2, Issue 1, Fall 94, 57–71, ©1994. Reproduced with permission from the American Society of Quality Control.

days of the U.S. See, for example, chapter 1 of Wadsworth, Stephens, and Godfrey (1986) and chapter 1 of Banks (1989). In pharmacy one could argue that the desire for quality can be traced back to at least Sumerian times, 3000 years before Christ, since clay tablets exist providing formulations and processes for pharmaceutical products. The ancient Egyptians, 1500 years later, were using defined drugs and drug products, some of which are still recognizably in use today. The *Papyrus Ebers*, for example, found wrapped around a mummy dated to 1600 B.C., contains formulations and standards for over 700 different drugs of organic and inorganic origin. By 1495 the *Florentine Pharmacopoeia* had been published in Europe as the first of the modern compendia in which descriptions and standards for drugs and drug products were mandatory. The United States Pharmacopeia of 1820 drew attention to the importance of the pharmacopoeia being kept up to date in order to keep pace with developing technologies. This statement is quoted with approval in the 1995 United States Pharmacopeia, now in the 23rd edition and 2391 pages long. This particular book of standards is republished every five years, with appendices in between editions as required to achieve this object of staying up to date. Since 1990 discussions have been initiated with a view to harmonizing the United States, European, and Japanese pharmacopoeias, a movement for unification of pharmaceutical standards that has a history stretching back five thousand years.

Quality/Quality Control/Quality Assurance

Many definitions of quality are offered by authors, organizations, dictionaries, and the like. Among these is,

> *Quality, as applied to the products turned out by industry, means the characteristic or group or combinations of characteristics which distinguishes one article from another, or the goods of one manufacturer from those of his competitors, or one grade of product from a certain factory from another grade turned out by the same factory* (Radford 1922).

Shewhart (1931) argues that quality is often best described as "qualities" (i.e., multiple characteristics); that it is quantifiable, from this perspective, but that there is both an "objective" and "subjective" side to quality.

From Crosby (1979) we get,

> *Quality is conformance to requirements or specifications.*

A different slant is offered by Juran (1988):

> *Quality is fitness for use.*

A definition, earlier expounded by the *ANSI/ASQC Standard A3* (1978) has been incorporated in *ISO 8402* (1994) as,

> *Quality: The totality of characteristics of an entity that bear on its ability to satisfy stated and implied needs.*

Obviously all of these definitions, and others, have their place in an overall system of quality that considers different viewpoints, perspectives, and aspects of the whole—from products to services; from customers, to employees, to operations; and even on to conservation and the environment and society in general. In recognition of the many aspects of quality to describe the wide spectrum of applicability, the author has developed a broad and comprehensive definition of quality, for which an earlier version[2] appeared in Wadsworth, Stephens, and Godfrey (1986). It defines quality from several important aspects and illustrates quality's ever-expanding influence on all aspects of our lives. This is shown in Table 3.1.

Attention to quality and recognition that quality—as a discipline— offers sound business strategies are evolutionary and revolutionary forces that have made major contributions to our current way of life.

As for quality, QC and quality assurance (QA) have undergone stages in the evolutionary process of refinement and application. The descriptive adjectives preceding the term(s), in themselves, reflect an evolution, such as *statistical quality control (SQC), total quality control (TQC), companywide quality control,* and so on. Basic definitions also vary, such as the following:

> *Quality control is the regulatory process through which we measure actual quality performance, compare it with standards, and act on the difference. This is due to* Juran (1988).

> **Quality Control:** *The operational techniques and the activities which sustain a quality of product or service that will satisfy given needs; also the use of such techniques and activities* (ANSI/ASQC Standard A3 1978).

> **Quality Control:** *The operational techniques and activities that are used to fulfill requirements for quality* (ISO 8402 1994). *[Two clarifying NOTES are also provided, the second of which recognizes the interrelationship of quality control and quality assurance.]*

On the other hand, a common definition for QA is as follows:

> *Quality Assurance is a system of activities whose purpose is to provide an assurance that the overall quality control is in fact being done effectively.*

---

[2]The earlier version, itself, is an adaptation and expansion from a paper by Dr. Farouk M. Fawzi (1978).

Table 3.1. Quality's Ever-Expanding Influence

Fundamental Aspects of Quality
Quality of design (specifications, standards, grades)

Measurable Aspects of Quality
Quality of conformance (ability to meet specifications, standards, requirements)

Consumer (Marketable) Aspects of Quality
Fitness for use (performance, reliability, life, price, availability, on-time delivery)

Operational Aspects of Quality
Quality of management, operations, employees, system (maximum output—productivity, minimum waste and cost—improvement, optimum storage, and delivery)

Conservational Aspects of Quality
Optimum use of resources (materials, machines, money, land, energy, people)

Environmental Aspects of Quality
Ecology, clean air, unpolluted waters, potable water, tolerable noise levels, safe waste, open space, landscape, recreation, beauty

Human Aspects of Quality
Quality of life, health, education, culture, society, freedom, ethical and moral values

Governmental Aspects of Quality
Quality of rule, law, justice, taxation, social security, defense, rights and privileges

---

And by *ISO 8402* (1994), that is similar to that of *ANSI/ASQC Standard A3* (1978), QA is defined as,

> **Quality Assurance:** *All the planned and systematic activities implemented within the **quality system** and demonstrated as needed, to provide adequate confidence that an entity will fulfill requirements for quality. [With three clarifying NOTES pertaining to internal and external quality assurance, the interrelationship between*

QC and QA, and the importance that "requirements" must match
the needs of the users.]

It is interesting to note that distinctions between the terms *quality
control* and *quality assurance* are not emphasized in Japan, as they are in
the U.S. and other Western cultures. This is primarily because the term
control in Japan does not bear any negative connotation (as it does to
some in the West), such as coercive supervision. The Japanese word for
"control" is *kanri*. It is said (see, for example, Lillrank and Kano 1989)
that *kanri* can be translated as "management" or "administration."
Hence, the term *quality control* in Japan implies a systematic and fact-
based management of quality. This view is also held, incidentally, by
many practitioners in the West; and strong distinctions are not made
between QC and QA.

Statistical Quality (Process) Control/Total Quality Control

Deming (1971) places considerable emphasis on the statistical aspects
of QC—following in the footsteps of Shewhart—applied soundly
and on an overall or total basis, with contribution of the following definition,

> *The Statistical Control of Quality is application of statistical princi-
> ples and techniques in all stages of design, production, maintenance
> and service, directed toward the economic satisfaction of demand.*

A further, interesting definition of SQC is provided by the Western
Electric (1956) *Statistical Quality Control Handbook*, as follows:

Statistical Quality Control

Statistical: With the help of numbers, or data,

Quality   : We study the characteristics of our process

Control   : In order to make it behave the way we want it to behave

Statistical thinking, methodology, and techniques still (and should
always) play an extremely important role in a total quality program. It
is an invaluable discipline for studying processes, achieving and maintaining control, and realizing breakthroughs for continuous improvement.

A significant evolution from the work of Shewhart (1931) is the
recognition of two major types of quality problems defined by Juran as
sporadic and chronic and by Deming as special and common, respectively. Whereas Shewhart's developments concentrated on the

sporadic or special causes, Juran and Deming have successfully placed emphasis on breakthroughs accomplished by addressing the chronic or common causes of quality problems. These concepts are summarized in Figure 3.1.

Feigenbaum (1961) ushered in a new development in the evolution of QC with his *Total Quality Control*. His definition, which emphasizes the managerial and system aspects, is, as follows:

> *Total quality control is an effective system for integrating the quality development (planning), quality maintenance (control), and quality improvement efforts of the various groups in an organization so as to enable production and service at the most economical levels which allow for full customer satisfaction.* (parenthetical additions by this author)

Is it safe to say that this work is *underutilized* and *undercredited* for what is now being called "Total Quality Management" or "Total Quality"? One place where it is not so underutilized or undercredited is in Japan. The Japanese continue to use the term *Total Quality Control* to describe their extremely effective quality programs, in spite of new buzz words in the discipline. Nomenclature is not as important as **what is being done!** And TQC as expounded by Feigenbaum, and expanded by the additional body of knowledge from mainly Japanese innovations is essentially the Total Quality we speak about today.

For a cogent discussion of the Japanese concept of TQC and how it is applied, see Lillrank and Kano (1989), pages 30–40.

Quality Systems/Quality Management

An advanced stage of the QC evolution is expressed by the term and concept of *quality system*. This concept emphasizes the identification,

---

Figure 3.1. Types of Causes of Poor Quality

1. Special or Sporadic Causes

Affect Specific Operator—corrected by the
operator with MANAGEMENT direction/support.

2. Common or Chronic Causes

Affect Groups of Operators, are faults of system—
corrected by MANAGEMENT with operator
cooperation/instruction.

---

elaboration, and implementation of a comprehensive and thorough set of elements to be included in the QC/QA effort.

A definition of the term as provided by *ANSI/ASQC Standard A3* (1978) is, as follows:

> **Quality System:** *The collective plans, activities and events that are provided to ensure that a product, process, or service will satisfy given needs.*

ISO 8402 (1994) gives the following definition:

> **Quality System:** *The organizational structure, procedures, processes and resources needed to implement quality management [with three qualifying/clarifying NOTES].*

Quality management as defined in *ISO 8402* (1994), with considerable expansion of the definition from the earlier edition, is as follows:

> **Quality Management:** *All activities of the overall management function that determine the* **quality policy,** *objectives and responsibilities, and implement them by means such as* **quality planning, quality control, quality assurance, and quality improvement** *within the quality system.*

Like quality, QC **is** exercising an expanding influence on the creation and management of products; services; enterprises; and national and international programs of commerce, trade, and cooperation. Early stages have helped to influence and expand the later stages—which, in turn, have contributed to understanding and expanding the earlier stages as they are continually modified within the overall system. A good example of this is the renewed attention of the operator under a system of employee involvement—which can be likened to the re-creation of craftsmanship at the mass production level. Similar to that for quality, the author has developed a diagram presenting the evolution and expansion of QC, for which an earlier version also appeared in Wadsworth, Stephens, and Godfrey (1986). It summarizes QC through multiple stages of evolution and progress and illustrates QC's ever-expanding influence on business strategy and results. This is shown in Table 3.2.

During the evolutionary process of development, of particular importance is the move up from detection (inspection QC) to correction to prevention (SQC/TQC/QA) to employee involvement to continuous quality improvement to customer satisfaction (yea, even *delight* via innovation) (TQ). This progress of ever-beneficial movement is shown in Figure 3.2. Three levels below defect *detection* are also shown to emphasize from where we have come—and from which large parts of the present world must still come, unfortunately.

Table 3.2. QC's Evolution and Development

Operator Quality Control
From craftsmanship to participation, involvement, and empowerment

Supervisor Quality Control
Brought in by mass production and the Industrial Revolution

Inspection Quality Control
- Relief for or check on the operator and supervisor
- Quantity (production) versus quality (inspection)
- Emphasis on *detection and correction*

Statistical Quality Control
- Development and implementation of statistical techniques for quality
- Facts replace opinions
- Scientific approach to process and product studies and control
- Emphasis on *improvement and control*

Total Quality Control/Assurance—Companywide Quality Control (Japan)
- Involvement of all departments, divisions, organizations
- Integration of quality and productivity
- Participation of all personnel from CEO's to operators
- Emphasis on *planning and prevention/improvement*

Quality Systems
- Structured elements/requirements—process, factory, corporation, association, national, international
- Emphasis on *comprehensive and thorough programs based on essential elements*

Total (Strategic) Quality
- Quality as business strategy
- Total quality management/systems
- Emphasis on *top management leadership, customer satisfaction, employment involvement, continuous improvement, and profound knowledge*

Figure 3.2. Beneficial progress in the evolution of quality control.

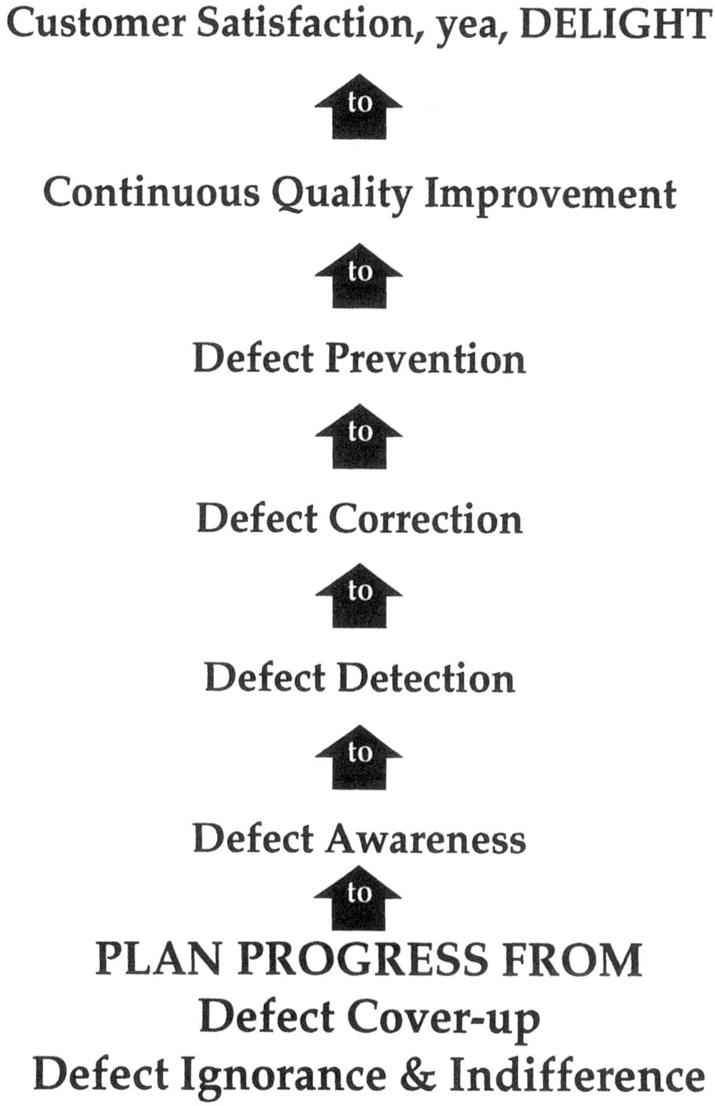

Customer Satisfaction, yea, DELIGHT

Continuous Quality Improvement

Defect Prevention

Defect Correction

Defect Detection

Defect Awareness

PLAN PROGRESS FROM
Defect Cover-up
Defect Ignorance & Indifference

The Evolution of the *ISO 9000 Series*

With the 1994 revisions of the *ISO 9000 Series*, including their widespread adoption and implementation, these international standards on *Quality Systems* continue to generate a lot of attention. This is deserving, but needs understanding as to what the *ISO 9000 Series* is and what it is **not**. Such an understanding will contribute significantly to the correct implementation and use of these standards and guidelines on quality systems and management—to meet customer demands for a quality system **and** to establish a *total quality* system that has direct and long-term beneficial results for the enterprise itself.

First, it is important to realize that quality systems or even standards for quality systems were **not** invented by *ISO 9000*. Inputs to the evolution of the *ISO 9000 Series* are reflected in Figure 3.3, which portrays the historical and widespread influences in the development of *ISO 9000*. A historical perspective of this development, with dates associated with the inputs, is given in Appendix A for greater detail.

Quality System Developments in Japan

Next, it is equally important to understand that the *ISO 9000 Series* was **not invented in Japan**. Many terms (with related concepts, methodology, and techniques) of the modern disciplines of quality have been borrowed from Japanese innovative and successful applications. Such terms as *kaizen* (improvement), *hoshin-kanri* (management by policy), *kanban* (visible record or order ticket—as an integral part of the kanban system of management by policy and just-in-time (JIT) production management), *jishu kanri* (self-management), *poka-yoke* (foolproofing/prevention), *seiri* (organization), *seiton* (neatness), *seiso* (cleaning),

---

Figure 3.3. Inputs to the evolution of *ISO 9000*.

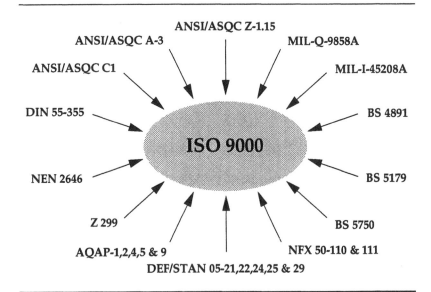

seiketsu (standardization), and *shitsuke* (discipline), with the latter five being referred to as the 5 Ss (Osada 1991). In addition to these, as further examples, additional terms (also with related concepts, methodology, and techniques) that took on English references directly include: *QCC* (quality control circles), *CWQC* (companywide quality control), *QFD* (quality function deployment), *CE* (cause and effect) *Diagrams*, *JIT* (just-in-time production/inventory management system), TPM (total productive maintenance), etc. To these must be added Deming's 14 Points for Management, emanating from the Japanese applications.

It is instructive to note that most of these concepts are not an integral part of *the ISO 9000 Series* and that **none** of these terms, concepts, and techniques represent quality system assessment, certification, registration, and/or accreditation—and especially as third-party programs. The system so successful in achieving quality in Japanese products—including innovative breakthroughs in design quality; low maintenance quality; competitive pricing via economic measures, such as scrap, rework, and waste elimination and avoidance; variation reduction; employee participation and involvement via quality improvement teams; planning and control for prevention; direct top management involvement and leadership; customer satisfaction; and continual improvement—was **not based** on quality system certification, registration, and/or accreditation! And this is the system that has attracted the attention of corporate management and the quality profession in the Western world—often with provocative exhortations by Western consultants, such as Deming and Juran. It is also the system that has been assimilated and implemented with equally beneficial results by many of the world's leading corporations and enterprises. It has gone under many names, some of which are mentioned later in this chapter. The above referenced inputs to the evolution and development of quality systems are reflected in Figure 3.4.

It **is a quality system** that has evolved with early principles and techniques transferred to Japan from the West and with many successful additions and innovations introduced by Japan for transfer back to the West. But it has **not** involved third-party quality system certification, registration, and accreditation—as presently being promoted with national and international fervor in conjunction with the *ISO 9000 Series*. And while certain positive aspects of international standardization of quality systems will be addressed as per *ISO 9000* below, one must pause and reflect whether we would not be better off promoting, teaching, and implementing the evolved quality system that includes the Japanese innovations rather than so much emphasis on third-party certification/registration of quality systems. In fact, enterprises and corporations implementing such quality systems would not need certification or registration—the results would speak for

Figure 3.4. Inputs to the evolution and development of quality systems.

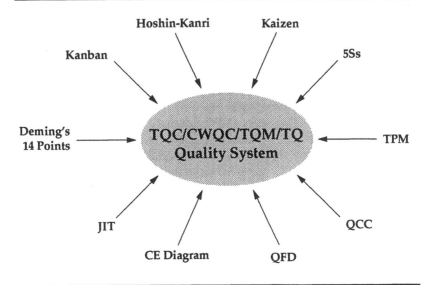

themselves—as is presently the case with well-recognized (by consumers) world class producers.

As a further historical perspective, the Japanese showed little or no interest in the early developments and implementation of the *ISO 9000 Series*, per se—they simply didn't need it; their quality systems were equal, and in most cases far superior (and recognized as such), to *ISO 9000*. One must recall or take note that the numerous tables showing national adoptions (in growing numbers) of *ISO 9000* either as "identical" or "equivalent" did **not** include Japan. In fact, the *ISO 9000 Series* was only adopted as the *JIS Z 9900 Series* in the Japanese national standards system in October 1991. The motivation for this was not based primarily on introducing and/or improving quality systems in enterprises and corporations (though they recognize this as one benefit for those enterprises lagging earlier developments—note discussion below), but more on reasons dictated by international harmony, trade, cooperation, and the sheer business opportunities related to third-party registration under *ISO 9000* (and its expected improved successors).

In an interim *Report of Special Committee on JIS Marking System* (Japanese Industrial Standards Committee (JISC)), the following items are mentioned as factors for consideration of adopting *ISO 9000* (parenthetical additions, bolding, and underlining are those of the author):

- Acceptance of the results of certification (registration) based on the *ISO 9000 Series* abroad in the JIS marking system (for **product** certification).

- Adoption of the *ISO 9000 Series* as JIS to start the private certification (registration) scheme by third-party bodies (in Japan).

- Internationalization of the JIS factory examination (for **product** certification) and active utilization of both domestic and foreign private inspection (assessment) bodies in response to the *ISO 9000 Series* movements.

- Japanese companies are beginning to recognize the merits of the *ISO 9000 Series*, such as the clear positioning of responsibility for quality control, by introducing the *Series* in their companies.

- Requests for establishment of a new scheme conforming with global movements are increasing for exporting to EC countries that are establishing unified criteria based on the *ISO 9000 Series*—or coping with the procurement policies of the governments and companies in foreign countries that are based on the *Series*.

- Necessity of adoption of the *Series* as JIS has been indicated, from a viewpoint of equalizing the bases of many existing certification, assessment, and registration schemes.

- Japanese companies wishing to receive factory assessment and registration based on the *ISO 9000 Series* (at the present time) have no alternative than to receive it from assessment bodies in foreign countries either directly or through the private assessment bodies (being proliferated in Japan without uniform national accreditation).

- Taking into consideration international trends and domestic company's demands, it is considered necessary to establish an accreditation body as soon as possible and to start the assessment and registration scheme based on the *ISO 9000 Series*.

- It is, therefore, essential for our country, too, to harmonize the *ISO 9000 Series* assessment and registration scheme from an international perspective, and to create the criteria for auditors according to international standards and guidelines.

- For accreditation of foreign accreditation bodies, it is desirable that the results of accreditation are to be respected between the two governments according to their mutual accreditation agreement.

- The JIS marking examination involves an examination of QC for <u>designated products</u>, and does not examine QC and QA concerning products other than the designated ones; therefore, it is not considered that JIS mark factories will easily be internationally regarded as factories registered under the basis of the *ISO 9000 Series* (this is perhaps more an example of Japanese humility).

- However, <u>since the *ISO 9000 Series* text is not considered the best in our country</u> and the ISO/TC 176 is proceeding with review work of the *ISO 9000 Series*, proposals should be actively made to ISO/TC 176 from the Japanese Industrial Standards Committee concerning any parts of which it considers revision to be desirable.

Other papers that should be considered in this respect are Hayashi (1991), Morita (1991), and Gomi (1991). Hayashi points out three merits of the ISO 9000 Series:

1. Gives manufacturers specific targets for their QC activities, leading to the improvement of the quality of their products.

2. The state of QC activities can be assessed and registered fairly, impartially, and neutrally by the third-party assessment body (Hayashi is perhaps being optimistic here) in line with objective check items *common to the world*.

3. The result of the assessment and registration becomes a "common passport to international markets," and there is a possibility that such schemes will grow up to be a measure to avoid unnecessary repetitive examination conducted in different countries.

Hayashi also presents some points (which he refers to as three traps) to be avoided when introducing the *ISO 9000 Series*:

1. Lack of unity in implementing the *ISO 9000 Series*, and a lack of harmonization of assessment by assessment bodies.

2. Perfunctory/bureaucratic examinations with assessment bodies requesting submissions of massive volumes of papers useless to the manufacturer, thus diminishing the meaning of the assessment of their quality systems.

3. Overconfidence of manufacturers in having passed the examination, *that no TQC is anymore necessary*.

ISO 9000 QUALITY SYSTEM STANDARDS AND TOTAL QUALITY

In terms of application the *ISO 9000 Series* of standards is well illustrated by the diagram presented in the pamphlet *Quality 9000* by ISO and shown in Figure 3.5. A description of each of the basic documents is presented in Appendix B for reference and completeness.

The *ISO 9000 Series* is a set of standards. Thus, it has both the advantages and disadvantages of standards, in general. As a standard, it is subject to periodic review and revision. The first cycle of that process has just been realized with the 1994 revisions, referred to as "phase I." More extensive, "phase II" revisions are already under consideration. See the paper by Tsiakals (1994). The standards need to be understood adequately in order to assure correct and beneficial implementation— together with other elements of a total quality system that are not a part of the *Series*. It is extremely important for enterprise/corporate managers to understand that the *ISO 9000 Series* is not intended as a standard on total quality. A total quality system must go beyond *ISO 9000*. This is discussed subsequently, in greater detail.

On the positive side its strengths lie in the structure that sets forth a uniform, consistent set of procedures, elements, and **requirements**

Figure 3.5. *ISO 9000 Series* **application schematic.**

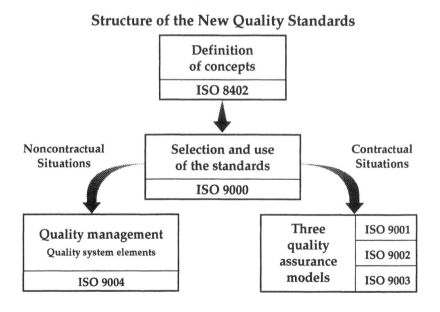

Structure of the New Quality Standards

that can be applied universally, albeit within limitations of interpretation and individual implementation. It provides a **basis** for designing, implementing, evaluating (assessing), specifying, and certifying (registering) a QA system. With widespread adoption (now a reality), it provides a common language for international trade with respect to the QA disciplines. It promotes (requires) a sound, well-documented contractual relationship between customer and supplier. Hence, it aims to establish a common understanding between these parties, based on agreed requirements.

For further understanding, the *ISO 9000 Series* is generic—and in two significant aspects. It is not *product or process specific*. The QA system (including the related quality manual—now a requirement) will, of necessity, have to contain specific subsystems related directly to the processes and the products to which it is being applied.

It is not even *quality system specific*. It does not specify a fixed system (beyond the requirements enumerated) for every enterprise (this, of course, is not a criticism). It provides considerable flexibility to the enterprise to design and specify (document) its own system within the framework of the requirements, and then directs attention to evaluating conformance to that system. It places considerable attention on documentation (perhaps with too much emphasis on conformance rather than on adequacy and/or effectiveness).

On the negative side, as for many standards passing through debate, review, negotiation, and consensus, *ISO 9000* represents a "least common denominator" in its coverage of the quality management/assurance/system disciplines. It would be good for everyone to understand this clearly, together with the understanding that it is not a standard on total quality, and realize that the actual quality system that is optimum for a given enterprise or corporation may go well beyond the requirements, elements, and procedures of the *ISO 9000 Series*. Note the remarks of Hayashi (1991) mentioned above with respect to, *"that no TQC is anymore necessary."* Many authors voice this caution (some as direct criticism). A number are included in the reference section. It **is** encouraging that many companies implementing quality systems **do, in fact,** go beyond the requirements of the particular standard used and incorporate other elements of total quality.

As mentioned above, there is too much emphasis on conformance rather than on adequacy and/or effectiveness. Meeting the requirements is the principal concern. Short-term corrective action is emphasized rather than long-term improvement. Sayle (1988) addresses these and other concerns with respect to *ISO 9000*. The *Series* is also believed by many to contain a weak quality audit program. Sayle (1992) also discusses this aspect while drawing contrasts with management audits.

Returning to the discussion above (Quality System Developments in Japan), there are many aspects of total quality systems not incorporated in *ISO 9000*. And, incidentally, the name one gives to these systems is not as important as the content. We see references to TQC, TQM (total quality management), IQM (integrated quality management, see APO 1990), SQM (strategic quality management), TQ (see Hutchins 1992), re-engineering, whole system architecture, and so on. And at this juncture we pause from the dialogue to enter a historical note. The quality sciences have always been plagued by problems of semantics and the NIH (not invented here) syndrome. Young (and/or new) proponents of TQM, for example, often are ignorant of, or ignore, the fact that programs and systems with previous names as simple as quality control or total quality control (and existent as much as 25–35 years ago, including those developed in Japan) included such concepts and methodologies as project-by-project continuous improvement with a prevention orientation and quality teams; customer needs assessments and satisfaction programs; quality as a strategic business component including its contribution to costs and cost reduction; design quality and innovation; and so on. This is not to say that important strides in refining and exposing these concepts to a wider audience have not been made in recent years; they have. But overzealous proponents of certain concepts have shown tendencies to idealize and ignore the conditions and necessities calling for a full range of tools (including statistical and others) for the *total* job of achieving quality and its related benefits.

Now with respect to understanding that *ISO 9000* is **not** a standard for total quality, in *ISO 9000* the following are either missing or less than adequately covered for a total quality system:

- Quality cost analysis and applications (other than as an element in ISO 9004-1 and a committee draft as ISO 10014)

- Top management involvement and leadership—top management driven quality councils and quality as a business strategy encompassing planning, control, and improvement

- Project-by-project improvement, pursued with revolutionary rates of improvement

- **Joy and pride** in work—and employee participation, involvement, and empowerment via project teams and quality circles

- Variation reduction, statistical process control, process capability, and process management

- Production/inventory management systems such as JIT with TQC

- The concept of single-sourcing, long-term cooperative supplier partnerships including product and quality system assistance (rather than mere assessment) based on trust and experience

- Innovation of products and processes

- Customer satisfaction including programs for surveying customer preferences and customer retention

- Benchmarking of competitors as well as best in class products, processes, and systems

- Deming's 14 points for management and the recognition of the necessity for transformation

Juran (1994) lists the following exclusions from *ISO 9000* as essentials to attain world-class quality:

- Personal leadership by upper managers

- Training the hierarchy in managing for quality

- Quality goals in the business plan

- A revolutionary rate of quality improvement

- Participation and empowerment of the workforce.

Now what has been said here must be properly interpreted and/or understood. It is not so much a criticism of *ISO 9000* as an emphasis to the exhortation not to limit one's quality program/system to that of *ISO 9000* alone. *ISO 9000* is intended as a set of standards on QA systems, **not** for Total Quality systems!

In fact, alternative (or additional) resources for total quality/management systems assessments and criteria for designing and implementing such systems are the various quality awards. Among these are the Deming Prize, the Malcolm Baldrige Award, and the European Quality Award. The latter, in particular, encourages self-appraisal and assigns 50 percent of the criteria to **results** in terms of people satisfaction (9 percent), customer satisfaction (20 percent), impact on society (6 percent), and business results (15 percent). See Conti (1991). See also the European Foundation for Quality Management (EFQM) with respect to the following: *The European Quality Award*, June 1992; and *Total Quality Management—The European Model for Self-Appraisal, 1992, Guidelines for Identifying and Addressing Total Quality Issues*.

A further resource for quality system requirements is that of FDA's current good manufacturing practices (cGMPs). While these apply on a mandatory basis to applicable organizations, they are, nevertheless, available to any organization to study and use. Of significance here is the move toward the ISO 9000 series quality assurance system standards in the proposed rules published in the *Federal Register*, Vol. 58, No. 224 on November 23, 1993. Harmonization of cGMPs with ISO 9001 (for example) is being sought to promote competitiveness, improve the quality of manufacture, enhance safer and more effective medical devices, and promote a more comprehensive quality assurance system than presently offered by the cGMP requirements. Thus, an improved set of requirements is expected by merging the product, facility, and process specific requirements of the cGMPs with the more comprehensive quality assurance system requirements of ISO 9001.

With consideration of these additional resources, the resultant system should be better, more dynamic, more comprehensive, more effective, and more economical than that of *ISO 9000* alone, in its present form. This is illustrated by Figure 3.6 with *ISO 9000* shown only as a basic foundation for a total quality system that for completeness includes other techniques and procedures as shown and mentioned above. Future revisions to the *ISO 9000 Series*, within the framework of its intended purpose, will address many of its inherent weaknesses. For example, the weaknesses of the "life-cycle model" versus a "process model" are being addressed by the phase II developments. With respect to its role within a total quality system, the phase II design specification prepared by the U.S. delegates to the ISO/TC 176 working group should be consulted, as given by Tsiakals (1994). The paper by Peach (1994) should also be studied.

ISO 9000 Assessment, Certification, and Registration

As explained in *ISO 9000-1*, in the *Scope* of *ISO 9001 to 9003*, and as illustrated in Figure 3.5, major parts of the *ISO 9000 Series* are intended for contractual use. This has its roots in the documents leading up to *ISO 9000*, including *MIL-Q-9858A*. *ISO 9004-1* is intended to assist in developing and implementing TQM systems.

Associated with these two major parts of the *ISO 9000 Series* are various assessment and certification/registration programs. It should be apparent from the above that the *ISO 9000 Series* is intended to facilitate two-party contractual arrangements for assurances of quality. The customer will want to have assurance that the supplier is, in fact, carrying out a comprehensive QA program in compliance with one of the *ISO 9000 Series* QA models as specified in the contract. To gain that assurance, the customer will arrange and carry out an external audit (assessment) of the supplier's QA system. Such an assessment is referred to as a "second-party audit or assessment." The *ISO 9000 Series*

Figure 3.6. *ISO 9000* only a minimum foundation for total quality systems.

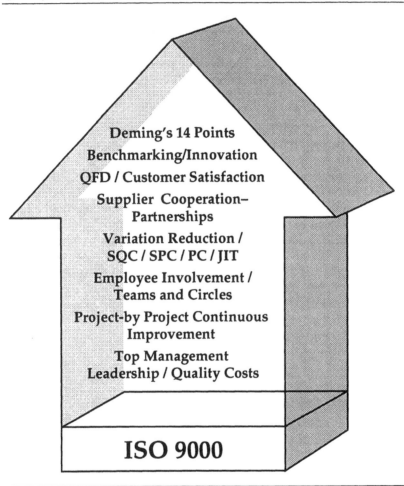

Deming's 14 Points

Benchmarking/Innovation

QFD / Customer Satisfaction

Supplier Cooperation–
Partnerships

Variation Reduction /
SQC / SPC / PC / JIT

Employee Involvement /
Teams and Circles

Project-by Project Continuous
Improvement

Top Management
Leadership / Quality Costs

ISO 9000

may also be used on a first-party basis (self-assessment, design and implementation of systems) to provide in-house management with assurances of quality. These were the original intentions of the standards, including their predecessors.

Additionally, extensive programs and efforts are now under way to use the *ISO 9000 Series* of standards on a third-party basis for assessment (on behalf of either of two parties) and eventual certification. These programs include the certification (and registration) of assessors and lead assessors for carrying out the assessment and certification activities. They also include the accreditation of bodies qualified to carry out third-party certification.

Third-party certification programs, especially with respect to product certification have been in existence for many years. One of the earliest is that of the British Standards Institution (BSI). This product certification mark, known as the "Kitemark" due to its shape, has been in existence since 1903, shortly after the establishment of BSI as a national standards body (see Stephens 1979). Hence, it is no wonder that third-party certification programs for quality systems have been established early and have grown rapidly in the United Kingdom.

In order to give some credibility to certification bodies in the UK, with respect to quality system certification, in particular, the Secretary of State for Trade and Industry agreed to accredit the certification bodies whose certificates could be relied on with respect to consistency, thoroughness, and competence. Hence, in 1982 the National Accreditation Council for Certification Bodies (NACCB), funded by the Department of Trade and Industry (DTI), was established, with BSI appointed to provide the secretariat.

Subsequently in 1985, the Association of Certification Bodies (ACB) was formed as a forum where certification bodies could collectively formulate and agree on their policies. The objective of the Association is to encourage improvement in the quality of goods and services manufactured or supplied in the U.K. through the promotion of independent certification to specific standards.

Efforts have been underway to standardize and harmonize the accreditation procedure. The European standard, EN 45012, *General Criteria for Certification Bodies Operating Quality System Certification*, was issued in September 1989 in support of these efforts. The accreditation bodies in Europe have come together to form the European Accreditation of Certification (EAC).

In the United States the Registrar Accreditation Board (RAB) was incorporated in November 1989 as an independent, wholly owned subsidiary of the American Society for Quality Control (ASQC) (Lofgren 1990). The American National Standards Institute (ANSI, the U.S. member of ISO) and RAB/ASQC jointly operate the American National Accreditation Program for Registrars of Quality Systems (*ISO 9000 News* 1992, 15–17). ANSI and RAB have also concluded a Memorandum of Understanding (MoU) with counterparts in the Netherlands (RVC), Australia and New Zealand (JAS–ANZ) and the United Kingdom (NACCB) to "achieve mutual recognition of each other's accreditation of quality system registrars." Then in January 1993, ANSI and RAB joined with these three counterparts, as well as organizations from Japan, Canada, and Mexico, to form the International Accreditation Forum (IAF), with ANSI as secretariat.

The ISO, itself, has now joined this effort. Its Council approved the setting up of a special group to propose the membership rules, organizational structure, and financing mechanisms for a worldwide system aimed at ensuring the international recognition of ISO 9000 certificates (*ISO 9000 News*, July/August 1993, 1–3; September/October 1993, 1, 3 & 4).

Another significant development in the UK was the formulation of the National Registration Scheme for Assessors of Quality Systems, administered by the Registration Board for Assessors (RBA). This was an element of the national quality campaign following the issue of the government's white paper, *Standards, Quality and International Competitiveness* in 1982. At the same time that NACCB was set up, the RBA was also sponsored by DTI to be managed by an independent Board of Management with the Institute of Quality Assurance (IQA) appointed to provide the necessary secretariat. The function of the RBA is to support the National Quality Initiative by setting qualifications and experience standards for the registration of Assessors of Quality Systems.

To qualify for registration by the RBA, individuals must be able to score a specified number of "credits" for a combination of academic qualifications, work experience, and experience in carrying out assessments of quality systems to the *ISO 9000 Series* or equivalent quality standards (BS 5750, for example, in the UK). In addition, it is a mandatory requirement that candidates must have successfully completed one of the training courses registered by the RBA. Hence, RBA/IQA operates another scheme, Registration of an Assessor Training Course.

In the U.S. the RAB operates a Certification Program for Auditors of Quality Systems and an Accreditation of Lead Assessor Courses. An extensive bibliography on certification, registration, and accreditation schemes is given in the references.

As with the quality system standards themselves, these programs of certification and registration also have both positive and negative aspects. Positively, third-party programs have as a basis the minimization of multiple audits carried out on given suppliers by multiple customers, that otherwise may involve considerable outlay of time, effort, manpower, and cost. The principle is that if a qualified, competent, reputable, and reliable auditing organization (registrar) carries out a thorough audit leading to approval (certification and registration) of a supplier's quality system, then this approval can serve all potential customers of this supplier's products and/or services for which the approved quality system applies. Hence, the initiation for the audit/assessment may come from the supplier itself as part of its marketing strategy.

On the negative side certification, registration, and accreditation programs must be watched very carefully and influenced by the international community to the extent that they do not become barriers to trade or, even to a lesser degree, economic barriers (with exorbitant fees for assessment/certification/accreditation). Small businesses have unique problems with these programs and have expressed concern. Developing countries, in particular, have also voiced their concern that they not be held ransom by demands to meet *ISO 9000* when no infrastructure exists locally to be assessed and certified. Certain aspects of third-party certification (registration) are being viewed as *technical colonization*. Additionally, the certification and accreditation process must be implemented very carefully to diminish or avoid business opportunists with no long-term interests in quality. Sayle (1992) mentions a deplorable practice of issuance of *conditional certificates*.

Much of the earlier reference, above, to the JISC *Report* with factors relating to the adoption of *ISO 9000* by the Japanese, have to do with assessment and certification (or registration). Hayashi (1991) mentions merits and traps associated with assessment and certification. Hutchins (1992) is especially critical with respect to the practice of third-party certification diluting the effectiveness of quality systems. He opines that "third party accreditation has probably cost the United Kingdom ten years of leadership in quality."

Reliance on third-party assessment and certification often represents an irresponsible delegation of responsibility (and leadership) on the part of top management. Hutchins (1992) refers to it as "one of the worst examples of delegation equals abdication."

A particular *bad scenario* is when companies feel *forced* to comply— that can lead to minimum attention being paid to the quality system that is implemented. This is often in response to initial or several assessment failures, simply plugging holes or filling gaps to satisfy the assessment body, perhaps with hostility and a desire to fake or cheat the system. This, in turn, often results in a fragmented quality system that no one is devoted to maintaining or even using, relegating quality to a police role rather than as a significant aid to business strategy and results.

Another UK quality professional has spoken out concerning third-party quality system certification. Burgess (1993) relates the following (with parenthetical additions for clarification):

Academicians (of the International Academy of Quality, IAQ, to whom he was writing) will know that the initiation of this new phenomenon (quality system certification), using the systems given in the ISO 9000 series, can be levelled at the United Kingdom. What they may not know is that since the first assessment of quality systems by third parties almost 20 years ago, the issue has now gotten out of hand. Many U.K. companies have developed quality systems for certification purposes only and, equally, many purchasers call up system certification as a condition of contract. Whilst this seems quite logical at first, it has led, in the United Kingdom and other countries, to the development of quality management for the wrong reasons. Commercialism is now leading to a lack of credibility for many of the activities associated with such quality systems and with registration.

At the World Quality Congress in Helsinki, I was able at the IAQ seminar to give some statistics relating to the U.K. situation. For example, there are now well over 20,000 companies registered to ISO 9000 (or its U.K. equivalent). There are 4,300 registered assessors (auditors) and you can imagine the variables that this produces. Further, I identified some 40 registration bodies, of which 27 are accredited quite properly by our accreditation body, the NACCB. Of course, the consultant community has grown in parallel, encouraged by government funding schemes. Academicians can imagine the different interpretations put upon the subject of quality in so many hands, not least the supplier himself!

All this (sometimes misguided) activity has led to concerns about value, short-term rather than long-term improvement, unfairness, different levels of achievement, and sometimes about the whole concept of system certification.

To many, certification has become an end in itself! Originally intended as a mark of distinction, it is in danger of becoming a shallow approach to quality achievement.

SUMMARY

ISO 9000 Series on Quality Systems for Quality Assurance

No single standard (or set of standards) has had more universal or worldwide results in increasing the awareness of quality than has the *ISO 9000 Series*—with its direct linkage to the unified market of the European Union (though mandatory compliance as a requirement of the EU is not an expected reality; see, for example, Zuckerman 1994). But one has to wonder what portion of the awareness created corresponds to the bad scenario mentioned above.

The national and international communities of quality professionals and related organizations have a serious responsibility (with related opportunity) to see that the *ISO 9000 Series* is properly used and promoted to create the correct and beneficial awareness and improvement of quality—in its broadest sense.

Every country, trade association, corporation, and enterprise that wishes to compete in international trade must give serious consideration to the use of the *ISO 9000 Series* on a first-party, second-party, and/or third-party basis for establishing and demonstrating QA of their products or services based on their quality management and systems, at least as a minimum program that may be extended to a more comprehensive system of total quality.

The publication of the *ISO 9000 Series* in 1987, and its revision in 1994, together with the accompanying terminology standard (*ISO-8402*) has brought harmonization on an international scale and has supported the growing impact of quality as a factor in international trade. The *ISO 9000 Series* has quickly been adopted by many nations and regional bodies and is rapidly supplanting prior national and industry-based standards (and otherwise filling existing gaps in QA system standards). It is a basis for promoting and disseminating quality management/assurance/systems globally. But, as detailed above, it is principally a standard for QA. It should be supplemented with additional elements of total quality systems for maximum benefits.

Total (Strategic) Quality

Many of the tenets of total quality are contained in the preceding sections, including *ISO 9000*. That is, the *ISO 9000 Series* is a subset of total quality. As alluded to earlier, there is much more—in content, approach, and benefits.

In the evolution to total quality a solid foundation is laid in defining quality as well as in describing stages of QC development to quality systems and total quality. The *ISO 9000 Series* is described as quality management, quality assurance, quality system guidelines and standards—all foundational to total quality. Reference is made to national quality awards and current Good Manufacturing Practices (cGMPs) as parts of a total quality system.

In the discussion pertaining to developments in Japan and the critique of *ISO 9000*, reference is made to many items in the *body of knowledge* of total quality that are missing in the *ISO 9000 Series*.

Hence, much of total quality is covered in earlier portions of the chapter. However, what sets total (strategic) quality apart from its earlier evolutionary entities is the emphasis and approach. Building on the strengths of earlier aspects and with recognition of impacting business strategy, total (strategic) quality is much more comprehensive. It places emphasis on:

- The customer and customer satisfaction

- Continuous improvement with innovation for all processes, services, and products

- Recognition of the value of the employee with employee involvement, empowerment, team participation, shared values and decisions, education and training

- Fact-based decision making, variability reduction, fast response with recognition of the internal customer, and the triple role of supplier/processor/customer at all processes

- Longer-term partnership relations with suppliers

- Competitive comparisons via benchmarking, including "best in class" on noncompetitive, but significant processes

- Productivity, cost reduction, and profitability enhancements

- Strong leadership from management with challenging statements of mission, vision, policy, values, and goals communicated throughout the organization.

This, together with all of the techniques, methodologies and elements of quality systems, constitutes total (strategic) quality.

Recent years have seen the proliferation of the term Total Quality Management and the TQM abbreviation. A word on semantics in the quality disciplines was given above. In terms of the description of the comprehensiveness of total (strategic) quality, management is an essential process. Total (strategic) quality management is defined as follows:

Total (Strategic) Quality Management is a process that integrates fundamental management art and techniques with the disciplines, principles, methodologies, and techniques of quality to develop and implement business strategies throughout the company (or other business entity) that emphasize and benefit from

- *A clear focus on customer needs, wants, and satisfaction*

- *Continuous improvement and innovation of all processes, services, and products*

- *Variability reduction, defect prevention, and fact-based decision making, with sound planning for quality*

- *Effective utilization, empowerment, and recognition of individuals under a team involvement/participation approach, including essential programs of education and training*

- *Integration of and mutual cooperation with suppliers*

- *Sensitivity to competitive comparisons*

- *Productivity and profitability enhancements*

- *Strong leadership by management at all levels with effective communication of policies, vision, mission, values, and goals*

While management, itself, may be considered a subset of total quality, it is key to the implementation and effectiveness of all it represents and all it can achieve. Encouraged and even goaded (by strong competitive performance) by the phenomenal Japanese successes, American (and other Western) management gradually have learned that it was not government subsidies or other external benefits that were giving Japanese companies a competitive edge on quality, productivity, and price. Eventually, it was seen that the disciplines of quality applied to business strategy with top management guidance, direct involvement, and strong leadership were paramount in permitting the Japanese to make significant inroads in market shares and domination of major industrial sectors. Until top management in key industries learned these lessons, total quality was not being utilized to its full potential. Even today, examples in the West are few. Dissemination is taking place and more examples are emerging, fortunately.

In order to implement the approach, with the emphases enumerated above, organization and management styles must undergo significant changes. Organizations must be flatter, more horizontal, with fewer hierarchial levels. Process ownership must be assigned to much lower levels. Even with hierarchal levels that will never disappear completely, facilitators of quality programs and cross-functional, multilevel, project improvement teams must have the facility to cut across many levels of management to accomplish their tasks. Communications between levels and across functions must be faster and more comprehensive. "Need to know" has taken on new meaning with greater employee involvement and participation.

In the paradigm shift to total quality, Juran (1989) mentions a "quality revolution" to respond to the quality crisis. Deming's reference to the necessity for "transformation" is now famous and also refers to the crisis facing American businesses and management, in particular. Of significant impact on the management of total quality are Deming's 14 points for management. These should be consulted in the development and implementation of any program (see Deming 1986).

CHAPTER APPENDICES

A. Historical Perspective to the *ISO 9000 Series*

As with many of the principles and methodologies of the quality sciences, quality systems and standards thereof have evolved. Early in the development was *MIL-Q-9858* issued on April 9, 1959, and *MIL-I-45208* of October 12, 1961. These were further developed as a companion two-part multilevel set of standards with the revision of December 16, 1963 as *MIL-Q-9858A*, Quality Program Requirements and *MIL-I-45208A*, Inspection System Requirements. Complementing these were *MIL-C-45662A*, Calibration System Requirements, February 9, 1962; *H-50*, Evaluation of a Contractor's Quality Program, April 23, 1965; *H-51*, Evaluation of a Contractor's Inspection System, January 3, 1967; and *H-52*, Evaluation of a Contractor's Calibration System, July 7, 1964.

An early nonmilitary standard was *ASQC Standard C1-1968*, Specification of General Requirements for a Quality Program. It was subsequently adopted as *ANSI Standard Z1.8-1971*, approved on November 18, 1971. It served a very useful purpose in procurement quality specifications. While not adopted as a standard, the *Aid: A Tested System for Achieving Quality Control, Technical Aids No. 91 for Small Manufacturers* (Small Business Administration), provided a fairly comprehensive set of essential elements of quality control. It was issued in January 1969.

MacDonald (1976a, 1976b, 1977) provides a list of quality system standards and discusses in some detail the British Standards *4891*, *5179 (Parts 1, 2, and 3)*; the NATO Documents *AQAP-1, 2, 4, 5, and 9 (Allied Quality Assurance Publications)*; and the British Defense Documents *DEF/STAN 05-21, 22, 24, 25, and 29* (Defense Standards). He traces the development of these standards as an evolution from *MIL-Q-9858A*. Following misgivings in many sectors of industry about the issuance of defense standards, covering quality management systems, as British standards, BSI developed *BS 4891*: A Guide to Quality Assurance that was issued in 1972. Following this was *BS 5179*: Guide to the Operation and Evaluation of Quality Assurance Systems, published in December 1974 and representing a three-part multilevel set of standards. It was the forerunner of *BS 5750*. At the time (1972–74) there were no comparable ASQC/ANSI standards for quality systems. The ANSI Committee Z1 on Quality Assurance was established in 1974 with ASQC as its secretariat. It was not until 1979 that ASQC was appointed secretariat to the U.S. TAG for ISO/TC 176 on Quality Management and Quality Assurance, shortly after the establishment of TC 176.

Further developments in the evolution of quality system standards took place over the period 1978–1979, in particular. *ASQC's Standard A–3 (1971)*, Glossary of General Terms Used in Quality Control was revised extensively in 1978 with change of title to Quality Systems Terminology with associated terms—eliminating other terms previously common to *A-1* and *A-2* (ASQC Standards).

A multilevel set of Quality Assurance Program Standards, *Z 299 Series*, was issued by Canada in 1978. Then in 1979, *BS 5750*, Quality Systems, was issued by BSI. It consisted of a three-part multilevel set of standards, and was based on *BS 5179*. The major changes were that (1) recommendations were changed to requirements; (2) parts 1 and 3 of *BS 5179* were changed to parts 3 and 1 of *BS 5750*, respectively; (3) references were made, where relevant, to *BS 5781*, Specification for Measurement and Calibration Systems; and (4) clauses dealing with *review, evaluation guidance*, and *significant questions* were excluded, with *BS 5179* and *BS 4891* still available for guidance purposes. *BS 5750 (1979)* was intended primarily for contractual purposes for a broad range of industries. It was a relatively brief document with less than three pages devoted to level 1 (design, manufacture, and installation) and less than one page devoted to level 3 (final inspection and test).

The American National Standard, Generic Guidelines for Quality Systems, *ANSI/ASQC Z-1.15-1979*, was issued on December 19, 1979, providing "guidelines for establishing, structuring, or evaluating product quality systems." It was intended to "encompass the quality functions at all stages and levels of the total product life cycle, from a product's conception through its development, manufacture, delivery, installation, and extended application by the ultimate user." It was not, therefore, a multilevel standard. It provided basic elements for structuring a quality system. An expanded set of basic elements based on *Z-1.15* is given in Wadsworth, Stephens, and Godfrey (1986). *ANSI/ASQC Z-1.15-1979* (with 17 pages) contained considerably more detail than *BS 5750 (1979)*. It was particularly useful in the development of *ISO 9004*.

The ISO/TC 176 was formed in 1979 and began work on standards for quality management, QA, and quality systems. The ISO is a worldwide federation of national standards bodies, comprising some 90 members, one in each country. The object of ISO, located in Geneva, Switzerland, is to promote the development of standardization and related activities in the world with a view to facilitating international exchange of goods and services, and to developing cooperation in the sphere of intellectual, scientific, technological, and economic activity. The results of ISO technical work are published as *international standards*. While international standardization was started in the electrotechnical field more than 80 years ago with the establishment of the International Electrotechnical Commission (IEC), ISO was created and began to function officially on February 23, 1947, with "the object to facilitate the international coordination and unification of industrial standards." The IEC remains intact and continues to cover the fields of electrical and electronic engineering standards. It shares the same building in Geneva with ISO.

A member body of ISO is the national body "most representative of standardization in its country." For the U.S. this is the American National Standards Institute (ANSI), 11 West 42nd Street, 13th floor, New York, NY 10036 (telephone: (212) 642-4900; telex: (212) 398–0023; telex: 424296 ansi ui; and telegram: standards, new york).

The technical work of the ISO is carried out through technical committees (TCs). Each technical committee may, in turn, establish subcommittees (SCs) and working groups (WGs) to cover different aspects of its work. By the end of 1991, there were 176 technical committees, 630 subcommittees, 1,827 working groups, and 18 ad hoc study groups. *Quality Management and Quality Assurance* (ISO/TC 176) has responsibility for the *ISO 9000 Series*. The Secretariat is the Standards Council of Canada (SCC).

In the QC discipline, there is yet another ISO technical committee, ISO/TC 69 on *Applications of Statistical Methods*.

The next major, and significant, phase in the evolutionary development of standards on quality systems was over the period of 1985–1987. In 1985 Canada revised its standard *Z 299 Series*, representing a four-part multilevel set of standards on Quality Assurance Programs. A discussion of these changes is provided by Sjoberg (1987). These standards are essentially contractual in nature, but can also be used in a noncontractual situation as a guide to establish, evaluate, and improve QA programs. These standards remain in effect today.

On June 15, 1986, ISO/TC 176 published Quality—Vocabulary, *ISO 8402*. It is comparable to *ANSI/ASQC A3-1987* on Quality Systems Terminology, presenting definitions and related discussion (via notes) for 22 items.

In March 1987 the ISO issued the *ISO 9000 Series*, prepared by ISO/TC 176. This was followed in June 1987 by the issuance of a revised set of *BS 5750* standards identical to *ISO 9000*. In November 1987 the Committee for European Standards (CEN) adopted the *Series* as their *EN 29000 Series*. This set in motion the rapid dissemination and adoption of the standards worldwide. The *ISO 9000 News* (the ISO newsletter on quality management standards), March 1992 issue, listed 45 countries with "identical" adoptions and 3 countries (China, Jamaica & Venezuela) with "equivalent" adoptions. The *ISO 9000 Handbook*, 2nd edition (Peach 1994) lists seventy-four countries that have adopted the *Series*. Thus, national adoptions are proceeding at a rapid pace.

National standards that contributed to the development and that were studied and used by the committee to write *ISO 9000* included *BS 5750* (the 1979 and 1987 versions), *BS 5179*, *BS 4891*, *ANSI/ASQC Z-1.15*, *MIL-Q-9858A*, *ANSI/ASQC C-1*, Canada's *Z 299*, *ANSI/ASME*

NQA-1, France's *AFNOR NFX 50-110 & 111*, Germany's *DIN 55-355*, and Netherlands' *NEN 2646*. The above referenced inputs to the evolution of *ISO 9000* are reflected in Figure 3.3 that portrays the historical and widespread influences in the development of *ISO 9000*. ASME is the American Society of Mechanical Engineers, a U.S. standards writing body. AFNOR is France's national standard's body (Association française de normalisation). DIN is Germany's national standard's body (Deutsches Institut für Normung).

The work of TC 176 continues. New standards in the 9000 series have been developed. These include:

- *ISO 9000-2 (1993)*, Quality Management and Quality Assurance Standards—Part 2: Generic Guidelines for the Application of ISO 9001, ISO 9002, and ISO 9003

- *ISO 9000-3 (1991, reissued in 1993)*, Quality Management and Quality Assurance Standards—Part 3: Guidelines for the Application of *ISO 9001* to the Development, Supply, and Maintenance of Software

- *ISO 9004-2 (1991, reissued in 1993)*, Quality Management and Quality System Elements—Part 2: Guidelines for Services

- *ISO 9004-3 (1993)*, Quality Management and Quality System Elements—Part 3: Guidelines for Processed Materials

- *ISO 9004-4 (1993)*, Quality Management and Quality System Elements—Part 4: Guidelines for Quality Improvement.

ISO/TC 176 has also started a new series for auditing. Standards completed include the following:

- *ISO 10011-1 (1990, reissued in 1993)*, Guidelines for Auditing Quality Systems—Part 1: Auditing

- *ISO 10011-2 (1991, reissued in 1993)*, Part 2: Qualification Criteria for Quality System Auditors

- *ISO 10011-3 (1991, reissued in 1993)*, Part 3: Management of Audit Programs. A further standard is *ISO 10012-1 (1992)*, Quality Assurance Requirements for Measuring Equipment—Part 1: Metrological Confirmation System for Measuring Equipment.

Revision of the basic 9000 series is now completed, including *ISO 8402*. This includes: *ISO 9000- 1, ISO 9001, ISO 9002, ISO 9003*, and *ISO 9004-1*.

ISO has published *Vision 2000, A Strategy for International Standards' Implementation in the Quality Arena During the 1990s*, a report of an ad hoc task force. This is further described by Marquardt et al. (1991). Four generic product categories are identified to help direct future standard's development:

1. Hardware

2. Software

3. Processed materials

4. Services

Four strategic goals have been set for further development:

1. Universal acceptance

2. Current compatibility (with parent documents and part supplements)

3. Forward compatibility (minimize revisions and with acceptance in existing documents)

4. Forward flexibility (combined supplements for the needs of industry/product categories and incorporation of useful supplements in revisions of parent documents).

B. A Brief Overview of the *ISO 9000 Series*

In terms of application the *ISO 9000 Series* of standards is well illustrated by the diagram presented in the pamphlet *Quality 9000* by ISO and shown in Figure 3.5. As mentioned above *ISO 8402* presents vocabulary or terms for quality systems. It thus serves as a reference base for terminology.

ISO 9000 is a series of five international standards (but as mentioned above with respect to new work completed and yet under way, various parts are being developed under these five basic standards) on quality management, QA, and quality systems. They deal with the structure, procedures, requirements, and the elements of quality management/assurance/systems.

The following individual standards make up the principal *Series*:

*ISO-9000-1: Quality Management and Quality Assurance Standards—
Guidelines for Selection and Use*

This standard consists of a general introduction, a set of definitions (referencing *ISO 8402* Quality—Vocabulary), the contractual and non-contractual situation, types of standards (9001 through 9004), selection of a quality assurance model (9001 through 9003), precontract assessment, tailoring and reviewing a contract, and a cross-reference list of quality system elements (between 9001 through 9004).

This standard provides the essentials of putting a management and QA policy into action. It clarifies the relationship between different quality concepts and specifies the rules for using the three models given in *ISO-9001, ISO-9002,* and *ISO-9003.* The standard introduces the notion of *degrees of demonstration* that is associated with the proof any client may require concerning the adequacy of the quality system and the conformity of the product with the specified requirements.

The three ISO Quality 9000 Models represent three distinct forms of functional or organizational capability suitable for two-party contractual purposes.

ISO-9001: Quality Systems—Model for Quality Assurance in Design/
Development, Production, Installation, and Servicing

Model 1 is for use when conformance to specified needs is to be assured by the supplier throughout the whole cycle from design to servicing. It is used when the contract (between supplier and purchaser, for example) specifically requires design effort and the product requirements are stated (or need to be stated) principally in performance terms. Model 1 represents the fullest requirements, involving 20 quality system elements at their most stringent level.

ISO-9002: Quality Systems—Model for Quality Assurance in Production,
Installation and Servicing.

Model 2 is more compact. It is for use when the specified requirements for products are stated in terms of an already-established design or specification. Only the supplier's capabilities in production, installation and servicing are to be demonstrated. All the quality system elements listed in *ISO-9001* are present at the same level, except for design control. All requirement clauses are now harmonized between the three QA models, with "placekeepers" in the clauses that do not apply. This is the case for "4.4 Design control" in ISO 9002.

ISO-9003: Quality Systems—Model for Quality Assurance in Final
Inspection and Test

Model 3 applies to situations where only the supplier's capabilities for inspection and tests (conducted on the product as supplied) can be (or must be) satisfactorily demonstrated. In this Model a reduced number of the quality system elements of *ISO-9001* are required. Four requirements, in particular, are given "placekeeper" clauses to align common requirement clauses between all of the QA models. For ISO 9003 these nonapplicable clauses are: design control, purchasing, process control, and servicing.

ISO-9004-1: Quality Management and Quality System Elements—Guidelines

This standard consists of a set of more than 90 quality system elements that should be considered when designing and implementing a quality system. It provides additional details for each of the broader categories of the 20 quality system elements referenced in *ISO-9001*, and the other system (model) standards. A manufacturer needs to understand an operation in sufficient detail so that only the appropriate elements are selected for each step of the operation. The object is to minimize the cost of the quality system while maximizing the benefits. *ISO-9004* is intended to serve as a guideline for this task.

REFERENCES

ANSI/ASQC Q9001-1994. *Quality systems—Model for quality assurance in design, development, production, installation, and servicing.* Milwaukee: American Society for Quality Control.

ANSI/ASQC Q9002-1994. *Quality systems—Model for quality assurance in production, installation, and servicing.* Milwaukee: American Society for Quality Control.

ANSI/ASQC Q9003-1994. *Quality systems—Model for quality assurance in final inspection and test.* Milwaukee: American Society for Quality Control.

ANSI/ASQC Q9004-1-1994. *Quality management and quality systems—Guidelines.* Milwaukee: American Society for Quality Control.

ASQC. 1978. ANSI/ASQC *Standard A3-1978—Quality systems terminology.* Milwaukee: American Society for Quality Control.

Asian Productivity Organization. 1990. *New waves in quality management—An integrated approach for product, process and human quality.* Workshop on Quality Management: An Integrated Approach, 4–8 September, 1989 (Taipei), Tokyo, Japan.

Banks, J. 1989. *Principles of quality control.* New York: John Wiley & Sons.

Burgess, N. 1993. Acn. Burgess addresses quality management certification. *CONTACT* (Newsletter of the International Academy for Quality), 53 (December): 5–6.

CEN/CENELEC, The Joint European Standards Institution. 1989. European standard EN 45012:1989E, general criteria for certification bodies operating quality system certification, September 1989, Brussels, Belgium (UDC 658.562.008.6).

Conti, T. 1991. Company quality assessments. *Total Quality Management* June: 167–172; August: 227–233.

Craig, R. J. 1991. Road map to ISO 9000 registration. In *ASQC Quality Congress Transactions* (pp. 926–930). Milwaukee: American Society for Quality Control.

Crosby, P. B. 1979. *Quality is free.* New York: McGraw-Hill.

Deming, W. E. 1971. Some statistical logic in the management of quality. In *Proceedings of the All India Conference on Quality Control* (pp. 98–119), 17 March, in New Delhi, India.

Deming, W. E. 1986. *Out of the crisis.* Cambridge, MA: Massachusetts Institute of Technology, Center for Advanced Engineering Study.

Dodge, H. F. 1928. A method of rating manufactured product. *Bell System Technical Journal* VII:350–368.

Eicher, L. D. 1992, The ISO 9000 Standards—An international phenomenon. *ISO Bulletin* 23 (7):3. (See other articles in this issue.)

EFQM. 1992. *Total quality management: The European model for self-appraisal: Guidelines for identifying and addressing total quality issues.* Jan. 1992, Eindhoven, Netherlands: European Foundation for Quality Management (ISBN 90-5236-035-9).

EFQM. 1992. *The European quality award.* The European Quality Award Secretariat. Brussels, Belgium: European Foundation for Quality Management.

Fawzi, F. 1978. Conservation of natural resources—A new role for quality control. *Transactions ICQC '78*, Tokyo, A2:15–17.

Feigenbaum, A. V. 1961. *Total quality control, engineering and management.* New York: McGraw-Hill.

Gomi, Y. 1991. Voluntary third party system for electronic components based on ISO 9000 Series. In *Transactions, seminar on achieving competitive quality through standardization and quality control*, MITI, JSA, UNIDO, SIRIM, 29–31 October, in Kuala Lumpur, Malaysia.

Hayashi, A. 1991. Japan's policy toward international assessment and registration system using ISO quality assurance standards. In *Transactions, seminar on achieving competitive quality through standardization and quality control*, MITI, JSA, UNIDO, SIRIM, 29–31 October, in Kuala Lumpur, Malaysia.

Hutchins, D. 1992. *Achieve total quality*, Cambridge, U.K.: Director Books.

ISO. 1990. *Vision 2000 A strategy for internal standards implementation in the quality arena during the 1990s.* Geneva, Switzerland: International Organization for Standardization.

ISO. 1994. *International Standard ISO 8402:1994 (E/F/R), Quality management and quality assurance—vocabulary.* Geneva, Switzerland: International Organization for Standardization.

ISO. 1994. *International Standard ISO 9000-1-1994, Quality management and quality assurance standard—Part 1: Guidelines for selection and use.* Geneva, Switzerland: International Organization for Standardization.

ISO. 1993. *International Standard ISO 9000-2:1993, Quality management and quality assurance standards—Part 2: Generic guidelines for the application of ISO 9001, ISO 9002 and ISO 9003.* Geneva, Switzerland: International Organization for Standardization.

ISO. 1991. *International Standard ISO 9000-3:1991, Quality management and quality assurance standards—Part 3: Guidelines for the application of ISO 9001 to the development, supply and maintenance of software.* Geneva, Switzerland: International Organization for Standardization.

ISO. 1993. *International Standard ISO 9000-4:1993, Quality management and quality assurance standards—Part 4: Guide to dependability programme management.* Geneva, Switzerland: International Organization for Standardization.

ISO. 1994. *International Standard ISO 9001:1994, Quality systems—Model for quality assurance in design, development, production, installation, and servicing.* Geneva, Switzerland: International Organization for Standardization.

ISO. 1994. *International Standard ISO 9002:1994, Quality systems—Model for quality assurance in production, installation, and servicing.* Geneva, Switzerland: International Organization for Standardization.

ISO. 1994. *International Standard ISO 9003:1994, Quality systems—Model for quality assurance in final inspection and test.* Geneva, Switzerland: International Organization for Standardization.

ISO. 1994. *International Standard ISO 9004-1:1994, Quality management and quality systems elements—Part 1: Guidelines.* Geneva, Switzerland: International Organization for Standardization.

ISO. 1991. *International Standard ISO 9004-2:1991, Quality management and quality system elements—Part 2: Guidelines for services.* Geneva, Switzerland: International Organization for Standardization.

ISO. 1993. *International Standard ISO 9004-3:1993, Quality management and quality system elements—Part 3: Guidelines for processed materials.* Geneva, Switzerland: International Organization for Standardization.

ISO. 1993. *International Standard ISO 9004-4:1993, Quality management and quality system elements—Part 4: Guidelines for quality improvement.* Geneva, Switzerland: International Organization for Standardization.

Japanese Industrial Standards Committee. 1991. *Report of special committee on JIS marking system* (interim). Special committee on JIS marking system. 17 May, Tokyo.

Juran, J. M., ed. 1988. *Quality control handbook,* 4th ed. New York: McGraw-Hill.

Juran, J. M. 1989. *Juran on leadership for quality.* Milwaukee: Quality Press, American Society for Quality Control.

Juran, J. M. 1994. The upcoming century of quality. *Keynote address to the ASQC Quality Congress.* Milwaukee: American Society for Quality Control.

Kalinosky, I. S. 1990. The total quality system—Going beyond ISO 9000. *Quality Progress* 23 (6):50–54.

Kume, H. 1992. The Japanese point of view on the ISO 9000 atandards In *Proceedings, 36th EOQ Annual Conference* (pp. 51–53), 15–19 June, in Brussels, Belgium.

Lam, T. C., and C. C. Liang. 1991. The need to implement ISO 9000 in Malaysia. In *Transactions, seminar on achieving competitive quality through standardization and quality control,* MITI, JSA, UNIDO, SIRIM, 29–31 October, in Kuala Lumpur, Malaysia.

Lillrank, P., and N. Kano. 1989. *Continuousimprovement.* Michigan Papers in Japanese Studies, No. 19. Ann Arbor: University of Michigan, Center for Japanese Studies.

Lofgren, G. Q. 1990. Accreditation of quality system registrars. *1990 ASQC Quality Congress Transactions—San Francisco* (pp. 979–982). Milwaukee: American Society for Quality Control.

MacDonald, B. A. 1976a. British Standard 4891: A guide to quality assurance. British Standard 5179 (Parts 1,2, & 3): Guide to the operation and evaluation of quality assurance systems. *Journal of Quality Technology* 8 (3).

MacDonald, B. A. 1976b. List of quality standards: Specifications and related documents. *Quality Progress* IX (9):30–35.

MacDonald, B. A. 1977. British Standard 4891: A guide to quality assurance, British Standard 5179 (Parts 1, 2, & 3): Guide to the operation and evaluation of quality assurance systems. *Quality Assurance* 3 (1):21–24.

Marquardt, D., J. Chove, K. E. Jensen, K. Petrick, J. Pyle, and D. Strahle. 1991. Vision 2000: The strategy for the ISO 9000 series standards in the '90s. *Quality Progress* 24 (5):25–31. (See also *EOQ Quality*, Vol. 2, 1991.)

Morita, C. 1991. Implementation of quality assurance activities (ISO 9000) in factories. In *Transactions, seminar on achieving competitive quality through standardization and quality control*, MITI, JSA, UNIDO, SIRIM, 29–31 October, in Kuala Lumpur, Malaysia.

Osada, T. 1991. *The 5S's: Five keys to a total quality environment*. Tokyo: Asian Productivity Organization.

Peach, R. W. 1990. ISO 9000 series—Quality management and quality assurance. In *ASQC Quality Congress Transactions* (pp. 968–974). Milwaukee: American Society for Quality Control.

Peach, R. W. 1994. Planning the journey from ISO 9000 to TQM. In *ASQC Quality Congress Transactions* (pp. 864–872). Milwaukee: American Society for Quality Control.

Peach, R. W., ed. 1994. *The ISO handbook*, 2nd ed. Fairfax, VA: CEEM Information Services.

Puri, S. C. 1991. Deming + ISO/9000, A Deadly Combination for Quality Revolution. In *ASQC Quality Congress Transactions* (pp. 938–943). Milwaukee: American Society for Quality Control.

Puri, S. C. 1992. The ABC's of implementing ISO/9000. In *ASQC Quality Congress Transactions* (pp. 1091–1097). Milwaukee: American Society for Quality Control.

Radford, G. S. 1922. *The control of quality in manufacturing*. Ronald Press Company.

Sawin, S. D., and S. Hutchens, Jr. 1991. ISO-9000 in operation. In *ASQC Quality Congress Transactions* (pp. 914–920). Milwaukee: American Society for Quality Control.

Sayle, A. J. 1988. ISO 9000—Progression or regression? *EOQC Quality* 1:9–13.

Sayle, A. J. 1992. Audits—The key to the future. *1st Annual Quality Audit Conference*, ASQC, 27–28 February, in St. Louis, MO.

Searstone, K. 1991. Total quality management: BS 5750 (ISO 9000, EN 29000). *Total Quality Management* 2 (3):249–253.

Shewhart, W. A. 1927. Quality control. *Bell System Technical Journal* VI.

Shewhart, W. A. 1931. *Economic control of quality of manufactured product.* New York: D. Van Nostrand Company, Inc. Republished by American Society for Quality Control in 1980.

Sjoberg, A. 1987. 1985 Revision of Z 299: Quality assurance program standards—Impact on effectiveness. In *Proceedings of the EOQC Annual Quality Conference* (pp. 283–296), 1–5 June, in Munich, Germany.

Stephens, K. S. 1979. *Preparing for standardization, certification, and quality control.* Tokyo: Asian Productivity Organization.

Stephens, K. S. 1994. ISO 9000 and total quality. *Quality Management Journal* 2 (1):57–71.

Stephenson, A. R. 1991a. Management systems for quality—A bonus for TQM? In *Proceedings, 3rd Conference of Asia Pacific Quality Control Organization* (Paper We A 01), 18–22 March, in Auckland, New Zealand.

Stephenson, A. R. 1991b. Auditor registration and the accreditation of certification bodies in the U.K. In *Proceedings, 3rd Conference of Asia Pacific Quality Control Organization* (Paper Th C 06), 18–22 March in Auckland, New Zealand.

Tsiakals, L. J. 1994. Revision of the ISO 9000 standards. In *ASQC Quality Congress Transactions* (pp. 873–881). Milwaukee: American Society for Quality Control.

Wadsworth, H. M., K. S. Stephens, and A. B. Godfrey. 1986. *Modern methods for quality control and improvement.* New York: John Wiley & Sons, Inc.

Western Electric Company. 1956. *Statistical quality control handbook.* Easton, PA: Mack Printing Company.

Zuckerman, A. 1994. EC drops ticking time bomb—It could prove lethal to the ISO 9000 community. *Industry Week* May 16:44–51.

4

Aseptic Processing of Healthcare Products— A Pending ISO Document

Michael S. Korczynski
James Lyda

INTERNATIONAL ORGANIZATION FOR STANDARDIZATION

The International Organization for Standardization (ISO) was founded in 1946 by 25 national standards associations. It is the largest of the international voluntary groups involved in technical cooperation within industries in every area of technology. Although ISO is nongovernmental, more than 70 percent of its members are the official standards bodies of various nations (i.e., governmental agencies or bodies incorporated by public law).

The ISO has 212 technical committees (TCs) and more than 2,800 subcommittees and working groups (WGs). At this writing at least 9,652 ISO Standards have been published. The productivity of the ISO continues to increase. A substantial proportion of its increased activity is attributable to harmonization among the nations belonging to the European Union (EU). Twelve countries are members of the EU and four may be added in 1995. The EU will represent 320 million people sharing institutions and policies. As a reference point, 1988 statistics indicate that the U.S. population was 246 million. Therefore, on the basis of population, the EU is approximately 20 percent larger than the United States and 39 percent larger than Japan. The EU contains 2,000 pharmaceutical firms. In 1992 more than 40 percent of the

81

new chemical entities launched worldwide originated in European laboratories.

The official U.S. member body of the ISO is the American National Standards Institute (ANSI), which appointed the Association for the Advancement of Medical Instrumentation (AAMI) to administer the Secretariat of TC 198 because the AAMI has generated many documents relating to sterilization technology that have become American national standards. Both AAMI and ANSI are accredited standards-generating bodies within the U.S.

When the U.S. is a member of an ISO Technical Committee such as TC 198, then U.S. nationals or their representatives can be appointed to a Technical Advisory Group (TAG). The TAG manages and must address U.S. interests before the Technical Committee and oversees U.S. participation in the Committee. Also, members of the TAG are instrumental in deciding what WGs will be formed under the umbrella of the parent ISO-TC.

The member nations of the ISO-TC 198 are voting nations and observer nonvoting members, as shown in Figure 4.1.

European Committee for Standardization (CEN) and ISO cooperation was acknowledged in the Lisbon Agreement of 1989. The most important provision of this agreement is that information from the ISO is shared with the CEN. In other words, the organizations have agreed to actively communicate. From a European viewpoint this means that the ISO has recognition and credibility. The Vienna Agreement (1991) acknowledged that ISO documents could be adopted by the CEN as CEN documents, especially where the subject matter has not been adopted in any manner by CEN technical documents.

The CEN presently includes 18 nations (12 voting) and provides harmonization for the nations of EU. CEN TC 204 addresses sterilization of medical devices, and five WGs are associated with the technical committee. Some of CEN TC 204's activities parallel activities within the ISO. For example, there are TCs within CEN 204 addressing ethylene oxide (EtO) sterilization, radiation, and steam. The EU will follow CEN documents and CEN can adopt what they consider the best portions of ISO documents.

The Institute for Environmental Sciences (IES), through ANSI, has been appointed Secretariat of ISO-TC 209, which will develop standards related to clean rooms and associated controlled environments. The IES is the same organization that wrote, under U.S. federal contract, the 209 series pertaining to particulate technology and clean rooms. Mr. Richard Mathews, current president of IES and CEO of Filtration Technology, Inc., is the chairman of ISO-TC 209.

Figure 4.1. ISO-TC 198 Membership

Voting Members (Participating members)	Nonvoting Members (Nonparticipating members)
Austria*	Argentina
Belgium*	Australia
Brazil	Colombia
Canada	Czech Republic
China	Finland*
Denmark*	India
Egypt	Israel
France*	Malaysia
Germany*	Poland
Ireland*	Saudi Arabia
Italy*	Slovakia
Japan	Switzerland*
Netherlands*	Tunisia
Norway*	Turkey
Philippines	Yugoslavia
South Africa	
Spain*	
Sweden*	
Thailand	
United Kingdom*	
United States of America	

*CEN Members

OBJECTIVE AND HISTORICAL CONTEXT

Shortly after the formation of ISO-TC 198, WG-5 of ISO-TC 172 envisaged a need for an ISO standard concerning the validation of aseptic filling processes for contact lens solutions. The need to expand the scope of activity beyond contact lens solutions became rapidly evident. Therefore, the initial proposal for work that was ratified by

ISO-TC 198 membership encompassed the aseptic processing of "sterile healthcare products."

Appointment and Role of Convener

The principal author (as an appointed member of the USTAG for ISO-TC 198) was appointed by the AAMI and ratified by the Central Secretariat of the ISO, and ISO-TC 198 member nations as convener of WG-9 pertaining to aseptic processing of healthcare products.

The convener must have technical knowledge related to the subject of the WG, and frequently must be able to negotiate a technical consensus among multinational participants on the WG. Further, the WG convener must report to the parent ISO-TC during plenary sessions, and describe technical process and issues that require potential resolution. Resolutions passed by the TC pertaining to any specific WG must be communicated to the WG by the convener and must be adhered to by the WG.

Role of the PDA

To add to the complexity of the ISO committee structure, when the U.S. becomes a member of an ISO-TC, a Subcommittee or Subtechnical advisory group (SubTAG) may be formed, composed of technical experts from the U.S. or representing U.S. interests. In the case of ISO-TC 198 WG-9 activities, a U.S. SubTAG for ISO-TC 198 WG-9 was formed. The chairperson of the U.S. SubTAG (principal author), in conjunction with the AAMI, recommended that a staff member of the Parenteral Drug Association (PDA, an international association for pharmaceutical science and technology) assist the convener in U.S. SubTAG administrative matters in lieu of an AAMI representative. The function of an administrator is to serve in a clerical and document processing capacity to facilitate the convener's workload. Mr. James Lyda, Vice-President of Internal and External Regulatory Affairs, PDA, served in this capacity. This was a new role for the PDA and presented an opportunity to be more active on a global basis relative to pending and future international guidelines and standards. Figure 4.2 is presented for clarification of the reporting structure within ISO-TC 198.

JUSTIFICATION

The U.S. Food and Drug Administration (U.S. FDA) proposed a requirement that aseptically filled drugs be terminally sterilized unless data indicated that the drug or the container was unstable to terminal

Figure 4.2. Overview of ISO-TC 198 Interactions

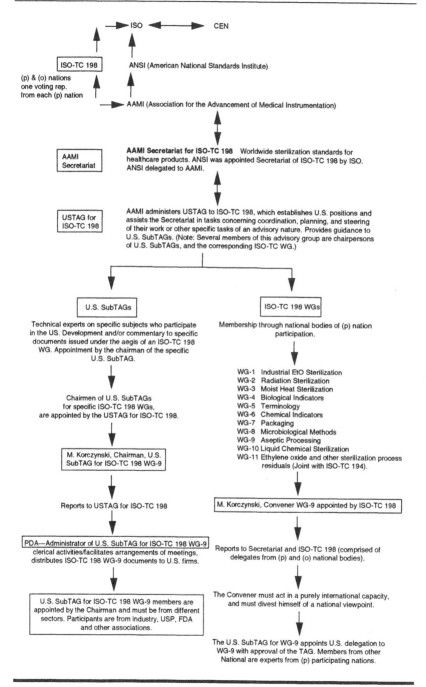

sterilization (1991). Aseptically processed drugs are recognized as necessary, but presently lack the defined assurance of sterility that exists for drugs that are moist heat sterilized in the final container. Therefore, both within the U.S. and abroad, there is a sense that new, emerging aseptic processing methods and the emphasis to terminally sterilize product will mandate requirements to improve the currently practiced or conventional aseptic processing of healthcare products.

SCOPE

The scope of the ISO aseptic processing document describes what the international standard is intended to include and its potential application. The scope of the ISO-TC 198 document can be paraphrased as follows:

Healthcare products that are labeled "sterile" have to be prepared using appropriate and validated methods. ISO-TC 198 has prepared standards for terminal sterilization of healthcare products by ethylene oxide (ISO 11135), irradiation (ISO 11137), moist heat (ISO 11134), and by liquid chemical sterilants (ISO-DIS 14160). However, for some healthcare products, aseptic processing can be used to produce a sterile product. There are two distinct situations in which aseptic processing is applied.

1. The aseptic preparation and filling of solutions.

2. The aseptic handling, transfer, and packaging of solid products that cannot be terminally sterilized in their final containers.

The processes, programs, and procedures relative to aseptic processing described within ISO-TC 198 WG-9's document pertain to such products.

The reader needs to be aware that an ISO document is a voluntary standard. This voluntary standard does not supersede or replace national regulatory requirements such as Good Manufacturing Practice (GMPs) and/or compendial requirements that a particular nation may be governed by or compelled to follow. Within this document the definition of healthcare products is a term encompassing medical devices, medicinal products (including pharmaceuticals and biologicals) and in vitro diagnostics. This is one of the terms that has been defined in the ISO-TC 198 WG-5 document that defines ISO-TC 198 sterilization terms.

A recent PDA survey (Agalloco and Akers, 1993) indicated that 80 percent of the U.S. sterile products filled are small-volume parenterals (SVPs) with a fill volume of less than 100 ml. About 75 percent of the SVPs are filled aseptically, and it is to the processing of these items that the standard primarily addresses. The number of aseptically filled healthcare products will undoubtedly increase as an eventual conversion from classically synthesized to biotechnology-derived therapeutic drugs occurs.

Important Areas of ISO Document

The aseptic processing document contains many requirement statements and describes guidelines for aseptic processing from the product batch stage to the filling of the finished product. A few of the major topics are *facility design, personnel training, microbiological environmental monitoring, and media filling procedures.*

Facility Design and Features

The document does not provide detailed design of rooms and air filtration systems. The CEN has a TC charged with the task of developing a clean room/contamination control standard. Also, ISO TC 209 is addressing this subject. The ISO-TC 198 document includes (1) location of aseptic processing area, and (2) facility design review, as requirements.

Batch (Manufacturing) Records

Batch records provide an audit trail for composition, formulation, and component specifications associated with the product and the final container.

The consensus of WG-9 is that one should have a strong data trail referencing specifications associated with the product or the process. The example, the actual environmental control data are not needed in the batch records, but the specifications and/or protocols that are followed for routine monitoring are needed. Actual data can be maintained within the quality assurance (QA) laboratory.

According to the U.S. FDA, batch records must describe the process testing and controls. Annual reports should contain information pertaining to product quality, including process problems, deviations, and complaints. The initial Drug Master File (DMF) for submission must document product process and analytical controls, and manufacturing controlled systems required for product sterility in detail.

Microbiological Environmental Control Program

The aseptic processing area (APA) shall be routinely monitored for the presence of microorganisms. Samples shall be collected from critical zones in which components and product are exposed to the environment, such as filling zones, filling rooms, and proximity to the product fill. These areas shall be monitored during each shift. Support areas outside the processing zone shall be routinely monitored, but may be monitored on a less frequent basis than critical areas.

Microbial Identification

The environmental control program shall include periodic identification of the recovered microorganisms (isolates). At least genus identification should be considered, with full speciation when corrective actions are necessary.

Frequently during discussions concerning environmental monitoring, it is asked whether the environment is examined for the presence of yeast and molds. It appears that many people conduct a yeast and mold monitoring program on a monthly basis. Regulatory people ask whether anaerobic monitoring is conducted. A recent U.S. PDA survey (Agalloco and Akers 1993) indicates that approximately 93 percent of companies conducting media fills use soybean-casein digest (SCD) broth or medium, which is not a conventional medium for strict anaerobes. Thirty percent of the companies in the PDA survey use an anaerobic organism to determine the growth support properties of the media-fill culture medium. This suggests that anaerobic monitoring during media-fills is not a general practice. It is generally considered that media fill monitoring for anaerobes is unnecessary unless environmental sampling demonstrates the presence of anaerobes.

Use of Quantitative and Semiquantitative Microbiological Air Sampling Techniques

Quantitative air sampling, using volumetric sampling methods, and semiquantitative air sampling (e.g., settling plates) shall be used to evaluate the microbiological quality of processing zones. Within the U.S. there appears to be a preferential use of volumetric air sampling such as slit-to-agar (STA) or centrifugal air sampling rather than wide scale usage of settling plates. The United Kingdom Medicines Control Agency (UKMCA) in past years has advocated the use of settling plates. Settling plates are now experiencing wider usage within Class 100 areas, in locations where it is difficult to place a dynamic air sampler. A perceived advantage of settling plates is that they permit continuous monitoring.

Originally, the U.S. Pharmacopeia (USP) proposed to classify clean room designations using a microbial action level of 0.03 microbes per cubic foot, or less than 1 microbe per cubic meter required for Class 100 areas. This would have required a sampler that could sample for at least 35.3 minutes to assure that 35.3 cubic feet (one cubic meter) of air had been sampled. This leads to issues of validation methodologies for the collection of air samples and the time required on a practical basis to collect such samples. Jensen et al. (1992) evaluated seven dynamic air samplers. Some samplers permitted the recovery of a wider range of particles (from submicron to 10 μm diameter) than other air samplers. Much discussion has occurred within the PDA and USP concerning means and ways to validate air samplers for the low-level detection of microbes related to the monitoring of clean rooms.

Initial Establishment of Sampling Sites

During initial facility line fills and prior to the completion of line validation, one might develop a grid pattern for the room or area. Sampling might be conducted daily for at least 3 weeks at numerous sampling sites. This permits the establishment of preliminary alert and action levels on a statistical basis. Alternatively, one may use suggested action levels for environmental sampling that appear in the literature. Sampling sites with the highest counts and those close to or in direct contact with the product should be included in the program. Based on historical data and the assessment of environmental counts, the firm should reserve the right to modify alert and action levels. This should be stated in the New Drug Application (NDA) or registration submissions.

Environmental Monitoring Sampling Locations

The sampling sites that are selected for each environmental control program should be at the discretion of each facility. Test site selection should include the following:

- Validation data

- Historical Data

- Potential for product quality impact

- Inaccessibility or difficulty in routine decontamination

- Microbial dispersion patterns (e.g., personnel traffic, air flow, etc.)

- Potential for microbial proliferation during production

In evaluating the results of environmental monitoring, the number of microorganisms obtained by various test methods should not be considered absolute values. The purpose of environmental monitoring is to be able to trend the microbial and particulate content of the manufacturing environment. A direct correlation between environmental counts from air and surfaces and product sterility cannot always be clearly demonstrated. Single environmental microbial excursions are not necessarily representative of out-of-control conditions. However, if a microorganism or microorganisms frequently occur or are detected in the environment, specific attempts should be made to determine the source of such microbes, and to identify them.

Alert and Action Levels

Development of Alert and Action Levels.   Alert and Action levels shall be developed for each sampling site within an APA.

Alert and action levels shall be established during the installation (IQ) and performance qualification (PQ) of APA operations. Action levels should be established for each area of environmental monitoring. Action levels should be based on statistical analysis of numerous sampling values and may be based on a percentile or a percentage factor. For example, historical data could be rank number ordered. Everything above a certain percentile, for example 95 or 98 percentile, could be considered an action level. Currently, no appropriate model exists in the literature concerning how to establish alert and action levels. This is a topic that could be addressed by a PDA TC.

Review of Data.   The results of each day's environmental monitoring shall be reviewed in a timely manner against the alert and action levels established for the facility. However, this is not a requirement to review data daily.

Environmental Monitoring Trend Analysis.   Environmental data shall be reviewed at regularly scheduled intervals and if performance problems are observed, an investigation must be initiated. Dr. Ron Tetzlaff (formerly of the FDA) advocated computer-generated environmental trending analysis. His instruction to field inspectors was to examine outlier data in order to determine how the firm responded to such data. He believes that computer analysis of environmental control data can show trends even when action levels are not exceeded (Tetzlaff 1992).

When action levels are exceeded repeatedly, investigative follow-up action is required. Included in such investigative action should be follow-up testing, designed to locate the source of the problem and demonstrate that the area is once again under control.

Also, written investigative procedures shall be available and include determination or review of the following:

- Data to be collected

- Extent of the problem

- Import on product or environmental control

- Need for product quarantine

- Whether environmental control has been attained

- Follow-up testing

- Notification of affected responsible personnel

A single overaction level event in a Class 100 area presents difficult follow-up actions and product release decisions. How does one react when 1 microbial count is found on a filling nozzle or 100 counts are found on the line conveyor before capping of the vials when all other environmental monitoring values are satisfactory? There appears to be an emerging trend to consider overaction level occurrences in Class 100 areas as absolute events that require resolution prior to the release of product that was manufactured during the sampling.

Media Fills

The media fill or simulated product filling portion of the ISO-TC 198 document is perhaps one of the most important sections of the document. This section also contains certain requirements that heretofore have not been consistently applied within, for example, the U.S. Within the U.S. during the last few years, regulatory inspectors have occasionally challenged firms relative to the adequacy of any positive control units occurring in a media fill run of only 3,000 units. The concern was that any positive units occurring in a media fill run of 3,000 would demonstrate less than 95 percent confidence that a 0.1 percent or less contamination rate existed in the media fill. Further, ISO-TC 198 WG-9 realized that many member countries on the working group were routinely filling in excess of 3,000 units during routine production media fills. If one accepts that product batches, especially with high speed filling lines, can include thousands of filled units, it is reasonable to conclude that the aseptic filling of 5,000–6,000 units will result in a more thorough evaluation of the process for a given media fill run. Also, with intensive training of line personnel, the removal of people from intervention during processing, placing the operation in barrier isolators, using form-fill-seal products, and decontaminating areas with hydrogen peroxide (H_2O_2) vapor are measures that move aseptic processing to further improvement. It is with these thoughts in mind that WG-9 members upgraded the requirements for media fills.

The use of statistics in assessing the potential contamination rate associated with a media fill run has been the subject of some discussion. Currently, the practice is to use a direct mathematical model to calculate the possible or projected rate of contamination in a media fill. Therefore, using the direct mathematical model, one could calculate that 6 positive units in a media fill of 6,000 units would still permit an acceptable contamination rate of 0.1 percent. Working Group 9 believes that it was time to refine the approach of calculating a possible rate of contamination in a media fill run. It selected the Poisson distribution model to assess the lower and upper 95 percent confidence limits for failed units in media fills of given batch sizes.

It was concluded that while the use of 3,000 media filled units is adequate for routine reevaluation of product filling lines, there will be cases where it would be more desirable to media fill greater than 3,000 units. Working Group 9 also believed that there were cases whereby a short duration media fill of only 3,000 units would not effectively include the majority of interventions and activities that could occur during large-scale production runs.

During the drafting of the document, our colleagues stated that many smaller firms are manually filling small batches during aseptic processing. This is especially true if clinical batch sizes are considered. Therefore, WG-9 included requirements for batch sizes less than 500 units. However, batch sizes that small still require a sizeable number of units in media fill during initial line qualification. This can be achieved by conducting at least 10 accumulative media fill runs. The repetitive nature of line qualification was thought to be important in qualifying extremely small batch operation because of the large amount of manual intervention frequently associated with the filling of clinical batches.

During initial line validation for larger batch sizes, one positive unit in any of the three required validation runs is considered an alert level. However, if in the series of three repetitive media fill runs, one positive unit occurs in each of two runs or two or more positive units occur in any one media fill run, the entire line validation must be repeated. For purposes of periodic reevaluation of filling lines, a table is provided in the document that identifies the lower acceptable 95 percent confidence interval value to assure that a 0.1 percent contamination rate has not been exceeded in the media fill run.

One of the most important elements of the media fill run is to run long enough to capture or include many of the planned and unplanned stoppages and interventions that will occur during actual product filling. The document also reviews the requirements for conducting periodic processing line reevaluations using media fills on a semi-annual basis.

In-Process Filtration

A documented filter evaluation program shall be established prior to filter acceptance and should include but not be limited to the following: *filter and filter equipment evaluation, physical integrity performance qualification, and procedures for filtration.*

The effectiveness of the filter or filter equipment shall be evaluated. Filters intended for retention of bacteria (sterilizing filters) shall be evaluated by an appropriate and defined bacterial challenge, or have evidence of such from manufacturers. Absorption or adsorption of drug or preservative shall be evaluated during the filter validation process.

In-process determinations of the physical integrity of a process filter shall be conducted. The ability of the filter-housing to maintain integrity in response to sterilization and gas or liquid flow (including pressure surges and flow variations) shall be determined. Also, post-filtration integrity evaluations shall be performed.

During the drafting of the aseptic process document, it became evident that a technical section pertaining to filtration could represent a substantive stand-alone document. This was also presented to ISO-TC 198 for consideration. Eventually, an independent document or an added detailed new section may be added to the current draft document pertaining to process filters and filtration.

Cleaning in Place

The ISO document contains guidance information on this subject, not requirements.

Documentation will vary based on equipment complexity. Critical cleaning steps, flowcharts, and piping diagrams for the identification of valves in cleaning procedures should be documented. The length of time between the end of processing in each cleaning step should be controlled. Drying of residuals will affect the efficiency of cleaning. Relative to analytical methods, the specificity and sensitivity of the analytical method used to detect residuals or contaminants should be determined. Thin layer chromatography assays are not recommended because of the lack of quantitation. Concerning sampling, swabbing is frequently used for sample areas that are physically difficult to reach and to remove dried material. However, one must be aware that certain adhesives or agents in swabs may impact the ability to remove the residual.

Once the cleaning process is validated, one can use an indirect test method such as conductivity testing, pH, or redox potential to determine the amount of residual. Rinsing of the equipment should persist until no traces of residual are found in the rinse.

A placebo batch may be tested for residual contamination. Limits established by the industry vary widely. A safety or toxicology assessment is made concerning the possible amount of contamination that could enter a dose of product delivered to the patient. A safety factor then is applied to that maximum level. The residual dose level (RDL) is the maximum amount of material allowed that would not result in patient pharmacological or toxicological effects. The RDL may be, for example, at least 100 times less than the minimum daily dose of the drug.

Lyophilization

Lyophilization is the removal of solvents by sublimation and is a drying method. It can be applied to filled products in ampoules and vials, as well as to bulk material. Lyophilization of bulk material requires additional processing, such as milling and sieving, before filling. The lyophilization of sterile products introduces an extra step in the aseptic process.

Media Filling for Validation of Lyophilized Product Lines

Aseptic processing operations that involve the handling of lyophilized products may result in an increased amount of manual manipulation and human intervention. Therefore, lyophilized products can present increased difficulties and challenges during media fills. In lyophilization operations product is moved through a series of transitions between Class 100 and Class 10,000 areas. Media fill tests should simulate the normal product environment transitional flow.

Phases of lyophilized product operations that should be incorporated into media fill runs include filling of product, loose capping, covering and transportation to a lyophilizer, uncovering, placement in the vacuum chamber, and seating of stoppers by compression of tray shelves or trays. During the process manual intervention may occur including the removal of tray tops or bottoms in the chamber when loading vials and ampoules, connecting thermocouples or thermoprobe junctions, and unloading the chamber.

Human intervention in lyophilization processes is increased if bulk product is lyophilized rather than the lyophilization of a unit vial or ampoule. In such cases entire trays of material are lyophilized and must be subjected to a milling process to render the product a homogeneous powder of uniform consistency prior to further aseptic processing. Such processing further exposes the product and can influence the sterilization assurance of the product.

Qualification or periodic reevaluation of lyophilization operations using media fills involves the use of various complicated tasks. Media fills may be done by filling sterile broth into vials or ampoules and then slightly capping them in anticipation of lyophilization. The filled broth solution should be transferred and loaded into the lyophilizer under normal product conditions. The draft ISO document states that the broth lyophilization cycle should include evacuation and release of vacuum, as well as exposure of the broth for the same period of time as the normal product. However, actual lyophilization or freezing of the solution should be avoided.

Sterility Testing of Finished Product

An investigation should be conducted when sterility testing positives occur. A correlation between types of microorganisms in the manufacturing environment sterility test room, and the sterility test positive isolate shall be sought. The recovery and identification of the same genus of microbe from the sterility test(s) and from the manufacturing environment would result in the rejection of the batch production lot. It should be noted that many firms are not retesting aseptically filled lots that have failed the initial sterility test, especially if a barrier isolator is used for sterility testing.

INTERPRETATION AND USE OF THE DOCUMENT

The requirements of final drug dosage forms generally appear in compendial monographs of many nations. The compendial monographs take precedence over requirements stated in the ISO document, and for this reason the scope section of the ISO document contains a disclaimer. An ISO document is a voluntary standard that each nation need not follow. Certain regions or trading federations, such as the European Community, need not adhere to the document.

What follows are predictions of how the ISO aseptic processing document will be used within the United States and abroad.

There are five FDA participants, each representing a different division within the FDA, on the US SubTAG that prepared the initial draft document. The fact that the FDA representatives participated on the US SubTAG to ISO-TC 198 WG-9 (composed of individuals from government, academia, industry, the USP, and other organizations) suggests that the FDA may, consider adapting portions of the ISO document into validation guidelines for aseptic processing.

The FDA has not adopted ISO 9000. However, ISO 9000 can be viewed as a framework for the more product-oriented GMPs and GLPs, the adverse reporting system, and other regulations; ISO 9000 does not conflict with FDA requirements.

Good Manufacturing Practices pertain to processing and product information. Good Laboratory Practices are developed specifically to assure the quality of data. The GMPs and GLPs are not intended to address overall quality management systems as found in the ISO 9000 series documents (Schwemer and Lynch 1993).

The USP has an interest in aseptic processing. Detailed information concerning requirements and guidelines are not in USP 23. However, the 1990–1995 USP Committee of Revision for Pharmaceutical Microbiology worked on a proposal to include new information on aseptic operations. The USP convened an open conference on water, microbiology, and particulates (Colorado Springs, July 1993). The consensus of the individuals present was that the USP proposal to use microbiological methods to classify clean rooms and develop cleanroom standards, should be used only for the periodic environmental monitoring of the APA. The consensus was that the classical method of particulate technology assessment should be used for the classification of clean rooms (Korczynski 1992).

EXISTING GUIDELINES ON ASEPTIC PROCESSING

No requirement guideline document on aseptic processing at present exists on the European continent. It is anticipated within the ISO that this aseptic processing document will be adopted by the EU nations as well as, perhaps, by the U.S. and Japan. It is also anticipated that other noncompendial nations may use the document for guidance. The ISO standards can eventually be adopted as American national standards through approval of the ANSI. Currently, very few aseptic processing guidelines exist in the U.S. although many articles and manuscripts exist concerning the subject.

The FDA did publish guidelines in 1987 relating to the aseptic processing of pharmaceutical products. (The FDA issued a proposed update of the 1987 document in 1991.) The FDA guidelines are less detailed than the ISO document. The most significant portions of the FDA guidelines were presented at several regional meetings held within the U.S. by the FDA. The purpose of the meetings was to describe the necessary validation requirements for submitting an NDA pertaining to moist heat sterilized or aseptically processed human or animal drug dosage forms. This information (FDA, 1993) was more explicit and detailed than the original FDA 1987 guideline on aseptic processing (FDA 1993). At this time there is no published American national standard on aseptic processing. However, some sections and proposed sections within the USP pertain to aseptic processing.

The only published European aseptic processing document that exists at this writing is from the International Pharmaceutical Federation (FIP) Committee on Microbial Purity (1991). This article was the fifth joint report of the section of official laboratories and drug control services in the section of industry pharmacists of the FIP. The two sections created a joint committee on microbial purity headed by Professor Dony, President of the National Hygiene Counsel of Belgium. Members were from Sweden, the UK, Germany, the Netherlands, Belgium, Poland, Switzerland, and France.

Provisions for Revision of the Standard

The state of the art of aseptic processing is in constant revision. Working Group 9 fully realized that new emerging technologies would be more widely used in the manufacturing of healthcare products. Within the text of the document, a statement appears that qualifies that the document address the validation and control of aseptic processing conducted in conventional cleanrooms. Many of the principles and programs described may also apply to aseptic processes conducted in barrier and isolator systems. However, requirements and guidance relating to barrier and isolation systems may need to be considered in the future when wider practical experience and knowledge are available. Therefore, WG-9 generally recognizes that additions to the document may occur in the future.

REFERENCES

Agalloco, J., and Akers, J. 1993. Current practices in the validation of aseptic processing 1992 (PDA survey). Technical report 17. *J. Paren. Sci. Tech.* 47:Supplement S1.

Food and Drug Administration, Center for Drugs and Biologics and Office of Regulatory Affairs. June, 1987. *Guideline on sterile drug products produced by aseptic processing.* HFN-320, FDA, 5600 Fishers Lane, Rockville, Maryland, 20857.

Food and Drug Administration. 1993. Information to be submitted to human and veterinary drug applications. Paper presented at the Sterilization Process Validation Conference, 14–15 January, in Northbrook, IL.

ISO-TC 198, N171. 1993. Aseptic processing of healthcare products. First committee draft (ISO-CD 13405).

IPF Committee on Microbial Purity. 1991 Validation and environmental monitoring of aseptic processing. *J. Paren. Sci. Tech.* 44 (5):272–292.

Jensen, P. A., W. F. Todd, G. N. Davis, and P. V. Scarpino. 1992. Evaluation of eight bioaerosol samplers challenged with aerosols of free bacteria. *Amer. Ind. Hygiene Assoc. J.* 53:660–667.

Korczynski, M. S. 1992. Response to the new USP proposal: Microbial evaluation and classification of cleanrooms. In *Proceedings of the PDA/IES joint conference on cleanrooms and microenvironments.* Institute for Environmental Sciences (IES), Mount Prospect, IL (ISBN # 1-877862-03-7).

Schwemer, W. L., and M. A. Lynch. 1993. ISO 9000 Policy implication for FDA. *J. Paren. Sci. Tech.* 47 (3):101–113.

Tetzlaff, R. F. 1992. Investigational trends: Cleanroom environmental monitoring. *J. Paren. Sci. Tech.* 46 (6):206–214.

ABBREVIATIONS AND GLOSSARY

AAMI   Association for the Advancement of Medical Instrumentation

Action level   Level of microbes in air, on product contact surfaces, or in product that when exceeded, requires specific action as a matter of policy

Alert level   Level of microbes in air, on product contact surfaces, or in product that when exceeded, alerts appropriate personnel that microbial count values may be encroaching upon action level values

ANSI   American National Standards Institute

APA   Aseptic processing area

CEN   European Committee for Standardization (= harmonization)

CEN 204   European Committee for Sterilization of Medicinal Devices

CEN-TC 243   European technical committee 243 working group 3, assigned a clean-room/WG-3 contamination control standard

CENELAC   European Committee for Electrotechnical Standardization

CEO   Chief executive office of a firm

DMF   Drug Master File (U.S. practice)

EU   European Union

FIP   International Pharmaceutical Federation

FTM   Fluid thioglycollate medium, a bacteriologic medium that supports the growth of anaerobes and microaerophiles

HEPA   High efficiency particulate air filter (n.): a system for the removal of fine particles (<3 μm diameter) from air

IES   Institute of Environmental Sciences

ISO   International Organization for Standardization

ISO 9000   International standards for quality management

ISO TC 198   ISO Technical Committee for development of worldwide standards for the sterilization of healthcare products

ISO TC 209   ISO Technical Committee for development of worldwide standards for the sterilization of healthcare products.

NDA   New Drug Application (filing for FDA approval to commercialize a specific drug or drug related product)

PDA   Parenteral Drug Association (Bethesda, MD): an international association for pharmaceutical science and technology

SCD   Soybean-casein digest, a broth medium commonly used to support the growth of aerobic and sometimes microaerophilic microorganisms; most often used in media fills and in sterility testing

Secretariat   The national standards body of a specific country that reports to ISO and has responsibility for the overall administration of the Technical Committee and its respective working groups

Sub-TAG   Technical experts comprised of nationals from the same national as the Secretariat that participate in the critique and review of documents generated by an ISO WG

TAG   Technical Advisory Group to the Secretariat

TC   Technical Committee

USP   United States Pharmacopeia, or U.S. Pharmacopeial Convention, Inc.

WG   Working Group comprised of multi-national participants

ACKNOWLEDGEMENT

The authors wish to thank the following individuals who are active participants in ISO-TC 198 WG-9 and the U.S. SubTAG for WG-9 activities.

ISO-TC 198 WG-9 Participation

W. Cobbett, Dr. Trevor Deeks, and D. P. Hargreaves (United Kingdom); Chuicki Ishiziki, Dr. Masayoshi Furuhashi, Yasuhiro Majori, Dr. Tsuguo Sasaki, Masayoshi Mishiyama, and Yoshito Hashimoto (Japan), Dr. Norman Franklin, Dr. Klais Haberer, G. Herfurth, and Georg Robling (Germany), Carol Mason, Margaret Muise, and P. Jefferson (Canada); Pauliene Mar and J. G. A. Mathot (the Netherlands); Matt Johansson (Sweden); Dr. Henri Van Rensburg (South Africa); Dr. M. DeMeo (Italy); James Lyda, William O'Connell, Joyce Adylett, Dr. Barry Garfinkle, and William Young (U.S.A.).

U.S. SubTAG for ISO-TC 198; WG-9 participation

James P. Agalloco, Dr. James Akers, Joyce Adylett, John Brau, Dr. Carl Bruch, Dr. Roger Dabbah, Pamela Deschenes, Dr. Barry Garfinkle, Beth Anne Grove, Dr. David Hussong, Sue Sutton-Jones, Dr. Patricia Leinbach, Mr. James Lyda, Dr. Anthony Parisi, William O'Connell, Dr. Raymond Shaw, Joseph Stojak, and William Young.

Authors' Note

1. It should be noted that the final published ISO aseptic processing document may in many areas differ from this published text since modifications continue to be made to the draft document by WG-9.

2. Any publication royalty income will be donated to the AAMI and the PDA to support the continuation of international standards participation and activties.

5

Validation of Aseptic Processes

Vijai Kumar
Ram Murty

The term *validation* is not defined under cGMPs (Federal Food, Drug, and Cosmetic Act; 21CFR211). However, many definitions have been offered and the one officially published, as a guidance, states: Validation is establishing documented evidence which provides a high degree of assurance that a specific process will consistently produce a product meeting its predetermined specifications and quality attributes (FDA 1987).

Several sections of the cGMP regulations state validation requirements in more specific terms. Excerpts from some of these sections are as follows:

Section 211.110, sampling and testing of in-process materials and drug products states:

. . . control procedures shall be established to monitor the output and VALIDATE the performance of those manufacturing processes that may be responsible for causing variability in the characteristics of in-process material and the drug product.

Section 211.113, control of microbiological contamination states,

appropriate written procedures, designed to prevent microbiological contamination of drug products purporting to be sterile, shall be established and followed. Such procedures shall include VALIDA-TION of any sterilization process.

The manufacturer of sterile products should evaluate all factors that affect product quality when designing and undertaking a process validation study. These factors may vary considerably with different products and manufacturing technologies and could include many operations. The product should be carefully defined during the research and development phase, in terms of its characteristics, such as physical, chemical performance, reliability, and stability. Acceptable ranges should be established for each characteristic and allowable variations should be expressed in measurable terms. The validity of acceptance criteria should be verified through testing and challenge of the product on a sound scientific basis during the development and initial production phase. Once a specification is demonstrated as acceptable it is important that all changes to the specifications be made in accordance with documented change control procedures.

ELEMENTS OF PROCESS VALIDATION

Prospective validation includes considerations that should be made before an entirely new product is introduced by a manufacturer or when there is a change in the manufacturing process that may affect the product's characteristics, such as identity and uniformity.

Concurrent validation means ongoing prospective validation studies prior to the start of full production. It can also mean gathering and evaluating data from current production batches retrospectively, or requalification and/or revalidation as a result of a significant change in equipment, facilities, batch size, or the manufacturing process.

Retrospective validation: In cases where a product may have been on the market without sufficient process validation, it may be possible to validate the adequacy of the process by examining the accumulated test data and records of the manufacturing process used. Retrospective validation can also be useful to augment initial prospective validation for new products or changed processes.

VALIDATION OF DRY HEAT PROCESSES

Dry heat is often the process of choice for sterilizing items that can tolerate relatively high temperatures and yet might not be adequately penetrated by steam or are damaged by moist heat. Dry heat sterilization processes are generally less complicated than steam, although higher temperatures and/or longer exposure times are required because microbial lethality associated with dry heat is much lower than that for saturated steam at the same temperature.

Validation of a dry heat sterilization process should include both physical and biological tests:

- Equipment installation qualification (IQ)

- Equipment operational qualification (OQ)

- Calibration of sensing, controlling, and monitoring devices

- Verification of thermodynamic characteristics of the sterilizer

- Engineering qualification of the process (PQ)

- Microbiological validation of the process

- Review of test data

- Final certification of documentation

Convection heating is the method of transferring heat in the dry heat sterilization process. The basic equation for convective heat transfer is as follows:

$$q/c = hcA\Delta T$$

where q/c is the rate of heat transfer by convection, BTU/hr, hc is the average unit thermal convective conductance, often called the surface coefficient or heat transfer coefficient, BTU/hr ft^2 °F; A is the area, ft^2; and ΔT is the difference between the surface temperature and the temperature of the fluid at some specified location, °F.

Installation Qualification

- Verify and document that equipment adheres or meets the original purchase specifications. Any exceptions and/or modifications must be supported and documented.

- All utility connections, such as pneumatic, electrical, and HVAC, meet design limits and any codes.

Operational Qualification

Actual operation of the various systems must be individually checked for proper operation. The following is a suggested list that must be checked during this phase of the verification:

- Electrical logic
- Cycle set point adjustability
- Gasket integrity (check for positive/negative pressure seal of all door gaskets)
- Door interlocks
- Vibration analysis (blower check for correct dynamic balancing)
- Blower rotation
- Blower rpm
- Heater elements
- Air balance
- Room balance (ΔP balance is positive from the sterile care to the prep area when one door is opened.)
- HEPA filter integrity
- Belt speed (in case of continuous convection equipment)

Equipment Calibration

There are two modes of equipment calibration, one by removing the piece from the sterilizer and the other calibrated in situ. A combination of both is recommended to achieve an accurate calibration. A metrology protocol must be developed and all calibration should be done in accordance with this protocol. The following items should be considered for calibration:

- Temperature records and thermocouples
- Temperature controllers (preferably in situ)
- Pressure gauges
- Cycle set point switches (preferably in situ)
- Timers
- Velometers

Verification of Thermodynamics of the Sterilizer

In dry heat sterilization an important parameter is the temperature uniformity in the sterilizer. This uniformity can be measured by obtaining the flow rates of the heating medium (i.e., air at the discharge of the sterilization chamber). The flow rate is equal to the product of the velocity and the cross-sectional area at the discharge source:

$$Q = AV$$

where Q is the volumetric flow, cfm; A is the cross-section area of the discharge, ft^2; and V is the velocity of air, ft/minute.

Once the airflow patterns are determined, thermocouples are placed in specific predetermined positions and heat distribution studies are performed on the empty chamber. Repeatability of temperature attainment and identification of cold spots are of most importance. Data obtained from empty sterilizer testing is used as a basis for all future flow-pattern studies. The number and type of tests necessary to demonstrate repeatability is determined from the evaluation of the results obtained.

Process Qualification

Heat-distribution and heat-penetration studies should be performed with thermocouples placed in load patterns that give the greatest amount of temperature information per run. These studies will determine minimum cycle times and the placement of microbial challenges for the microbial validation of the process.

Three replicate heat-distribution and heat-penetration microbiological challenge units (biological indicators and/or endotoxin challenges) are placed adjacent to the thermocouples. The microbiological challenge verifies the thermodynamic parameters of the established cycle.

Cycle parameters are set at the minimal cycle specifications to assure process efficiency. Mechanical repeatability in terms of air velocity, temperature consistency, and sensitivity and reliability of instrumentation must be verified and documented.

Data from the runs are collated and analyzed. The results are judged as acceptable or unacceptable based on the protocol criteria for acceptability.

VALIDATION OF HEAT STERILIZATION

Heat sterilization is a probability function dependent on heat exposure, the number of microorganisms, and the heat resistance of the microorganisms. It is recommended that steam sterilization processes provide a level of assurance of at least 1×10^{-6} probability of survival for terminally sterilized parenterals.

Two basic approaches are used by the industry in developing sterilization cycles. The first approach is to establish cycle parameters based on the number of microorganisms in the product (bioburden) and the heat resistance of those microorganisms. The combination of number and resistance will determine the amount of heat required to provide less than 10^{-6} probability of microbial survival in the product. The probability of survival approach recognizes that the microbial attributes of a product will determine the F_0 required for adequate sterilization. This approach is generally used when developing and validating sterilization cycles for each product.

The second method is an overkill sterilization process that is normally used when the material can withstand heat treatment without adverse effects. Use of the overkill approach will provide assurance of a 10^{-6} probability of survival regardless of the number of microbes and heat resistance. This method offers the advantage of eliminating the need for bioburden and resistance studies in developing sterilization cycles.

Probability of Survival Method

In order to establish that the materials are consistently exposed to sufficient heat, the following two studies should be performed:

1. Laboratory studies to determine the number (bioburden) and resistance of microorganisms associated with the product. These studies will determine the minimum temperature required to obtain a specified probability of a microorganism surviving the sterilization process (probability of survival).

2. Studies in manufacturing equipment to determine

    -   Heat-distribution studies to confirm provision of uniform heating.

    -   Heat-penetration studies to determine cold spots and loading patterns.

    -   The minimum lethality provided by the sterilization cycle through heat- penetration and biological challenge studies.

    -   The reproducibility studies for the sterilization cycle.

Overkill Approach

The overkill approach is utilized when the primary concern is assurance of sterilization and not heat degradation. When using the overkill approach, extremely high F_0 values are generally used. It is desirable that the sterilization cycle provide an F_0 that will result in at least a 12-log reduction of microorganisms having a D value of at least 1 minute. Bioburden and resistance data are not required to determine F_0 values when using this method. Rather, cycle parameters are adjusted to assure that the coldest point in the loading pattern received an F_0 that will provide at least a 12-log reduction of microorganisms having a D value of 1 minute. In order to establish consistent exposure the following studies should be performed:

- Calibration of biological indicators (BIs)

- Studies in manufacturing equipment to determine

  — Acceptability of equipment for providing uniform heating medium.

  — The coldest location within a loading pattern.

  — The minimum lethality provided by the cycle sufficient to provide at least 12-log reduction of microorganisms having a D value of 1 minute.

  — The reproducibility of the sterilization cycle.

D Value

The D value is a quantitative expression of the rate of killing microorganisms. It is the time required for a 90 percent reduction of the microbial population.

D values are experimentally determined by either a colony-count method or by a fraction-negative method. When survival data is determined by colony forming units, a semilogarithmic curve is generated by plotting the logarithm of the number of survivors versus heating time, as shown in Figure 5.1. A straight line is fitted to the data by linear regression according to the following equation, which assumes first order death rate kinetics:

$$y = a + bx$$

where y is the log10 number of survivors at time x; x is the heating time; a is the y intercept at time 0; and b is the slope of the line as determined by linear repression. The D value is the negative reciprocal of the slope.

The D value can be affected by the

- Type of microorganisms used as the biological indicator[1]

- Temperature of the sterilization process

---

[1]Biological indicators are live spore forms of microorganisms known to be the most-resistant organisms to the lethality of the sterilization process. For steam sterilization the most-resistant organisms are *Bacillus stearothermophilus*.

Figure 5.1. Thermal resistance plots of *D* versus temperature, showing slopes equivalent to Z = 10°C and Z = 20°C.

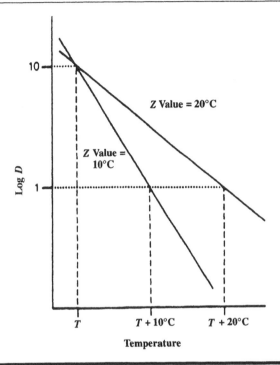

- Surface to which the microorganisms are exposed
- Formulation ingredients and characteristics

Z Value

The Z value is defined as the number of degrees (°F or °C) that are required to change the *D* value by a factor of ten. The Z value of a microorganism is a measure of how heat resistance changes with changes in temperature. It is the Z value that allows integration of the lethal effect of heat as the temperature changes during the heating and cooling phases of a sterilization cycle. The Z value may be determined by the following three-step procedure:

1. Determine the *D* value of an organism at a minimum of three different temperatures.

2. Plot a thermal death curve by plotting the logarithm of *D* on the ordinate of graph versus the temperature, as illustrated in Figure 5.1.

3. Draw a straight line through the data points. The Z value is the change in temperature for the D value to change by a factor of ten. The Z value is also equal to the negative reciprocal of the slope.

Figure 5.2 represents the thermal resistance plot for a Z value of 10°C accepted value for steam sterilization of *B. stearothermophilus* and *B. subtilis* spores. The Z value should be verified for BIs when the indicator is used to measure lethality during a validation cycle.

VALIDATION OF ASEPTIC FILTRATION PROCESS

Aseptic filtration is utilized to remove microbial contamination from the product. Filtration of product under aseptic conditions through a sterilized bacterial retentive membrane of 0.2 μm pore size is a means of rendering the product sterile. The objective of this protocol is to prove through a series of well-defined tests that this process produces sterile product with reasonable assurance. The following general acceptance criteria applies to this process:

Figure 5.2. Survivor curves showing the effect of decreasing the microbial load (A) from 10^6 to 10^2 on the time required to achieve a probability of nonsterility (B) of 10^{-6}.

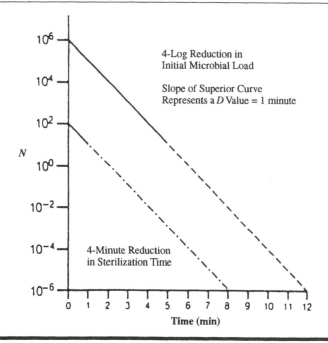

- The filter must qualify as a "sterilizing filter" defined as capable of producing a sterile effluent when challenged with a specific microorganism at a minimum concentration of 10^7 organisms per cm^2.

- The filter and the components of the filtration system must be compatible with the product. The drug product must meet all specifications for the product after being in contact with the materials.

- The filter must remain integral after undergoing sterilization at least as challenging as its standard sterilization process.

- The filter must maintain its integrity after undergoing filtration process with respect to product contact time, pressure differential, and temperature gradient if applicable.

- The filtration regimen must not affect the stability of the product.

Biological Challenge

- The bioburden of the product prior to the filtration process on at least three production batches should be determined and documented.

- The filter manufacturer's data on the biological removal capabilities of the filter should be reviewed and the reports and validation data should be included. An audit of the filter manufacturer should be undertaken to assure the credibility of the data.

Product Compatibility

Samples should be evaluated for product compatibility using the following tests when applicable.

- Appearance
- pH
- Preservative
- Assay
- UV-visible spectra
- Toxicity.

Effects of Sterilization on Filter Integrity

- Wet integrity tests using water for injection (WFI) should be performed using three different filters prior to sterilization.
- Filters should be sterilized, challenging them to sterilization time and temperature.
- Wet integrity tests should be performed using WFI poststerilization and results documented and evaluated.

Effects of Filtration on Filter Integrity

- Wet integrity tests should be performed on three different sterilized filters.
- Filters should be challenged with product contact times, pressure differentials, and temperature gradients at least as extreme as would occur during normal manufacturing conditions.

Product Stability

- Postfiltration stability tests should be conducted as per stability protocol on at least three production batches to determine if the filtration affects product stability.
- Compiled data should be compared with the acceptance criteria and conclusions are drawn as to the acceptability of the validation effort.

VALIDATION OF STEAM STERILIZATION PROCESS

Validation protocols should be prepared to demonstrate the effectiveness of the sterilization cycles to be routinely used for equipment; product loads should be validated using the maximum and minimum load configurations. The following tests are generally used:

- Operational qualification for empty sterilizer chamber temperature distribution and heat-up time should be performed in triplicate for each of the following two cycles: prevacuum with fast exhaust and slow exhaust.
- Load heat-penetration studies should be conducted on both maximum and minimum equipment load configurations to determine heat distribution throughout the chamber load.

- Cycle parameters that are established by heat-penetration studies should be validated for equipment loads. The validation studies should be conducted in triplicate on the two equipment loads that exhibit the longest time to achieve temperature and have the lowest accumulated F_0 values.

- Cycle parameters established for product loads should be validated by conducting the studies on *each* product loads in maximum and minimum configurations. Additional validation studies should be conducted on product loads exhibiting the longest time to achieve temperature and lowest accumulated F_0 values.

The reference test equipment should be in current calibration and, as applicable, traceable to the National Bureau of Standards (NBS). Reference standards indicating instruments range of use, accuracy, date of last calibration, due date, and traceability to NBS should be documented.

During the qualification cycle several thermocouples are positioned in the sterilizer, one of them adjacent to the temperature sensing probe for the sterilizer's recorder. The cycle is repeated for three consecutive runs.

Biological indicators (*B. stearothermophilus* with a *D* value ≥ 1.0 minute) are placed next to each thermocouple tips in the load along with additional BIs placed throughout the load. The exposed BIs should be removed immediately after the cool-down period and incubated for 7 days at 50–55°C along with two positive controls.

Equipment load cycle parameters should be established using PDA *Technical Monograph No. 1* as a guide, specifically Section III, "Overkill Approach to Sterilization Validation." Upon sterility testing no growth should be exhibited of the indicator organisms in the exposed BIs during the 7-day incubation period. Positive controls will show growth.

VALIDATION OF A FREEZE-DRYING PROCESS

Freeze-drying is a means of stabilizing a drug product by freezing it and then subliming the resultant ice, to leave a substantially dry, porous cake of the product. The steps for freeze-drying in vials are outlined below:

1. Trays of prefilled, partially stoppered vials are loaded onto the shelves of a presterilized freeze dryer. Tray bottoms are removed, leaving the vials directly on the shelves.

2. The product is frozen by applying refrigeration to the shelves and cooled below the product freezing temperature for a specified period of time developed for the product.

3. Refrigerant is then circulated through the condensers to lower the temperature below that of the product.

4. The chamber pressure is lowered to a predetermined set point and heat is applied to the shelves to cause the sublimation of ice. This step is called primary drying.

5. After completion of primary drying, the chamber pressure is further lowered to a second predetermined value and the shelf temperature is again raised. This step is called secondary drying.

6. Upon completion of secondary drying, the chamber pressure is returned to atmospheric pressure with filtered nitrogen and the vials are stoppered. The vials are then removed from the freeze dryer.

The validation protocol for the freeze-drying process should include the validation of freeze-drying equipment as well as the freeze-drying process. The freeze-drying equipment validation should include IQ and OQ.

1. Testing should include water load runs of approximately equal to or greater than the condenser capacity of the freeze dryer.

2. Shelf temperature studies must meet ±2–3°C from the mean shelf temperature across all shelves.

3. Leak rate determination performed and acceptance criteria should be established (typically leak rate <2 μm mercury/minute).

4. The steam sterilization process should include the following heat-distribution/penetration studies:

   - All thermocouples must maintain an average deviation of ±1.5°C across the chamber.

   - All thermocouples inside the chamber should maintain a minimum temperature of 121°C during the cycle.

   - A minimum F_0 of 12 times the D value at all thermocouple locations should be maintained.

   - Three consecutive runs meeting the above acceptance criteria are recommended to demonstrate reproducibility.

- The BI (spore strips) must show complete destruction during challenge evaluation.

The freeze-drying process validation can be performed using product placebo filled into the vials. The following process parameters should be monitored:

- Time required to cool product to eutectic point or below

- Length of primary drying (i.e., time required to heat all product to a given drying temperature)

- Temperature depression upon reduction from primary vacuum to secondary vacuum

- Second vacuum set point timer

- Product temperature spread throughout the cycle

- Shelf vs. product temperature differentials

- Freeze-dried placebo vials should be sampled from the top, middle, and bottom of the freeze-dryer chamber and tested for the following characteristics:

  — Fill volume

  — Weight variation

  — Moisture content

  — Appearance of cake

  — Reconstituted product

- Placebo runs should be performed for the minimum and maximum loads (maximum number of vials) and reproducibility should be demonstrated by replicate runs.

- The aseptic filling and freeze-drying process should be validated using media fills demonstrating no more than 0.10 percent contamination.

- Depending on the product, chamber cleaning validation should be carried out specifying acceptance criteria for samples based on assay level of detectability.

VALIDATION OF ASEPTIC FILLING PROCESS

Validation of an aseptic filling process is normally performed using a microbiological growth nutrient medium as a substitute for product to simulate the filling operation. The presterilized growth media is exposed to the same conditions as a normal product run, such as operators, equipment, surfaces, and room environment. The fill should include a minimum of one personnel shift change and a 15–30 minute simulated machine stoppage, when applicable. Actual production filling assemblies are employed onto the filling machine designated for a particular filling room. Environmental conditions should be monitored, such as temperature, relative humidity, viable particulates (surface and airborne) and nonviable air particulates.

The media fill consists of an absolute minimum of 3000 containers for incubation and physical examination. Following closure, the containers are incubated for 14 days. Each vial, ampule, or syringe is then visually inspected for contamination along with the bulk media that is used as the negative control.

Three consecutive successful trials are conducted to validate the process. A microbial contamination level at or below 0.10 percent is considered in compliance.

VALIDATION OF MEDIA STERILIZATION

Prior to using sterilized media for validating aseptic processes, it is of utmost importance to validate media sterilization. The objective of this protocol is to prove, using physical and microbiological controls, that the media sterilization cycle is sufficient to attain a 1×10^6 survivor probability and that the resulting media *will* support growth.

The following procedure is used:

1. The media is inoculated with *Bacillus stearothermophilus* in suspension with a population of at least 1×10^6 per ml.

2. Two thermocouples are placed in the media container, one in the center of the media, and the other in the middle near the side of the container. Load the media bottle in the autoclave.

3. A container with water (with as much liquid volume as the media) with the load probe (suspended in the center of the water) and with two similarly placed thermocouples is loaded into the autoclave.

4. The sterilization cycle is initiated with the starting of a multi-point recorder, recording the thermocouple temperatures at 1 or 2 minute intervals. In addition, record the time the sterilizer turned on, the start and end of sterilization cycle, and the autoclave settings.

5. After completion of the cycle, the media is sent to the microbiological laboratory for incubation for 7 days at 55–60°C.

6. At the end of incubation period, media is removed from the incubator and visually inspected for turbidity and/or sedimentation. If found, testing is performed to identify the organism.

REFERENCES

Agalloco, J. P. 1986. Validation of aseptic filling operations. In *Validation of aseptic pharmacuetical processes,* edited by F. J. Carleton, and J. P. Agalloco. New York: Marcel Dekker.

Akers, M. J., and N. R. Anderson. 1984. Validation of sterile products. In *Pharmaceutical process validation,* vol. 23, edited by B. T. Loftus, and R. A. Nash. New York: Marcel Dekker.

Avis, K. E., and M. J. Akers. 1986. Sterilization. In *Theory and practice of industrial pharmacy,* 3d ed., edited by L. Lachman, H. Lieberman, and J. Kanig. Philadelphia: Lea & Febiger.

Federal Food, Drug, and Cosmetic Act. Section 501(a) 2(b). Title 21, Code of Federal Regulations, Part 211. Washington, D.C.: U.S. Government Printing Office.

Food and Drug Administration, 1987. *Guideline on sterile drug products produced by aseptic processing.* Rockville, MD: Center for Drugs and Biologics.

Food and Drug Administration, 1987. *Guideline on general principles of process validation.* Rockville, MD: Center for Drug Evaluation and Research.

Parenteral Drug Association, 1981. *Validation of dry heat processes used for sterilization and depyrogenation.* Technical Report No. 3.

United States Pharmacopeia. 1990. *USP XXII/The National Formulary, NF XVII.* Rockville, MD: United States Pharmacopeia Convention.

6

Laboratory Techniques in Aseptic Manufacturing

Rahul C. Mehta
Ram Murty

The use of complex and advanced analytical techniques for the evaluation of the quality of drug substances and excipients used in parenteral formulations is becoming increasingly feasible. This chapter attempts to present an overview of some of the techniques. The contents are divided in four sections based on the nature of the analytical techniques. Techniques for the determination of physicochemical properties are summarized under Physical and Chemical Methods. Separation Methods include extraction and chromatography. Methods for identification and analysis using spectroscopies are included under Spectroscopic Methods and techniques based on immunological reactions are included under Immunological Methods. Two other groups of techniques, Biological Methods and Microbiological Methods, deserve special consideration, but are not included here because extensive overview and parenteral formulation-related information on these techniques is readily available from the current United States Pharmacopeia (USP).

The current Food and Drug Administration (FDA) guidelines for the *Inspections of Pharmaceutical Quality Control Laboratories* and the *Inspections of Microbiological Pharmaceutical Quality Control Laboratories* are appended to this chapter. These guidelines are intended to provide an overview of the requirements for maintaining an analytical laboratory under FDA compliance and as general reference for maintaining "quality" in an analytical laboratory.

PHYSICAL AND CHEMICAL METHODS

Physical and chemical methods are extensively used in pharmaceutical analyses of sterile products and constitute a variety of techniques that provide information about the structure, purity, and quantitation of active ingredients and excipients.

Determination of pH

Although pH is used to describe the degree of acidity or alkalinity of a solution, the exact value, defined as the negative log of hydrogen ion concentration, cannot be measured directly. As a result, pH measurements are made in relation to a *pH standard* solution. Several pH scales are recommended based on the standard solutions used (Kristensen et al. 1991). The USP provides five standard solutions with assigned pH values. The National Institute of Standards and Technology (NIST, formerly National Bureau of Standards) uses six primary standards and two secondary standards. The International Union of Pure and Applied Chemistry (IUPAC) recommends six primary standards and 0.05 moVkg potassium hydrogen phthalate as the reference value standard (RVS).

pH is measured using a setup of a *reference electrode* and an *indicator electrode* with pH sensitive glass. The two may be combined in a *combination electrode*. The selection of the electrodes is the most important step in obtaining reliable and reproducible pH values. The reference electrode contains a salt-bridge solution that may be saturated KCl, K_2SO_4, LiCl, or KCl/KNO_3. *Liquid junctions* in reference electrodes permit outflow of the salt-bridge solution to establish a galvanic cell, in combination with glass electrode, whose potential is dependent on the pH of the surrounding solution. A *porous plug* is the most common type of liquid junction. High temperature measurements require an Ag/AgCl reference electrode; high pH and high salt concentration require an alkali-resistant glass membrane. Measurements in emulsions and fatty solutions require electrodes with easily cleaned liquid junctions such as an *open* or *sleeve junction*. A salt-bridge solution containing LiCl is preferable for fatty solutions due to the greater solubility of LiCl (as compared to KCl) in many organic solutions. If chloride contamination of the sample is undesirable, a $Hg/HgSO_4$ electrode with K_2SO_4 solution may be used. Solutions with low ionic strength require high outflow of the salt-bridge solution and, hence, liquid junctions with *annular rings* are recommended. Very high ionic strength solutions also require high outflow of the salt-bridge solution and either open or sleeve junctions are required.

For reliable measurements calibration with two buffers is necessary under the temperature and stirring conditions that are similar to the sample measurements. Electrode maintenance, including replacement of the solution and cleaning, as prescribed by the electrode manufacturer is essential for accurate pH measurements.

Titrations

Titration techniques are most commonly encountered in standardization procedures. Primary forms of titration include those involving acid-base reactions, redox reactions, precipitation, and complexation. Direct titrations involve reactions between the titrate and an appropriate titrant. The titrate may be present in the test solution or may be liberated by some other reaction. Residual titrations involve addition of excess standard titrant followed by titration of the remaining titrant with another standard titrant.

Acid-base titrations are used for substances such as citrates, oxidized cellulose, polysorbates, and polyethylene glycol (PEG). Redox titrations result from the change in the valence of titrate resulting from an oxidation or a reduction reaction. Titrates are either easily oxidizable or easily reduced and include ascorbic acid, quaternary ammonium compounds, mannitol, cellulose esters, and procaine. Titrants include potassium permanganate, potassium iodate, and iodine. Precipitation titrations result in the formation of precipitates, usually of an insoluble silver compound generated by titration with silver nitrate. The most commonly used titrates include halogen-containing compounds. Complexation of ions with ethylenediaminetetraacetic acid (EDTA) is commonly used for the determination of calcium, aluminum, and magnesium. Substances insoluble in water or those that do not give a sharp end point in the presence of water can be titrated using nonaqueous solvents. Glacial acetic acid with perchloric acid may be used to neutralize amine salts. Weak organic acids can also be titrated in nonaqueous solvents by alkali salts of methanol or ethanol (e.g., sodium ethoxide titration of carboxylic acids).

End-point determinations in most of the reactions can be made by a suitable color indicator. Color indicators are chemicals that change color at or near the equivalence point of the titration. Accurate measurement of the equivalence points can be made electrochemically by a potentiometer. Most titrations involve a sharp change in the potential across two electrodes at the equivalence point. Systems are also available for automated titrations, where the end point is determined potentiometrically (*The Merchant* 1994).

Water/Moisture Content

Moisture determination is generally done by the reaction of water with an anhydrous solution of sulfur dioxide and iodine (*Karl Fischer reaction*). The determination can be done by direct titration with a visual end point or by coulometric determination of the end point. The Karl Fischer reagent is prepared by the addition of 100 ml pyridine containing 100 ml dry sulfur dioxide to a solution of 125 g iodine in 670 ml methanol and 170 ml pyridine. The reagent must be standardized 1 hour prior to use. One ml of fresh reagent is equivalent to 5 mg water. The Karl Fischer reagent is also available commercially. The direct titration method is suitable for the determination of 10–250 mg water in the test sample. *Residual titration* may be done if water is bound and is released slowly. For determination of very low moisture contents (0.1 percent to 0.0001 percent), *coulometric titration* is performed. In coulometric titrations iodine is generated by anodic oxidation of iodide solution. Excess iodine after reaction with water is detected coulometrically.

Thermal Methods

Thermal analysis measures changes in physical properties as a function of temperature. Typical properties measured include melting and boiling points; glass transition; the presence of impurities, hydrates, and solvates; and polymorphic forms. Thermogravimetric analysis (TGA) involves heating the sample at a predetermined rate and measuring the change in the weight. At specific temperatures evaporation or sublimation results in a loss of mass that is recorded by the instrument. The effluent from the TGA can be injected directly into a gas chromatographic system or a mass spectrometer to identify and quantify the nature of the volatile materials. Differential thermal analysis (DTA) measures the difference in the temperature between sample and a reference. Transitions result in heat changes that affect the temperature of the sample and are recorded as a function of temperature or time. Differential thermal analysis only measures the temperatures of transitions and not the amount of heat involved in the transitions, which is measured by differential scanning calorimetry (DSC). Thus, DSC measures the heat capacity and heat of transitions such as melting, vaporizing, and so on. These two methods, DSC and DTA, are extensively used in the characterization of polymers, including determination of melting, glass transition, and recrystallization temperatures. The magnitude of glass transition for partially crystalline polymers depends on the degree of crystallinity of the sample. Thermomechanical analysis (TMA) is used for determining the changes in the viscoelastic properties, shape, penetration properties, and the volume of polymers. Extensive discussion of the techniques and theory of thermal methods of analysis can be found in Ford and Timmins (1989) and the references there.

Viscosity

Viscosity is the resistance to relative motion by adjacent layers of liquid and is defined as the tangential force per unit area required to maintain a difference in velocity of 1 cm/sec between two parallel layers of liquid 1 cm apart. Measurements of viscosity are usually made using standard solutions of known viscosity for the calibration of viscosity-measuring instruments. Flow viscometry is the most common method used for viscosity measurements. A known volume of liquid is allowed to flow through a capillary at a precisely controlled temperature and the time required to flow is recorded. The capillary is calibrated using a liquid of known viscosity. The time required for the liquid to flow through the capillary is proportional to its viscosity. Ostwald and Ubbelohde viscometers are available with known "viscometric constants" obtained by calibration using a liquid of known viscosity and specific gravity. A rotational viscometer uses a rotating spindle completely immersed in the liquid. The instrument measures the resistance to rotation of the spindle as a measure of viscosity. The Brookfield viscometer is an example of this type of instrument. Automated viscometers of the Ostwald and Brookfield type are available from a number of manufacturers. Capillary viscometers have optical sensors to record the time required for the liquid meniscus to pass through the starting and ending lines on the viscometer tubes.

Osmometry

Osmometry determines the tonicity of the injectable solutions—to determine if the solutions are isotonic with blood serum. All large-volume parenterals (LVPs), small-volume parenterals (SVPs) designed for rapid intravenous injections, and ophthalmic solutions must be isotonic. The most common type of technique for the measurement of osmolality is the freezing point depression method. The freezing point of water is depressed to -0.52°C due to the presence of solutes in blood. Thus, a concentration of any substance, in water, that has a freezing point of -0.52°C is considered isotonic. Hypotonic solutions have a freezing point above -0.52°C and hypertonic solutions have a freezing point below -0.52°C. Highly sensitive and automated instruments are available for the measurement of osmolality.

The sodium chloride equivalent method is used for adjusting the tonicity of solutions. A sodium chloride equivalent is the weight of sodium chloride that will produce the same osmotic effect as 1 gm of drug prepared as an isotonic solution. Values for sodium chloride equivalents for many drugs are available from the literature (Siegel 1990). The sodium chloride equivalents of all the substances in the formulation are added together and a sufficient amount of additional sodium chloride is added to bring up the total sodium chloride to 0.9 percent.

Optical Rotation

Many organic substances have the property of rotating plane-polarized light passing through them to a measurable extent. The measure of the rotation can be used for qualitative and quantitative analysis. Specific rotation, *a*, of a liquid is defined as the angular rotation in degrees caused by passage of light through a 10 cm length. It is usually accompanied by the temperature and the wavelength of measurement. Most optical rotation assays in USP use a wavelength of the D line of sodium (589.0 nm and 589.6 nm)

Particle Counting and Sizing

The counting of particles in parenteral solutions is required by the USP. The visible limit of detection is 10–50 μm, depending on the technique used and the training of the person. The prescribed limit of particulate matter in parenteral solutions is not more than 10,000 particles ≥ 10 μm and not more than 1000 particles ≥ 25 μm. The most commonly used technique for counting and sizing particles is light blockage. The particles pass through a small orifice sensor through which a high intensity light beam is passing. The extent of light blockage determines the size of the particles and the number of times the light is blocked determines the particle count. Various channels of appropriate size range can be selected for data reporting. This technique is more sensitive and rugged than the Coulter technique where the reduction in conductance is used to measure the size of the particles passing through an orifice. Light scattering techniques can also be used to count and size particles up to 0.5 μm in diameter. Other techniques include counting particles collected on a filter after filtration of a required volume of solution.

Total Organic Carbon

The determination of total organic carbon (TOC) in water is becoming an important technique for routine quality control (QC) of pharmaceutical grade water. The principle of operation of TOC analysis is conversion of organic carbon to carbon dioxide (CO_2) followed by determination of CO_2 concentration. The oxidation of organic carbon is achieved by combination of ultraviolet (UV) irradiation; peroxydisulfate produces hydroxyl radicals that oxidize the organic carbon. The reaction can also take place with peroxysulfate at 100°C. Alternately, high temperature (680°C) combustion can also produce CO_2. The CO_2 is detected by nondispersive infrared spectrometry or conductivity. Many new systems are specifically designed for the in-line determination of TOC in water for injection (WFI) generation systems and have stable calibrations for long-term performance.

SEPARATION METHODS

Successful quantitation and identification·of analyte from a dosage form often requires separation from formulation additives, including other drugs. The analyte may be separated by *extraction* for further evaluation or by *chromatography*, where separation and analysis can be done simultaneously. Ionic substances may be separated and analyzed based on their charge properties using *electrophoresis;* substances with specific immunological properties may be analyzed by *immunological methods,* usually combined with another analytical technique (such as chromatography or spectrometry).

Extraction

Extraction of analyte for the purpose of further analysis using other techniques is discussed here. Extraction is often the most difficult step in the development of an analytical procedure. Successful extraction results not only in the removal of extraneous materials but also in concentration of the analyte, thus improving the specificity and sensitivity of quantitative analysis. Four major types of extractions are common in analysis:

1. Liquid/solid extraction, where the analyte is in solid form

2. Liquid/liquid extraction, where the analyte is in liquid form

3. Precipitation-extraction, where the extraneous materials are usually proteins in liquid samples

4. Solid phase (solid/liquid) extraction, where the analyte is in
    the liquid form

Liquid/Solid Extraction

Extraction of solid dosage forms usually involves liquid/solid extraction to isolate the analyte from matrix components. Solvent selection is the most critical aspect of any solvent-based extraction procedure. Extraction involves *solubilization* and/or *desorption* of analyte from the solid matrix. Multiple extraction always results in a larger fraction recovered when compared with single extraction of the same volume. Desorption of analyte may be aided by the addition of other compounds capable of displacing analyte from binding sites on the solid matrix. Usually higher temperatures increase desorption and recovery of solute. Another important parameter is the rate of extraction, which can be increased by particle size reduction, increased temperature, and chemical destruction of the matrix.

A common technique of extraction includes mixing and agitating the matrix with an extraction solvent followed by filtration, centrifugation, or decantation. *Soxlet extraction* provides automated multiple extractions using the same extraction solvent, resulting in a more concentrated analyte solution. Sterile devices such as *implants* and *microspheres* may be extracted using solid/liquid extraction techniques.

Liquid/Liquid Extraction

A system consisting of two solvents that are only partially miscible with each other can be used for the extraction of analyte from liquid samples. *Partition coefficient* of analyte and other substances in the two solvents and the ratio of the solvents determines the efficiency of extraction. An *internal standard* is usually required for liquid/liquid extraction to compensate for volume changes due to partial miscibility of the two phases and loss of phases due to multiple manipulations. *Temperature* effects on the partition coefficients must be taken into account during method development. Solvent pairs resulting in increased partitioning of analyte into the extraction solvent at higher temperatures are preferable. Other substances dissolved in solvents often change the apparent partitioning of analyte without changing the partition coefficient. *Ionization, complex formation, ion-pairing,* and so on may increase the extraction of analyte when appropriately used. The addition of appropriate salts to the aqueous phase during extraction results in the "salting out" of organic compounds, with increased extraction by organic phase. The effectiveness follows the *lyotropic series*—cations: $Mg^{+2}> Ca^{+2}> Sr^{+2}> Ba^{+2}> Li^+> Na^+>K^+>Rb^+>Cs^+$;

anions: citrate>tartrate>SO_4^{-2}>Cl^->NO_3^->ClO_3^->I^-. *pH* is another important parameter for the extraction process. A pH at which the analyte is in its most extractable form by the extraction solvent should be selected. Usually pH providing 95–97 percent unionized analyte is suitable for extraction, which can be easily calculated if the *pK$_a$* of analyte is known. If buffers are used in extraction, the selection of ionic strength must include consideration for salting out of analyte or other compounds. The selection of extraction solvent must include consideration for *technical parameters* including UV cutoff, volatility, flash point, odor, toxicity, and cost.

Methods for liquid/liquid extraction involve agitation followed by separation of the two phases, either by sedimentation or freezing of aqueous phase and centrifugation. Emulsion of the two phases may form if the analyte solution contains surfactants or colloids. Gentle agitation and use of phases with greater difference in density may avoid emulsion formation. The addition of salts, alcohols, or silicon defoaming agents may break emulsions if already formed. Liquid/liquid extraction may be employed for the purification of formulations containing a large amount of excipients, ophthalmic ointments and creams, nonaquous injections containing oils, and serum samples.

Precipitation-Extraction

Precipitation-extraction is commonly used for serum analysis where a large amount of extraneous proteins must be removed to isolate and concentrate analyte (Markell et al. 1991). Organic modifiers (such as acetone or acetonitrile), acids (such as trichloroacetic, perchloric, and phosphotungstic), or a combination of the two (such as alcoholic HCl) are used to precipitate proteins. The samples are then centrifuged to remove precipitates and evaporated to concentrate the analyte.

Solid-Phase Extraction

Solid-phase extraction (SPE) involves extraction of analyte from solution by a solid matrix followed by isolation of the analyte from the matrix using a process resembling column chromatography. Thus, the process includes a combination of liquid/solid and solid/liquid extraction techniques. It is the most commonly used technique for extraction and concentration of analyte from a wide variety of solutions, including biological specimens. Solid-phase extraction provides a high degree of specificity in sample preparation and has been used for many years in ion-exchange cleanup and concentration. The use of phases other than ion-exchange resins for sample retention has enormously increased the potential utility of SPE in sample preparation, cleanup, and concentration. A typical SPE column is a syringe barrel filled with solid extraction phase-containing frits at both ends. The column is conditioned using an appropriate solution and then sample containing analyte is passed through the column. The extraction phase retains the analyte and other similar impurities that are then removed from the extraction phase by rinsing with an appropriate solution. Finally, the analyte is eluted using another solution and collected for analysis or further concentrated. Typically, the samples are passed through the column at 1–2 ml/min and analyte is recoved in 0.2–0.5 ml solution. The dual cleanup and concentration properties of SPE have made the technique very popular for the extraction of biological samples and for the enrichment of trace contaminations in large-volume samples (Markell et al. 1991).

Extraction matrices are available in a variety of forms including C_{18}, C_8, C_2; cyano and phenyl on bonded silica; and cation- and anion-exchange resins. Amino functionalities are available for reaction with user ligands, such as antibodies or substrates. The retention and elution of analytes and impurities follow the same principles as in chromatography for each of these phases (e.g., a hydrophobic compound may be isolated from an aqueous solution by retention on a C_{18} matrix, followed by washing of the matrix with aqueous solution to remove impurities, and, finally, elution of analyte in a small volume of organic solution). Most commercial suppliers of SPE devices provide an extraction guide to aid in sample preparation and the choice of phases and solvents.

Recent developments in membrane sciences have attempted to overcome some of the disadvantages of conventional column-based SPE, such as low cross-sectional area resulting in low flow rates that limit the utility of SPE in extensive trace enrichment. Membrane-based SPE cartridges resemble conventional membrane filters, where the active functionalities are bound to the pores of the membrane. Membranes of polytetrafluoroethene (PTFE) and polyvinyl chloride (PVC) are commonly impregnated with bonded silica or resins (e.g., Empore by Varian, Bio-Rex by Bio-Rad). Membranes with limited types of chemically linked functionalities are also available (Memsep by Millipore). The cartridges and disks have comparable packing capacities and differ only in geometry, but the disks have added advantages of higher flow rates and faster mass transfer (for retention/elution of analyte) (Markell et al. 1991).

High Performance Liquid Chromatography

Chromatography is a separation technique in which separation is achieved by partitioning of compounds between a stationary phase and a mobile phase. Partitioning between two phases may be achieved by several mechanisms, based on ion exchange, molecular size, specific affinity, and partitioning between two liquid phases. High performance liquid chromatography (HPLC) uses small particles for the stationary phase, which results in high column efficiencies and high pressures. Following separation, the eluant is monitored for peaks of separated compounds using detectors that provide qualitative and quantitative analysis. Table 6.1 summarizes the HPLC techniques, their characteristics and applications.

Instrumentation

The instrumentation for HPLC is primarily modular, allowing addition, upgrading, and substitution of various components based on

Table 6.1. Summary of HPLC Techniques

Chromatography	Stationary Phase	Mobile Phase	Samples
Reversed-phased	High density organosilanes	Aqueous with *organic* modifiers	Ionic or soluble in nonpolar organic solvents
Hydrophobic interaction	Low density organosilanes	Aqueous with *salt* modifiers	Ionic, usually proteins
Normal-phase *Hydrophilic-interaction*	CN, amino, nitrile on silica	Organic	Soluble in polar organic solvents
Size-exclusion *Gel-permeation/ Filtration*	Polymeric gels or silica	Aqueous or organic	Macro-molecules
Ion-exchange	NH_4^+ or sulfonyl on polymer beads	Aqueous buffers	Ionic
Affinity	Ligands specific for analyte	Aqueous with modifiers	With affinity for stationary phase ligands

intended application. The basic components include a high pressure pump capable of delivering pulseless flow of mobile phase, a sample injector, a column, a detector, and a recorder/computer. A *single or double piston reciprocating pump* with appropriate pressure sensors and pulse reducers is the most commonly used pump system in HPLC. This pump uses an inlet and outlet check valve to maintain the direction of mobile phase flow. The piston used for pumping comes in direct contact with the mobile phase in most pumps and, hence, the pumps must be selected based on the compatibility of the piston and seals with the intended mobile phases. A *gradient pump system* allows the mixing of two or more mobile phases at a predetermined rate, allowing maximum flexibility in method development for complex separation. Gradients may be formed on the high pressure side of the pumps, where multiple pumps are used to deliver individual solvents. Newer pumps allow the mixing of solvents on the low pressure side of the pump and use only one pump to achieve the gradient. These are the most popular gradient systems in use. An injection-type *sample inlet device* is used for the introduction of samples onto the column. The most common type of device uses a six-port two-position sampling valve with a sample loop. Sample injection may be manual or with the aid of an *autoinjector* or *autosampler*. Automated injectors use an array of small vials that are accessed by a needle either serially or randomly to draw samples for injection onto the column. *Columns* are the heart of HPLC systems. Typical columns are 15 cm or 25 cm long, with an internal diameter of 4–4.6 mm. Packing type depends on the intended application and is described with each type of chromatography. The majority of columns are constructed from stainless steel tubes with end fittings as inlets and outlets. Cartridge columns contain a stainless steel tube with packing and embedded frits that can be placed in reusable holders with end fittings. Advances in particle and filling technologies have made it possible to make columns with smaller internal diameters. These columns (*minibore* or *microbore*) provide significant mobile phase saving, without compromising the separation, which is cost-effective and environment-friendly, especially due to the extensive need for using organic solvents in mobile phases. After separation of the components, the mobile phase is passed through a *detector* that detects and quantitates the amount of substance passing though. The type of detector depends on the components of interest and the mobile phase composition. Absorption detectors such as *UV-Vis and fluorescence* are extensively used for organic compounds containing chromophores. Other detectors include those measuring *refractive index, electrochemical reactions,* and *radioactivity*. Specialized detectors, such as *mass spectrometers*, are also increasingly utilized for quantitation and identification of analytes. Often detectors are connected in series to increase the detection capability. Detectors also measure the total solid contents by evaporation of the mobile phase and detection of particles by light scattering *(evaporative light-scatering detectors, ELSD)*. The mobile phase used with ELSD must contain only volatile ingredients *(Bulletin #264 1994)*. *Fourier-transform infrared spectroscopy (FT–IR)* has been successfully used as a liquid chromatography (LC) detector by collecting the column output on a germanium disk that can be placed in a special sample holder for FT–IR analysis on conventional FT–IR spectrometers *(LC Transform 1994)*.

Reversed-Phased High Performance Liquid Chromatography

Reversed-phased HPLC (RP–HPLC) is the most commonly used technique for the analysis of pharmaceutical products. The stationary phase, which is chemically bonded to an inert support, is hydrophobic and the mobile phase is hydrophilic relative to the stationary phase. The stationary phase is available in different hydrophobicities such as octadecyl (C_{18}) and octyl (C_8). The support material (usually silica) varies in size from 3 μm to 10 μm, shaped as spherical or irregular and phase density (percent silanol converted to C_{18} or C_8) ranging from 5 percent to 20 percent. Small spherical particles with high density provide the most efficiency, but are also the most expensive. Silica with free silanol groups tend to produce broadening of the chromatographic band, thus reducing efficiency and sensitivity. End-capped columns tend to minimize these effects by converting free silanol groups to triethylsilane. The bonded phases are stable over a pH range of 2.5 to 7.5 in aqueous solution and in most common organic solvents.

The solvents most commonly used for RP–HPLC include aqueous buffers with methanol and/or acetonitrile as organic modifiers. Suitable selection of pH and organic modifier concentrations are sufficient for the separation and analysis of most common small molecular weight water soluble substances with some degree of hydrophobicity. Compounds with ionic character under the condition of analysis can also be separated using ion-pair techniques. An ion pair between two oppositely charged molecules favors the organic phase and, hence, the method is suitable for RP–HPLC. Alkyl sulfonates with high pH mobile phases are generally used for basic compounds and quarternary amines with low pH mobile phases are used for acidic compounds.

Micellar chromatography is a special form of RP–HPLC, where surfactants above their critical micelle concentration are used in the mobile phase. The micelle provides hydrophobic sites for interaction of the compounds; retention and selectivity are varied by changing the concentration and type of surfactant, respectively (Armstrong 1985).

In addition to its extensive compendial uses, RP–HPLC is extensively used in the clinical analysis of biological samples (Anderson 1993).

Hydrophobic Interaction Chromatography

The primary applications of hydrophobic interaction chromatography (HIC) (Ingraham 1991) are based on the surface hydrophobicities of proteins in their native form. These columns have stationary phases that are less hydrophobic than those used in RP–HPLC and include hydroxypropyl, benzyl, phenyl, isopropyl, and pentyl. The density in an HIC column is also lower, about 10 percent of that used in the RP–HPLC stationary phase. The mobile phase is generally neutral with a high salt concentration; reverse salt gradients are used to alter surface hydrophobicities, which results in separation. Salts with greater molal surface tension increment cause greater retention of proteins. The most commonly used salt for HIC is $(NH_4)_2SO_4$, due to its high solubility, very low UV absorbance, low microbial growth, and availability in high purity grade. Initial salt concentrations of 1 M to 3 M are commonly used. Temperature plays an important role in HIC since hydrophobic interaction increases with temperature. Temperatures of 15°C to 50°C are used. The primary applications are in the analysis of proteins and peptides in native form.

Normal Phase Chromatography

Normal phase chromatography uses a hydrophilic stationary phase and an organic mobile phase. This technique is used only for selected steroids and some fatty acid derivatives. A primary disadvantage of this method is the use of high concentrations of organic solvents, such as methylene chloride, heptane, hexane, butyl chloride, and isooctane, which are not only expensive to use but also pose greater toxicological and environmental concerns. The solvents are used in combination with each other and are often saturated with water or made acidic using acetic acid. Normal phase chromatography is rarely used in routine analysis.

Size-Exclusion Chromatography

Size-exclusion chromatography (SEC) is primarily used for separation and quantitation of macromolecules. Aqueous SEC (Aq-SEC) is used for proteins, peptides, and other water soluble macromolecules; organic SEC (Org-SEC) is used for synthetic polymers and resins. Sample viscosity is of concern due to high molecular weight substances; higher viscosity results in loss of resolution. In general, for substances with molecular weight of 50,000, sample concentration of 0.25 percent is reasonable whereas for samples with molecular weight of 2.5 million, only 0.01 percent can be used. Larger sample sizes may be used if column temperature is increased. Size-exclusion columns are graded according to the exclusion pore size, with smaller pore size columns having a lower molecular weight limit of separation. Single columns usually have a molecular weight range of 1.5 to 2.5 orders of magnitude. Multiple columns in series are commonly used to increase resolution in SEC. Similar pore size multiple columns increase resolution, whereas columns with different pore sizes increase the molecular weight range. Pore size and molecular weight ranges are available from manufacturer catalogs.

Aq-SEC depends on the size of macromolecule that can be varied based on the ionic strength of the mobile phase. Higher salt concentration causes protein folding, resulting in smaller molecular size. Salt gradients are commonly used for separation. Ionic interactions between the analyte and column materials must be removed by use of the appropriate pH based on the pK_a of the analyte. The addition of detergents for large protein separation may cause multisubunit proteins to divide into individual subunits. Sodium dodecyl sulfate (SDS) is an example of a detergent commonly used for large protein separation. Proteins and peptides with known molecular weights are used as standards.

Org-SEC requires mobile phase selection based on compatibility with the column, solubility of the sample, and temperature of operation. Calibration is usually required to prepare a standard curve of molecular weight vs. retention time. Polystyrene of narrow molecular weight is commonly used for the calibration of Org-SEC techniques.

Ion-Exchange Chromatography

Ion-exchange chromatography (IXC) is one of the oldest chromatographic techniques. It employs polar ionizable groups, such as sulfonic acid and quarternary amines, that attract substances with opposite charge at the appropriate mobile phase pH. For proteins and peptides IXC offers milder conditions for higher recovery of bioactivity as compared to other chromatographic techniques. Anion exchange is used for some acidic proteins. Cation exchange is more widely used than anion exchange for peptides and proteins due to the ease of ionization of basic residues such as Lys, Arg, and His at pH $\approx$ 3.0. Also, strong cation exchange columns are more suitable than weak cation exchange columns. Primary factors affecting separation in IXC are the number and polarity of the charged residues at the pH of mobile phase. The addition of organic modifiers improves selectivity for compounds with similar charges by a combination of ion exchange and hydrophobic effects on the structure of the molecule. The combination of ion-exchange and reversed-phased columns has been used in the separation of complex protein and peptide mixtures (Albert and Andrews 1988).

Affinity Chromatography

High performance affinity chromatography uses a specific ligand that binds to the substance of interest. The ligand is usually an enzyme substrate or antigen. The immobilized ligand only binds to its substrate and retains while purifying it from other nonbinding substances. This technique is described in detail under *Immunological Methods of Analysis.*

Gas Chromatography

Gas chromatography (GC) separates volatile or volatilizable substances between a solid (GSC) or a liquid (GLC) stationary phase and a gaseous mobile phase. Gas-liquid chromatography is more selective and versatile due to the availability of a large number of liquids that can be adsorbed on a suitable stationary support to form a column. The layout of GC equipment is similar to HPLC, with a carrier gas supply, injector, heated column, and detector. Helium and N_2 are the most commonly used carrier gases. Nitrogen provides better separation efficiency, whereas He provides faster separation. The selection of the gas also depends on the type of detector used. A *sample injector* vaporizes the sample and delivers the vapors to the column. The temperature of the injector is usually about 50°C higher than the column temperature. Very small sample volumes can be introduced in the column using a *sample splitter* that injects a reproducible fraction of the injected volume. Solid samples can be "injected" using sealed metal capsules that are pierced to volatilize samples in a closed chamber. Columns for GC are made primarily from glass and stainless steel. The majority of columns are packed with a solid stationary phase or with a liquid-coated solid phase. Column efficiency improves with small and narrow size range material. The selectivity of the column depends on the differences in the solubility of the sample components in the stationary phase. The selection of a suitable stationary phase may be made using *McReynold's constants,* which are directly related to the retention times of specific test compounds (Souter 1991). *Capillary columns* are used for the separation of very close peaks. The columns are either "wall coated," where the stationary phase is coated directly on the column walls, or "support coated," which are similar to conventional packed columns.

Detectors used for GC include *thermal conductivity* detectors, which measure the thermal conductivity of eluting substances and can be used at high temperatures (up to 450°C) with He, H_2, and N_2 and are universal in detection. A *flame ionization* detector (FID) ionizes the sample with H_2 flame and the increased electrical conductivity is detected. It can be used for almost all organic compounds and in the presence of water with He and N_2 as carrier gases. The FID provides the best stability of all the GC detectors. An *electron capture* detector (ECD) ionizes carrier gas, usually N_2 or Ar with $\approx$ 10 percent CH_4, providing baseline conductivity. Eluting electron-adsorbing molecules capture some of the electrons and reduce the conductance that is detected. The ECD is the most sensitive detector for compounds responding to it.

The sample for GC must be volatile under the condition of separation. Samples with low or very high volatility can be chemically derivatized. Derivatization can also be performed to improve the separation of close peaks, increase sensitivity, or to protect thermolabile compounds. Silylation using trimethylsilane is one of the commonly used derivatization techniques for compounds containing carboxylic acid, amine, alcohol, and phenyl functionalities. Other silyl compounds have also been used for derivatization. The primary use of GC is in quantitative analysis and in the determination of volatile impurities.

Supercritical Fluid Extraction/Chromatography

Supercritical fluid is fluid above its critical temperature and pressure, where its solvent capacity is similar to that in a liquid stage at very low density and viscosity and the diffusion coefficients are 10 to 100 times higher. The increased diffusion coefficients in supercritical fluid extraction (SFE) permit rapid and more efficient extraction of components from solid matrices. Supercritical fluid chromatography (SFC) is derived from GC and HPLC. The most commonly used mobile phase is CO_2, with or without a small amount of solvent modifiers. Other solvents include nitrous oxide, pentane, and other hydrocarbons. Most polar solvents have very high critical temperatures for use in pharmaceutical analysis. The mobile phase is pumped in a liquid or gas state and its temperature and pressure are then increased to above the critical point. The injector and column are placed in an oven. The injector is similar to an HPLC injector. Commonly used columns are open tubular capillary columns providing less pressure drop, which can be significant with packed columns due to the high compressibility of critical fluids. The stationary phases are similar to those used in capillary GC columns. Supercritical fluid chromatography can use GC detectors if allowed to expand into a vapor state and can use HPLC detectors if liquid is kept under pressure.

Electrophoresis

Electrophoresis is migration of charged molecules under the influence of an external electrical field. A support medium is usually used for the separation. Paper and cellulose acetate membranes are the simplest form of support media. The most commonly used technique, however, is *polyacrylamide gel electrophoresis (PAGE)*. The sample is applied on a plate coated with the gel and electrophoresis is conducted. For separation of proteins, SDS is added to the gel to provide uniform shape and consistent size of protein molecules. The separated bands can be derivatized with specific reagents or can be *blotted* on a suitable medium, such as nitrocellulose or nylon, for detection. For nucleic acids, probes of complementary sequence are used for detection in a *Northern Blot* (for DNA) or a *Southern Blot* (for RNA). Proteins can be detected by reaction with specific antibody probes in a *Western Blot*. The use of antibodies for detection, identification, and quantitation of antigens after electrophoretic separation is known as *immunoelectrophoresis*. The use of pH gradients in an electrophoretic medium causes proteins and other amphiphiles to change charge as they migrate under the influence of current. At isoelectric pH the proteins have no net charge and hence migration stops. This technique is known as *isoelectric focusing* (IEF) and it separates proteins based on their pI (isoelectric point).

Capillary electrophoresis (CE), which uses capillaries to hold support media for electrophoresis, is becoming an important tool for the rapid and reproducible separation and quantitation of proteins and peptides (Albin et al. 1993). *Capillary zone electrophoresis (CZE)* is emerging as a powerful technique due to the high efficiency, the small sample requirement, and the availability of several automated systems similar to HPLC systems. Several approaches are used to prevent the adsorption of proteins to the silanol groups that are commonly seen in CZE. These include the use of high pH buffers to produce a negative charge on the protein, low pH buffers to prevent ionization of silanol groups on the capillary, high ionic strength buffers, buffer additives to neutralize the silanol charge, and coating of the capillary walls with appropriate agents (Turner 1991). Other applications of CE include capillary isoelectric focusing (CIEF), capillary gel electrophoresis (CGE), and micellar electrokinetic capillary chromatography (MECC) (Schoneich et al. 1993).

SPECTROPHOTOMETRIC METHODS

Spectrophotometric methods are the most extensively used techniques in the identification and quantitative analysis of sterile products. These techniques depend on the ability of chemical entities to either absorb, scatter, or otherwise modify the properties of a beam of electromagnetic radiation or induce changes in the properties of the molecules. Spectroscopic methods involve the measurement of changes in the properties of the electromagnetic radiations that are either transmitted, emitted, or scattered after passage through a chemical substance. Table 6.2 lists the spectrometric techniques with the various regions of the electromagnetic spectrum utilized and the applications in pharmaceutical analysis.

Table 6.2. Summary of Spectrometric Techniques

Spectrometric Technique	Spectral Region	Wavelengths	Applications[a]
UV-visible	UV, visible	190 nm–900 nm	Q, I
NIR	Near-IR	900 nm–2500 nm	Q
IR	Fundamental IR	2500 nm–25 μm	I, Q
Atomic Absorption	UV, visible	190 nm–900 nm	Q, I
Flame Photometry	UV, visible		
Fluorescence	UV, visible		
Raman	UV, vis, NIR, IR	190 nm–25 μm	
NMR	Radio-frequency	1 m–3000 m	I, Q

[a]Q: Quantitation; I: Identification

Ultraviolet-Visible Spectrometry

Absorption in the UV and visible region of the spectrum arises from electronic transitions within the molecules. Absorption of chemical substances in the UV-Vis region is often strong, resulting in the best sensitivity, among various spectrometric methods, for the quantitative analysis of many organic compunds. Essentially, all compounds containing conjugated double bonds—aromatic ring, halogen, thioether, amino, hydroxy, and ether—show absorbance in the UV-Vis region. The spectra are usually recorded as absorbance vs. wavelength. The lack of specificity may create interference during analysis that can usually be minimized by appropriate seperation procedures.

Analysis

The *wavelength of maximum absorption* (λ_{max}) depends on the type of electronic transitions resulting from absorption. The λ_{max} thus depends on the chromophore, the solvent, and the geometry of the chromophore (e.g. *cis* vs. *trans* isomers). The λ_{max} can be easily determined from a spectrum of wavelength vs. absorbance in the selected solvent.

Molar absorptivity (ε) *and* $E_{1cm}^{1\%}$ are absorbance of either a known molar concentration or a 1 percent concentration of solute in a suitable solvent measured in a 1 cm path-length cuvette. True ε and $E_{1cm}^{1\%}$ are constants, but should not be used in quantitative measurements due to instrumental deviations from true values.

Extraction of analyte for the removal of interfering excipients was a common procedure used for quantitative analysis using UV-Vis spectrometry. Although it is still used somewhat, most separations of analytes are done by chromatographic processes, resulting in extraction and analysis in a single step without subsequent spectrometry. Some chromatographic systems equipped with photodiode array detectors combine separation and spectral analysis in a single step and are useful in identifying components in a mixture.

Solvents with an UV cutoff lower than the λ_{max} of the substance being analyzed are suitable for analysis. *UV cutoff* is the maximum wavelength at which solvents show absorbance that is high enough to interfere with analysis; the values are provided by the manufacturers for solvents commonly used in UV spectrometry. Aqueous buffers and many common organic solvents are suitable for most analysis.

Standard solutions of substances to be analyzed with known concentrations should be measured at the same time as the samples being analyzed. The unknown concentrations (C_u) may be obtained either from a plot of absorbance vs. concentration or from a single standard using the formula $C_u = (A_u / A_s) C_s$, where A_u is the absorbance of the unknown, A_s is the absorbance of the standard, and C_s is the concentration of the standard. Standard curves are linear in accordance with Beer's Law, which states that absorbance $(A) = a \cdot b \cdot c$, where a is the absorptivity of the solute at a particular wavelength, b is the length of the path of the beam through the solution, and c is the concentration of the solute. Substances present in equilibrium with several other forms, such as protonated vs. nonprotonated, dimers vs. complexes, may significantly deviate from Beer's Law. Measurements in such cases may be made at wavelengths other than the λ_{max} (such as isosebestic wavelength for two closely related species) or by carefully buffering the system to minimize changes in the concentrations of the species.

Instruments

A wide variety of spectrophotometers are available for UV, visible, and UV-Vis ranges of wavelengths, which are either single beam or double beam type instruments. In *single beam spectrophotometers* the transmittance of the blank solution is adjusted to 100 percent prior to measurement at each wavelength. These spectrophotometers are relatively simple in design, inexpensive, and suitable for analysis if the wavelengths of measurements are not changed very frequently. *Double beam spectrophotometers* are the most commonly used instruments in UV-Vis spectrometric analysis. One beam of light each (from the same source) is passed through a reference cell and a sample cell and simultaneous measurements of absorbance are made. This eliminates the need to adjust the transmittance for each wavelength and, hence, permits easy recording of spectral scans.

Diode-array spectrophotometers use an array of diodes, each capable of measuring one or two wavelengths for detection of absorbance. The detection is very fast, usually the entire spectrum can be scanned in a few milliseconds and, hence, multiple scans can be averaged for each measurement. While most spectrophotometers use closed optical and sample compartments, many diode-array spectrophotometers can be used with open sample compartments due to scatter and background correction. These are, however, among the most expensive spectrophotometers. Most double beam and diode-array spectrophotometers are available with multiple sample holders and automated analysis, using either computer-controlled or built-in analysis protocols. Measurement intervals, wavelengths, and sample positions can be programmed for extended automated analysis as required in some stability and release studies.

Infrared Spectrometry

Absorption and emission in the IR region arises from transitions between molecular rotational and vibrational energy levels. Molecules have a large number of vibrational and rotational energy levels, the combinations of which are highly structure specific. This results in spectra that are useful in the identification of molecules. Infrared spectra are usually recorded in absorbance vs. frequency (cm^{-1}) of electromagnetic radiation. Infrared methods have lower sensitivity, but much greater specificity than UV or visible spectrometry. Individual functional groups in the molecules are responsible for the bands in an IR spectrum. The bands in the high frequency region of the spectrum are usually due to *stretching vibrations*, whereas the bands in the lower frequency region are usually attributed to *bending vibrations* of functional groups. The correlation between the band frequency and the structure of the underlying functional group is central to understanding and interpreting IR spectra.

Analysis

It is a common practice to compare IR spectra of unknown samples with reference spectra for the purpose of identification. *Reference spectra* may be obtained from spectral libraries or from the spectroscopy of reference standards. To improve the signal-to-noise ratio, the frequency range should be narrow and scanning should be slow. The selected region should be free from interference by other components in the sample. If a suitable band cannot be found, extraction of interfering substance must be performed. Bands that are moderately sharp are best for analysis, as intensities of very sharp bands have greater sensitivity to operator and instrumental parameters.

The intensity of bands corresponds with the concentration of the molecule and follows Beer's Law; hence, IR spectrometry may be used for *quantitative analysis*. However, deviation form Beer's Law is more common in IR spectrometry than UV and visible spectrometries due to greater sensitivities of absorptivities to intermolecular interactions.

Adequate *sample preparation* for IR spectrometry is important to obtain meaningful identification and quantitation. Spectra in solution form are preferred for quantitative analysis where the path length is known and the sample is homogeneous. Two other most commonly used methods are suspensions in mineral oil (*Nujol mulls*) and pellets of KBr. Mulls are prepared by grinding solid substances with mineral oil (Nujol), hexachlorobutadiene, perfluorokerosene, or other heavy liquid to a preferred particle size of less than 2 μm. The mulls are pressed between two salt plates and used for analysis. *KBr pellets* are prepared from solid substances by grinding a known amount with a known amount of spectral quality KBr powder in an agate mortar. The mixture is then placed in cylindrical die, evacuated, and pressed into a pellet using a 20,000 to 100,000 psig pressure. The pellet must be transparent, without any air bubbles or surface irregularities. The die with the pellet is used for obtaining the spectrum. The path length (KBr pellet thickness) and ratio of analyte to KBr must be known if no internal standard is used during pellet preparation. Both the mulls and KBr pellet introduce variability due to particle size. Mulling oils can contribute bands in spectra and water in KBr and oils may also contribute bands that must be considered in structure elucidation of the sample.

Solutions and neat liquids can be directly used for IR spectrometry by placing them in salt cells, provided they do not affect the salt cell. The solvents must be transparent in the spectral region of interest. The path length is obtained by scanning the empty cell; the sample is then placed for quantitative analysis. Intensities (absorbance) of C=O, N–H or O–H bands can be used for quantitation in a manner similar to UV and visible spectrometry. Infrared spectrometry, however, can be used for the simultaneous quantitation of a mixture of compounds with different frequencies for vibration of a particular functional group, such as carbonyl in a mixture of aspirin, phenacetin, and caffeine.

The latest techniques for sample preparation involve the use of a *microporous polyethylene* substrate that absorbs and holds the analyte on *disposable sample cards*. These are suitable for liquids, semisolids, and pastes as well as samples containing water.

Instrumentation

The basic design of IR spectrophotometers is similar to UV-Vis spectophotometers. The radiation source is usually a heated rod or wire providing the large range of wavelengths required for IR spectrometry. Most instruments are *double beam spectrophotometers* to minimize the effects of fluctuation in the source output. The selection of wavelength requires special crystals made from inorganic salts, as glass and quartz are not transparent in the IR region of the electromagnetic spectrum; however, gratings are used in most newer spectrophotometers for selecting wavelengths. Appropriate selection of crystal material for other optics depends on the spectral region of interest. Materials used for crystals range from salts (such as KBr, NaCl, AgBr, and CsI) to artificial sapphire, selenium, germanium, and diamond. The detectors are usually very sensitive thermocouples or Golay cells based on the expansion of gas upon collecting radiation. The intensity of radiation in conventional dispersive IR spectrophotometers is very small, resulting in limited signal-to-noise ratio at the detector. High resolution requires smaller slit widths, further limiting the intensity at the detector.

Fourier transform IR (FT–IR) spectrophotometers overcome most of the sensitivity and resolution limitations of dispersive IR spectrophotometers. These spectrophotometers utilize two radiation beams with slightly different path lengths, which upon combination produce a finite degree of interference. This radiation beam contains all the frequencies that are detected simultaneously at the detector. The resulting interferogram is transformed to a spectrum by Fourier transform. Fourier transform IR thus provides simultaneous measurement of all frequencies, improving the signal-to-noise ratio, permitting higher resolution, and reducing the time required for measurements. The mathematical transformation requires the use of a computer, which is also used to average multiple scans to further improve the signal-to-noise ratio. Fourier transform IR instruments are more expensive than dispersive spectrophotometers and are not very suitable for quantitative analysis due to the variable intensities resulting from the generation of beam interference. Fourier transform IR, however, is better than dispersive IR for qualitative estimation and identification of minute changes in the microenvironment of functional groups.

Infrared microspectroscopy appears to be a promising technique for the qualitative analysis of microscopic sample sizes. Often the results obtained are better than macrospectroscopy under suitable conditions. With further developments, IR microspectroscopy may emerge to be a very good quantitative tool (Katon and Sommer 1992).

Near-Infrared Spectometry

The transitions between molecular rotational and vibrational states upon adsorption of IR radiation result in fundamental bands in the IR region. Each fundamental band gives rise to a series of overtone bands at frequencies that are multiples of the fundamental bands. These bands are increasingly weaker and are obsereved in the near-infrared (NIR) spectral region. The combination of fundamental and overtone bands may also give rise to combination bands, detectable in the NIR spectral region. Multiple bands from a single functional group result in complicated band overlaps. Most bands arise from H-containing functional groups.

Analysis

The wavelength range for the measurement of NIR spectrometry is more flexible than IR or UV spectrometry due to the presence of two to three overtone and combination bands for each fundamental mid IR band. For example, OH bonds in water can be analyzed at 1.94 μm, 1.45 μm, 1.19 μm, 0.97 μm, or 0.76 μm wavelengths. Lower wavelength bands have lower intensity; hence, the selection of the appropriate band must include the required sensitivity as one of the parameters. The absorption follows Beer's Law; hence, quantitative measurements can be made using absorbance values.

Sample preparation in NIR does not require any special techniques. Solutions and neat liquids can be introduced into the spectrometer using cuvettes similar to UV spectrometers. Solvents without hydrogens are usually transparant in NIR. Due to very low levels of absorbance by most substances, path lengths extending to several meters can be used in NIR analysis to obtain the maximum absorbance of analyte. Path lengths of 1 cm or less are usually suitable for most analysis.

Solid samples can be used for NIR analysis using *diffuse reflectance spectrometry*, where instead of the transmitted radiation, radiation reflected from the surface of the sample is collected for detection. Sample cups are packed with particles of uniform size or are flattened between glass plates before analysis.

Instruments

The basic instrument design for a NIR spectrophotometer is similar to a UV spectrophotometer. The radiation source is usually a high temperature incandescent lamp. Commercial NIR instruments for absorption spectroscopy are often combined with a UV-Vis spectrophotometer. Instruments with only NIR spectrometry are often equipped to handle diffuse reflectance work. On-line instruments are available for many specialized operations, such as dosage form content uniformity analysis.

Applications

The primary pharmaceutical applications of NIR are in the qualitative analysis of samples; however, qualitative and quantitative methods have been developed (Corti et al. 1991). Other potential applications of NIR include noninvasive on-line QC techniques, which can be adapted for content uniformity of vials and ampoules. Equipment is available for content analysis from lyophilized parenterals and LVPs from commercial sources (NIR Systems, Silver Spring, MD).

Atomic Absorption Spectrometry

Radiation absorbed by free atoms is the source of spectra in atomic absorption (AA) spectrometry. The wavelength of absorption is generally near the resonance lines of the atoms, usually in the UV or visible range of the spectrum.

Analysis

The *wavelength of excitation* selected should be close to the resonance lines of the atoms of interest. *Organic solvents* enhance the intensity of absorption and, hence, are better suited than aqueous solutions. Large quantities of *ions of analyte* tend to interfere with the detection of atomic absorption, as ionic absorption occurs at wavelengths other than atomic absorption. Excess of ions of elements other than the analyte usually improves the analyte signal-to-noise ratio. *Sample handling* in AA involves atomization of the sample solution into an open flame followed by excitation of the free atoms using an appropriate radiation source.

Instrumentation

The AA spectrometer includes a hollow cathode tube as the radiation source, a burner with oxyacetylene or nitrous oxide–acetylene flame on which the sample is atomized, and other optical components similar to a UV spectrophotometer. New atomization techniques have been introduced to improve the detection limits of some elements. Electrothermal atomization with AA spectrometry is emerging as a powerful technique for studying trace elements (Li et al. 1990).

Flame Emission Spectrometry (Flame Photometry)

Flame emission spectrometry utilizes the measurement of emission at characteristic wavelengths for group IA and IIA metals. A gas-air flame is used as the source of excitation in traditional flame photometers and emission is measured using an optical system similar to the AA spectrophotometer. Samples in solution are atomized and introduced into the flame. Standard solutions are required to obtain a baseline with respect to solution viscosity and other atomization properties. This technique is primarily used for the measurement of Na, K, and Ca in solutions. Primary limitations of conventional flame emission spectrometry are low sensitivity and possibility of chemical interactions from oxide formation.

Plasma Emission Spectrometry

Plasma emission spectrometry is based on the principles of flame emission spectrometry, except that conventional atomization methods are replaced by electrically generated plasma that is heated to 9000 K. This produces a stable source of excited atoms and permits very sensitive and simultaneous analysis of multiple elements at ng/ml levels. Two major sources of plasma include *direct current* (DC) *argon plasma* and *inductively coupled argon plasma* (ICAP). Direct current argon plasma uses electricity to generate plasma that is then heated and an aerosolized sample is introduced. Inductively coupled argon plasma uses a magnetic field to sustain the plasma into which the aerosolized samples are introduced for analysis. Samples in solid, liquid or gaseous form may be used with plasma emission spectrometry. Inductively coupled argon plasma has been used as a plasma emission detector in combination with a variety of separation techniques, including HPLC, GC, and SFC. Plasma emission spectrometry is extensively used in other areas of science, such as geology, but is proving to be useful in biomedical analysis, such as Cr in deionized water and Fe, Cu, Ca, and Al in serum samples.

Fluorescence Spectrometry

Energy absorbed by molecules upon excitation is released (*relaxed*) in a variety of ways, one of which is by photon emission. Fluorescence is the photon emission from an excited singlet state of molecules. Fluorescence spectroscopy measures the intensity of fluorescence that occurs at a wavelength of 20–30 nm higher than the wavelength of the absorbed radiation. Any compound generating fluorescence or with functional groups that can be derivatized with fluorescent/fluorogenic compounds can be measured using fluoroscence spectrometry. The sensitivity for detection using fluorescence spectrometry is highest among all absorption spectrometries and the detection responses are linear over three to four orders of magnitude.

Analysis

Compounds with conjugated double bonds such as *aromatics* and *polyenes*, are good candidates for fluorescence spectrometry. Nonfluorescent compounds with appropriate functional groups can be chemically derivatized using fluorescent or fluorogenic compounds. These *fluorescence derivatization* techniques have been extensively used in detection of compounds after separation by HPLC and other liquid chromatographies.

Aromatic rings with ortho/para electrophilic substitutions may increase the fluorescence, whereas meta electrophilic substitutions decrease or eliminate fluorescence. The combination of ortho/para and meta *electrophilic substitutions* does not affect the fluorescence. Substitution of *heavy atoms* generally decreases the fluorescence. Lower *temperature* and higher *viscosity* increase the fluorescence. A shift towards larger wavelengths is seen with increasing *solvent polarity* and if hydrogen bonding also increases, the fluorescence decreases.

Solvents or other solutes with heavy atoms and inorganic anions also decrease fluorescence. At *pH* values near the pK_a of functional groups in molecules, shifts in the spectra are seen (Ireland and Wyatt 1976). The pK_a for the ground state and the excited state may differ by five units. *Dissolved oxygen* in solution may also reduce fluorescence and, hence, deoxygenation may be required. *Supercritical fluids* have been recently evaluated for fluorescence spectrometry (Betts and Bright 1990). Other factors affecting fluorescence include the *inner filter effect*, resulting from absorption of radiation by molecules, scattering, and fluorescence from impurities.

Instrumentation

The basic design of the spectrofluorometer includes an excitation source, usually a mercury or a xenon lamp, an excitation filter or monochromator, an emission filter or monochromator that is at right angles to the excitation source, and a photodetector similar to UV spectrophotometers. The cell is usually a cuvette with all four sides clear so that emission can be measured at right angles to the excitation source. Advanced techniques in fluorescence spectrometry include frequency-domain fluorescence spectrometry and laser-based and fiber optics–based instruments (Warner and McGown 1992).

Raman Spectrometry

The spectrometric techniques reviewed above are based on the absorption of radiations by molecules that change the radiation properties. Most nonabsorptive or scattering interactions are measured by techniques based on no change in the wavelength of the scattered radiation as seen in turbidimetry, nephelometry, and polarography. Scattering interactions, however, do result in change in the wavelength of a small fraction of scattered light. This shift in the wavelength is called *Raman shift* and the technique to measure the magnitude and intensity is called Raman spectrometry. Structural information about molecules can be generated using Raman spectrometry, in a way similar to IR spectrometry. The primary advantage of Raman spectrometry, however, is its ability to handle *aqueous samples*, a significant advantage over IR spectrometry. Raman and IR spectrometries, nonetheless, are complementary to each other since many transitions not visible in IR may be seen using Raman. Infrared radiations have been extensively used for Raman spectrometry and, in fact, *FT–Raman* accessories for many commercial FT–IR spectrometers are now available. Fourier transform–Raman spectrometry has been extensively investigated for the interaction of proteins and peptides with biological and polymeric substances (Watts et al. 1991), which may be applied to many parenteral containers and closures and drug delivery systems (Almond et al. 1990). *Resonance-enhanced Raman* spectrometry significantly increases the sensitivity of Raman spectrometry in the UV range of electromagnetic spectrum (Asher 1993).

Nuclear Magnetic Resonance Spectrometry

Nuclear magnetic resonance (NMR) spectroscopy involves absorption of energy by atomic nuclei from radiofrequency electromagnetic radiation in the presence of a strong external magnetic field, which results in the transition of nuclei to a higher energy state and the corresponding generation of NMR spectra. Tetramethylsilane (TMS) is used as a standard for proton NMR. Since all the protons are equivalent, only one peak results. Protons from other compounds produce resonate peaks that are shifted from the TMS peak, which is assigned an arbitrary value of zero. This phenomenon is called *chemical shift*, which depends on the magnetic field strength and the environment of the protons producing the resonate peak. To eliminate the dependence of chemical shift on field strength, it is represented in dimensionless units of parts per million (ppm). Thus, an NMR spectrum is plotted as magnetic field strength (or resonate absorption frequency) vs. chemical shift. The nuclei that can be used for NMR studies include ^{1}H, ^{13}C, ^{31}P, ^{19}F, and ^{11}B.

Analysis

The interpretation of NMR spectra is complicated due to the fact that protons (or atoms) show different chemical shifts depending on their position in the molecule. Each group of equivalent protons can further undergo *spin-spin coupling* with protons on neighboring groups that are not separated by more than three bonds. This is due to the tendency of the electron to pair its spin with the nearest protons. This complexity is also the reason for the extreme selectivity of NMR spectroscopy in structure elucidation. A detailed decription of NMR principles and its biological applications is extensively reviewed (Pettehrew 1990). In general, the magnitude of the chemical shift establishes the nature of the proton and the pattern of spin-spin coupling determines the interactions and environment of the proton. Complex spectra usually require computer-assisted analysis or the use of reference standard materials.

The magnitude of the intensity of a group of protons (singlets or multiples) is obtained by integrating the peak areas and it is directly proportional to the number of protons. This is the basis for *quantitative*

NMR spectrometry. Since NMR is an absolute technique, standards of known purity with a chemical structure identical to the analyte are not required. Any substance of known purity capable of generating a suitable NMR signal may be used as a standard. The amount of analyte in an unknown sample can be obtained from the NMR spectra of unknown analyte sample and a standard substance using the following formula:

$$\text{Amount of Analyte} = \left(\frac{A_u}{N_u}\right) \cdot \left(\frac{N_s}{A_s}\right) \cdot \left(\frac{M_u}{M_s}\right) \cdot (\text{Amount of Standard})$$

where A_u and A_s are the areas under the sample and standard peaks, N_u and N_s are the number of protons (or nuclei) contributing to the area, and M_u and M_s are the molecular weights of the analyte in the unknown sample and the standard. For best results strong peaks of the analyte should be chosen for analysis, the standard should include a single peak in the vicinity of the peak of interest, and results from several scans should be averaged.

Other isotopes used for NMR include ^{13}C, which is used for the structure elucidation of complex compounds. Carbon-13 has a natural abundance of only 1.1 percent of ^{12}C and, hence, larger samples are required. Coupling between ^{1}H and ^{13}C can be removed by *spin-spin decoupling* techniques. Fluorine-19 and ^{31}P can also be used for NMR.

Sample preparation for most NMR spectrometers involves dissolving the compound in a proton-free vehicle or vehicles with the protons replaced by deuterium (such as D_2O). Solid-state NMR is now feasible using new technologies where solid samples can be analyzed.

Instrumentation

The basic instrumentation for NMR includes a magnet system, a radio-frequency (RF) oscillator, and a RF receiver. The magnet must provide uniform magnetic fields across the entire width of the sample. The field strength of the magnet is varied by a few milligauss using sweep coils. The uniformity of field may be achieved by placing the sample in the exact center of the magnetic field or by spinning the sample tubes. Sample spinning produces additional lines in the spectra that must be identified and must not interfere with the integration of other lines in quantitative spectrometry. The strength of the NMR equipment is rated in terms of mHz, based on the frequency of the RF oscillator, as 60 mHz, 200 mHz, 400 mHz, and so on. The change in the voltage generated by nuclei is received by the RF receiver and is amplified on several orders of magnitude. The RF receiver is at 90° to the RF oscillator.

The excitation of nuclei by absorption of RF is followed by dissipation of the excess energy and return to equilibrium state. This process is called *relaxation*. Relaxation may occur by conversion of energy to heat as in *spin-lattice relaxation* or by transfer of energy to other nuclear spins as in *spin-spin relaxation*. The return to equilibrium state is the Fourier transform of the NMR spectrum and it can be recorded using *pulsed FT–NMR* technique. Pulse NMR involves the application of a short burst of high strength and controlled duration to the sample followed by measurement of the nuclear relaxation. This technique results in quicker and more sensitive NMR spectrometry.

Another significant progress in NMR includes *solid-state NMR* which, as the name suggests, is capable of analyzing solids. This along with *magic angle spinning* and *multidimensional NMR* should prove very useful in the quantitative and qualitative determination of pharmaceuticals and sterile dosage forms (Haw 1992).

Mass Spectrometry

Ionization of molecules by bombardment of electrons produces ionic fragments, the amounts of which are plotted against their mass/charge (m/e) ratio to produce a mass spectrum. The patterns of fragmentation are characteristics for every compound; thus, mass spectrometry (MS) is a highly selective and sensitive technique. The sample sizes are very small, usually in the μg to ng range, and patterns of fragmentation are more-or-less unchanged in the presence of other components. Successful mass spectrometry, however, requires extensive and skilled data analysis; it is also expensive to set up and maintain a mass spectrometer.

Analysis

Samples in liquid form are vaporized under vacuum, whereas solid samples are volatilized (heating to below their decomposition temperature under very high vacuum). The vapors are then ionized and accelerated toward an ion collector through an analyzer tube where they are separated according to charge and mass. The fragmentation of molecules follows known patterns. The site of cleavage depends on the chemical structure and can be identified from standards and mass spectra of other known compounds (Bursey and Nibberning 1990).

Instrumentation

A mass spectrometer includes a source of ions, a mass analyzer, and an ion detector. The resolution of a mass spectrometer depends on the techniques of ion generation, ion focusing, and mass determination. Methods of sample ionization include electron bombardment; chemical means such as protonation/deprotonation, thermospray, and electrospray ionization; and laser desorption ionization. These methods provide ions from a wide variety of sample forms, including solutions, ionized substances, and effluents from liquid chromatographic systems. Ions from high molecular weight materials, such as proteins and polymers, can be obtained using some of these novel ionization techniques (McEwen and Larsen 1990). A *quadruple mass spectrometer* uses a different technique for ion focusing, whereby four charged poles govern the ion *m/e* that can pass through at any given time. Rapid sweep in the voltage and frequency results in the scanning of all *m/e* of interest. This produces greater sensitivity and resolution. A *time-of-flight mass spectrometer* provides a constant energy to all the ions, which, thereby, acquire different velocities. The time required for the ions to reach the detector is then inversely proportional to the mass of the ion. Complete mass spectra can be collected over 2000 times every second, making it possible to follow kinetic reactions with rapid rate constants and detection of effluents from GC and LC systems.

Applications

Mass spectrometry is generally used only for development work and in the initial synthesis/isolation of peptides and proteins. Routine analysis methods using MS are not yet feasible due to the high cost of installing and operating a MS facility.

IMMUNOLOGICAL METHODS

The interaction between antigens and bodies has been extensively used in analytical laboratories to identify, separate, and purify a variety of molecules. Many immunological methods depended on the precipitation that occurs during antigen-antibody reactions. Others depend solely on the binding between the two. *Immunoelectrophoresis* combines electrophoresis for separation with recognition and precipitation of antigen by antibody. *Immunoaffinity chromatography* involves the presence of antigen or antibody as a part of the stationary phase which retains the antigen-antibody counterpart on the column and permits easy separation from noninteracting molecules. *Immunoassays* are routinely used for quantitative analysis of substances to very low solution concentrations. An *enzyme-linked immunosorbent assay (ELISA)* uses the indirect method of amplification-quantitation. Antibody specific for the analyte is linked with an enzyme, typically horseradish peroxidase. After binding the antibody with analyte from the matrix of interest, a substrate for the enzyme is added. The substrate converts to products that can be quantitated by spectrophotometric methods. A *double antibody ELISA* includes a primary antibody without linked enzyme and a secondary antibody with linked enzyme. The secondary antibody is usually specific for the Fc part of the primary antibody and, thus, binds to the primary antibody-analyte complex. The advantage of this method is that only one enzyme-linked antibody is needed for all the primary antibodies produced in the same species of animal. A *radioimmunoassay (RIA)* uses a radiolabeled tracer that competes for the binding site on the antibody with the analyte. The amount of radiolabeled tracer bound to the antibody is inversely proportional to the amount of analyte present. After the binding excess tracer is washed off and the antigen-antibody complex is precipitated for radioactivity measurements. *Solid-phase RIA* uses tubes coated with antibody and, hence, no precipitation is required.

REFERENCES

Albert, A. J., and P. C. Andrews. 1988. Cation exchange chromatography of peptides on poly(20sulfoethyl aspartamide)-silica. *J. Chromatog.* 443:85–96.

Albin, M., P. D. Grossman, and S. E. Moring. 1993. Sensitivity enhancement for capillary electrophoresis. *Anal. Chem.* 63:489A–497A.

Almond, M. J., C. A. Yates, R. H. Orrin, and D. A. Rice. 1990. Fourier-transform Raman spectroscopy—A tool for inorganic, organometallic and solid state chemists? *Spectrochim. Acta. Part A* 46A:177–186.

Anderson, D. J. 1993. Coatings. *Anal. Chem.* 65:1R–11R.

Armstrong, D. W. 1985. Micelles in separation—A practical and theoretical review. *Sep. Purif. Methods* 14:213–304.

Asher, S. A. 1993. UV resonance Raman spectroscopy for analytical, physical and biophysical chemistry, Part 1. *Anal. Chem.* 65:59A–66A.

Betts, T. A., and F. V. Bright. 1990. Instrumentation for steady-state and dynamic fluorescence and absorbency studies in supercritical media. *Appl. Spectrosc.* 44:1196–1202.

Bulletin #264. 1994. Deerfield, IL: Alltech Associates Inc.

Bursey, M., and N. Nibberning. *Mass spectrometry reviews.* New York: John Wiley and Sons, Inc.

Corti, P., E. Dreassi, G. Ceramelli, S. Lonardi, R. Viviani, and S. Gravina. 1991. Near infrared reflectance spectroscopy—Applied to pharmaceutical quality-control—Identification and assay of cephalosporins. *Analusis* 19:198–204.

Ford, J. L., and P. Timmins. 1989. *Pharmaceutical thermal analysis: Techniques and applications.* Ellis Horwood.

Haw, J. F. 1992. Nuclear magnetic resonance spectroscopy. *Anal. Chem.* 64:243R–254R.

Ingraham, R. H. 1991. Hydrophobic interaction chromatography of proteins. In *High-performance liquid chromatography of peptides and proteins: Separation, analysis and conformation,* edited by C. T. Mant, and R. S. Hodges. Boca Raton, FL: CRC Press.

Ireland, J. F., and P. A. H. Wyatt. 1976. Acid-base properties of electronically excited-states of organic molecules. *Adv. Phys. Org. Chem.* 12:131–221.

Katon, J. E., and A. J. Sommer. 1992. IR microscopy—Routine IR sampling methods extended to the microscopic domain. *Anal. Chem.* 64:931A–940A.

Kristensen, H. B., A. Salomon, and G. Kokholm. 1991. International pH scale and certification of pH. *Anal. Chem.* 63:885A–891A.

LC Transform. Bulletin #BN4. 1994. Marlborough, MA: Lab Connections Inc.

Li, Z., G. R. Carnick, J. Schickli, S. McIntosh, and W. Slavin. 1990. Using the sodium sulfate interference for lead to test the fork platform design. *At. Spectrosc.* 11:216–221.

Markell, C., D. F. Hagan, and V. A. Bunnelle. 1991. New technologies in solid-phase extraction. *LC–GC* 9:332–337.

McEwen, C. N., and B. S. Larsen. 1990. *Mass spectrometry of biological materials.* New York: Marcel Dekker, Inc.

Patonay, G., and M. D. Antoine. 1991. Near-infrared fluorogenic labels: New approaches to an old problem. *Anal. Chem.* 63:321A–327A.

Pettehrew, J. W. 1990. *NMR: Principles and applications to biomedical research.* New York: Springer-Verlag.

Schoneich, C., S. K. Kwok, G. S. Wilson, S. R. Rabel, J. F. Stobaugh, T. D. Williams, and D. G. Vander Velde. 1993. Separation and analysis of peptides and proteins. *Anal. Chem.* 65:67R–84R.

Siegel, F. P. 1990. Tonicity, osmoticity, osmolality and osmolarity. In *Remington's pharmaceutical sciences*, 18th ed., edited by A. R. Gennaro. Easton, PA: Mack Publishing Company.

Souter, R. W., 1991. Gas-liquid chromatography. In *Modern methods of pharmaceutical analysis*, vol. II, edited by R. E. Schirrner. Boca Raton, FL: CRC Press.

The Merchant, chemistry ed., vol. 1. 1994. Westlake, OH: Radiometer America Inc.

Turner, K. A., 1991. New dimensions in capillary chromatography. *LC–GC* 9:350–356.

Warner, I. M., and L. B. McGown. 1992. Molecular fluorescence, phosphorescence and chemiluminescence spectrometry. *Anal. Chem.* 64:343R–352R.

Watts, P. J., A. Tudor, S. J. Church, P. J. Hendra, P. Truner, C. D. Melia, and M. C. Davies. 1991. Fourier-transform Raman spectroscopy for qualitative and quantitative characterization of sulfasalazine-containing microspheres. *Pharm. Res.* 8:1323–1328.

7

Aseptic Production of Radiopharmaceuticals

Per Oscar Bremer

A radiopharmaceutical is defined as a pharmaceutical that incorporates a radionuclide. A radionuclide may decay by emitting three different types of ionizing radiation: alpha, beta, and gamma radiation. Alpha decay is characterized by the emission of an alpha particle from the nucleus. This particle is a helium ion containing two protons and two neutrons. In beta decay a negatively charged particle with the same charge and mass as an electron is emitted. Gamma radiation is characterized as electromagnetic radiation. Particulate radiation is less penetrating in matter than electromagnetic radiation. Depending on the radiation characteristics of the nuclide, the radiopharmaceutical can either be used for diagnosis or for therapy. If a radiopharmaceutical containing a gamma-emitting nuclide is given to a patient, it is possible to monitor externally the in vivo distribution, accumulation, and excretion of the drug, by using advanced detection systems. Normally, these radiopharmaceuticals are given in such small quantities that the radiation dose received by the patient can be compared to that of a simple radiology investigation. It is important to note that radiation is a general property of all radiopharmaceuticals and, therefore, inevitably will give the patient a radiation dose. For therapeutic purposes radiopharmaceuticals containing a beta-emitting nuclide are mostly used. Here it is the damaging effect of the radiation that is the required property of the radiopharmaceutical. Cells are most vulnerable to radiation damage at the moment they are dividing; therefore, radiation is used to kill rapidly dividing cells, such as cancer cells. Beta radiation has only a short penetration distance in living tissue. If the radiopharmaceutical is accumulated exclusively in the target area, it is possible to give large amounts of high energy radiation without exposing other organs to high radiation doses. A radiopharmaceutical may be as uncomplicated as a simple solution of a salt of a radionuclide, such as sodium iodine or sodium pertechnetate. Usually radiopharmaceuticals contain at least two major components:

1. A radionuclide that provides the desired radiation characteristics

2. A chemical compound with structural or chemical properties that determine the in vivo distribution and physiological behavior of the radiopharmaceutical

Approximately 3000 radionuclides have been discovered so far, but only about 30 of them are routinely used in nuclear medicine. Most of these are artificial radionuclides, which are among the 2700 radionuclides that are produced by irradiation in nuclear reactors, cyclotrons, or large linear accelerators. In a nuclear reactor the core has a high flux of thermal neutrons. For the production of a new nuclide, a suitable target element is placed in a special irradiation channel in the reactor core. There the target nucleus will take up a thermal neutron, emit a gamma ray and produce a new nuclide of the same element as the target. Of the best known radionuclides that can be produced by neutron caption reactions are iodine 131 (^{131}I), chromium 51 (^{51}Cr), and molybdenum 99 (^{99}Mo). The nuclear reactor can also be used to produce radionuclides by a fission reaction. By using this type of reaction, a heavy nucleus is divided into two parts of approximately equal masses. Such a reaction can be obtained by placing a target of a heavy element in the reactor core for bombardment with thermal neutrons. Uranium 235 (^{235}U) is the most commonly used target material for fission reactions, and the process results in nuclides with atomic numbers in the range from 28 to 65. As the nuclides formed are of different elements, they can be separated by appropriate chemical and physical methods. Many radionuclides used in nuclear medicine to day are produced by fission (e.g., ^{131}I, xenon 133 (^{133}Xe), and ^{99}Mo). In particle accelerators, cyclotrons, and linear accelerators, the target material is bombarded by charged particles such as protons, deuterons, and alpha particles. To overcome the repulsion of the target nuclei, these particles must be given sufficient kinetic energy to penetrate the electric barrier. In a cyclotron this is done by accelerating the bombarding particles along a spiral path under vacuum by using alternating electromagnetic fields. Among the most used radionuclides produced in cyclotrons are gallium 67 (^{67}Ga), iodine 123 (^{123}I), thallium 201 (^{201}Tl), and indium 111 (^{111}In). Many short-lived radionuclides developed for use in positron emission tomography (PET) are also produced by bombardments in cyclotrons. Among these are carbon 11 (^{11}C), nitrogen 13 (^{13}N), oxygen 15 (^{15}O), and fluorine 18 (^{18}F).

More than 90 percent of the radiopharmaceuticals used today are parenteral solutions. The rest are mostly given by inhalation as gases or aerosols, or they are taken orally as capsules or in liquid form. To a large extent the final preparation of parenteral radiopharmaceuticals takes place in hospitals, either in the nuclear medicine department or in the hospital pharmacy.

In this monograph the manufacture and preparation aspects of aseptically produced radiopharmaceuticals will be described. Radiopharmacy and nuclear medicine are relative newcomers to the field of natural sciences. The first radiopharmaceutical to have a widespread clinical use was the radionuclide ^{131}I. It was introduced at the end of World War II to treat carcinoma of the thyroid and to study other abnormalities of the gland. The stunning success of this cancer treatment and the interesting story of the new drug gave nuclear medicine a flying start. The production of the radionuclide was a direct result of the construction of the first nuclear reactor, which again was part of the work to develop the atomic bomb. This radiopharmaceutical, given as an oral preparation, was soon given the nickname "Atomic cocktail" by the public. After administration the radioactive iodine was transported directly and exclusively to the thyroid gland, as this is the only place in the body where the element is used. The new miracle drug was looked upon as just the first of many "magic bullets" in which it was hoped that the nuclear sciences would give us in the years to come to cure a multitude of various cancer forms.

However, the following years showed that it was a difficult task to develop such specific-targeting radiopharmaceuticals for therapeutic use; and nuclear medicine has, therefore, been dominated by diagnostic investigations until now. In the beginning static studies of various organs were most important. But the rapid progress within data processing technology, the development of advanced detection systems, and the introduction of new radionuclides and radiopharmaceuticals in dynamic function studies have now often made nuclear medicine investigations the preferred choice over other imaging modalities (computerized axial tomography, nuclear magnetic resonance and radiology) Today it is possible with routine nuclear medicine techniques to perform studies of how regional brain perfusion is affected by external stimuli or to distinguish between viable, ischemic, and dead regions of the myocardium.

Radiopharmacy is now on the brink of entering a new important stage in its development. Progress in biotechnology has made it possible to produce sufficient amounts of monoclonal antibodies to be used as carrier molecules for specific tumor-targeting radiopharmaceuticals. By attaching small quantities of radionuclides to these molecules, they can be used for diagnosis. By attaching larger amounts of alpha- or beta-emitting nuclides, they can be used for therapy. Several such products are used routinely today. It has been much more difficult to develop new "magic bullets" than first anticipated, because of problems of in vitro and in vivo stability and the change of tumor specificity of the antibody after labeling. But the progress in antibody labeling has also opened up interest for other types of therapeutic radiopharmaceuticals. We have already seen the development of several new products for the treatment of pain in patients with skeletal metastasis from breast or prostate cancer. New radiopharmaceuticals are being developed with hormones or peptides as the targeting component. Further work on therapeutic radiopharmaceuticals will be an important task in the years to come. They will bring new problems both in production, distribution, and drug administration in the hospitals.

CHARACTERISTICS OF RADIOPHARMACEUTICALS

Several aspects make radiopharmaceuticals different from regular drugs:

- All handling of radiopharmaceuticals from production through patient administration requires special training in working with radioactive materials. The products must be handled in accordance with radiation protection requirements in order to protect personnel and the environment.

- Due to radioactive decay radiopharmaceuticals have a changing composition with time. The volume of a liquid preparation to be given to a patient will depend on the radioactive concentration at the time of administration.

- The physical half-life of a radionuclide is often so short that the final preparation of the radiopharmaceutical must take place immediately before administration. This is solved by providing the nuclear medicine departments with preparation kits (semimanufactured radiopharmaceuticals), which are combined with the eluate from a radionuclide generator to prepare the product to be given to the patient.

Among the aseptically produced parenteral radiopharmaceuticals several galenic drug forms are represented, such as solutions, suspensions and colloids. These products must be produced aseptically, either because of the drug form (colloids or suspensions) or because they contain starting materials (human serum albumin) that are damaged by a sterilization procedure during manufacture. Therefore special precautions will have to be taken during manufacture, quality control (QC), and final preparation of these products to ensure the efficacy and safety of these radiopharmaceuticals.

Good Radiopharmaceutical Practices (GRPs) are the guidelines for the manufacture, preparation, and handling of radiopharmaceuticals. They combine the principles of traditional Good Manufacturing Practice (GMP) for pharmaceuticals with aspects of radiation protection. The purpose is to safeguard the quality of the radiopharmaceutical up to the point of administration to the patient. The two most important aspects for patient protection incorporated in the guidelines are as follows:

1. No radiopharmaceuticals should be used unless the benefit in terms of better diagnosis and better treatment outweighs any risk induced by radiation or other properties of the product.

2. All exposure to radiation should be as low as reasonably achievable for patient and staff. Equipment and facilities should be of a satisfactory quality, ensuring the protection of personnel and the environment. The radiopharmaceutical should be of optimal quality (Nordic Council on Medicines 1989).

Four types of aseptically produced parenteral radiopharmaceuticals are most commonly used:

1. *Ready-for-use radiopharmaceuticals.* These radiopharmaceuticals contain radionuclides with a sufficiently long physical half-life to allow distribution to the hospital of a product in a ready-for-use form.

2. *Radiopharmaceuticals prepared from semimanufactured products.* Several radionuclides with short half-lives, in particular ^{99m}Tc (the meta stable form of technetium 99) with a half-life of 6 hours, are supplied to hospitals in the form of radionuclide generators. By using such a generator hospitals can have a fresh supply of the desired radionuclide available every day

and can make the final preparation by combining the sterile radionuclide solution with a sterile semimanufactured product (a preparation kit). Both the generator and the kit may have been produced aseptically.

3. *Radiopharmaceuticals prepared directly prior to patient administration.* The radionuclide to be used may have such a short half-life that the radiopharmaceutical must be administered immediately after production. The radionuclide may be produced in hospital cyclotrons or by using radionuclide generators. (The most common products in this group are used for PET; the radionuclides have a half-life from a few minutes up to 30 minutes.)

4. *Radiopharmaceuticals prepared from biological samples from the patient (autologous-labeled products).* Blood cells and plasma proteins from a patient may be labeled with a radionuclide before readministration to the same patient.

PRODUCTION ASPECTS

Aseptically produced radiopharmaceuticals are a complex and varied group of drugs. When compared with regular drugs, the most obvious difference is that a large portion of the radiopharmaceuticals are either produced entirely or go through the final step(s) of preparation in hospitals. But for this group of pharmaceuticals there is not a large difference in the production techniques applied or the production facilities required for manufacture. Even for the largest commercial manufacturer of radiopharmaceuticals, the batch volume for aseptic production of a parenteral solution would be small compared to a nonradioactive parenteral prepared in a regular hospital pharmacy. For radiopharmaceuticals batch sizes from 5 ml up to 300 ml would be normal. The description of the production aspects for these products is, therefore, relevant to both industrial and hospital manufacture.

General Requirements for Isotope Laboratories

Both small- and large-scale production of radiopharmaceuticals must take place on premises designed, constructed, and maintained to suit the operations to be carried out. Special emphasis must be put on the control of radiation hazard. National regulations with regard to the design and classification of radioisotope laboratories must be fulfilled. Such laboratories are normally classified according to the amount of the various radionuclides to be handled at any time and the radiotoxicity grading given to each radionuclide. When planning the layout of the laboratory, it is recommended to allocate separate working areas or contained units for the various procedures to avoid possible cross-contamination of radionuclides.

One of the most important factors in planning a radioisotope laboratory is the design of the ventilation system. Laboratories with medium or high grading must be designed with the purpose of protecting personnel from inhaling radioactive gases or particles. The system should be designed to provide a negative pressure at the actual working area relative to the surrounding environment. The system should have an appropriate number of air changes per hour and the replacement air should be filtered. Exhaust air to the environment should be monitored for radioactivity and it may be necessary to install active charcoal filters to absorb radioactive gases and small particles.

There are three basic principles for the reduction of radiation doses to the operator:

1. *Time.* The shorter the time of exposure to radiation, the lower the dose to the operator. Good training and experienced personnel are essentials for running a safe production of radiopharmaceuticals.

2. *Distance.* The radiation dose decreases with a factor equal to the square root of the distance from the radiation source. The operator's distance from the source can be increased by using forceps, tongs, or manipulators in handling the radioactive material.

3. *Shielding.* The radiation dose can be reduced by placing shielding material between the source and the operator. For protection against gamma radiation, walls made of heavy concrete or lead bricks are used in radioisotope laboratories.

Production Premises for Radiopharmaceuticals

Premises must be designed with two important requirements in mind. The product should not be contaminated by the operator, and the operator and the environment should be protected from contamination by the radioactive product. However, the hygiene requirements for the production of pharmaceuticals are very similar to the requirements existing for radiation protection for work with radioactive substances. It is, therefore, an interesting task to combine these requirements to cover both aspects of production. The first similarity is met already at the entrance to such premises, where both sets of regulations indicate the establishment of a physical barrier. Personnel crossing the barrier will have to dress in proper, protective garments and gloves must be worn. By establishing a physical barrier, the access of unauthorized personnel to the premises is controlled.

The general requirements for the design of a production laboratory for radiopharmaceuticals should be the same as for nonradioactive pharmaceuticals with regard to materials and paints used for benches, walls, floors, and ceilings. Special care must be taken to choose materials with favorable absorption properties for spill of radioactive materials and that allow easy decontamination. It is a part of radiation protection requirements that this type of laboratory should have a sink, so that fast decontamination procedures can be carried out on the spot. To avoid the spread of contamination, it must be possible to operate the taps without using hands.

The production of a radiopharmaceutical will normally take place within a contained box unit, which is designated for the production of only one particular product to avoid contamination of different products and radionuclides. The box may be constructed as an airtight plastic box on a metal frame. The box is placed on concrete fundaments on a reinforced floor. The box is provided with funnels in the floor of the box, which are connected to tubes that lead down to separate, shielded storage containers for dry and wet radioactive waste under the box or in the basement (Figure 7.1). If the box is used for the production of a radiopharmaceutical incorporating a beta-emitting radionuclide, no additional shielding is needed to the plastic walls. For gamma-emitting radionuclides lead brick walls of 5 cm to 15 cm are built around the box for radiation protection. The thickness of the wall will depend on the energy of the gamma emission and the quantity of radioactivity to be handled (Figure 7.2). The lead bricks used to construct the wall have a special design. When they are stacked on top of each other, they will interlock. This is important to avoid cracks in the wall, through which radiation can escape (Figure 7.3). Radioactive material is introduced into the production box either through a maintenance corridor at the back of the box or through specially designed transport hatches fitted in the lead shielding in the front of the box. These hatches are normally constructed with an inner, tight-fitting plastic door with a rubber seal and an outer lead plug of the same thickness as the wall. The hatch must not be larger in area than to assure a negative pressure in the box when the lead plug and the plastic door have been opened. The inward airflow velocity is recommended to be of minimum 0.5 m/s when the hatch is open.

Most contamination incidents in isotope laboratories are caused by some weakness in the transport and transfer systems for the

Figure 7.1. Box unit for the production of beta-emitting radiopharmaceuticals. (Photo courtesy of Institutt for Energiteknikk)

radioactive material within the laboratory. The transport of larger quantities of radioactive materials should be done by using shielded transport containers, which are constructed to fit all transfer hatches in the production boxes. (The height of the working area and the height of the hatches above the floor should be standardized throughout the

Figure 7.2. Shielded box unit for production of gamma-emitting radiopharmaceuticals. (Photo courtesy of Institutt for Energiteknikk)

laboratory.) The transport container, fitted to a fourwheel carriage, may contain a drawer designed to contain irradiated target material or batch flasks. By placing the transport container close to the open hatch of the production box and opening a sliding door, the drawer with the radioactive material can be pushed by a long handle through the hatch and into the box (Figure 7.4).

Inside the box the material is either handled by remote control equipment or by manipulator tongs that are incorporated in the wall. The tong is fitted in the wall as part of a large tungsten sphere, which acts as a ballbearing and thereby allows more flexibility for the movement of the tong inside the box (Figure 7.5). Lead glass windows are fitted in the lead brick wall to allow the operator to overlook the area of operation inside the production unit. The lead glass is transparent and has good shielding properties. Because of the advantageous refraction property of the glass, just one or two windows will be needed in each cell. The thickness of the lead glass window is normally double the thickness of the lead brick wall in which it is installed. Thus the

Figure 7.3. Interlocking lead bricks. (Photo courtesy of Institutt for Energiteknikk)

window gives almost the same radiation protection as the wall, as the glass contains approximately 50 percent lead.

Parenteral radiopharmaceuticals produced at such premises are terminally sterilized in the final container either by autoclaving or by membrane filtration into the vial. The same type of premises will often be used for the production of oral solutions, and capsules and for the preparation of solutions of radionuclides that will be regarded as a starting material in the production of other radiopharmaceuticals. This is, for example, the case for the production of ^{131}I solution, which is used as a starting material for labeling proteins, peptides, and monoclonal antibodies.

Waste management is a very important element in the operation of isotope laboratories. The key factor is to reduce the amount of radioactive waste to a minimum. National legislation will vary considerably and influence the requirements that must be set for the handling of waste material. Effluents from nonradioactive parts of laboratories and water used for cooling systems in the radioactive section may normally by connected to the ordinary sewage system. All other waste, both liquid and solid waste, must be monitored carefully before release. Liquid waste is lead directly from production boxes to storage

Figure 7.4. Transfer of radioactive material from transport container to a box unit for production of radiopharmaceuticals. (Photo courtesy of Institutt for Energiteknikk)

tanks. The waste is often divided into separate storage tanks according to the physical half-life of the radionuclides. In this way small quantities of short-lived radionuclides may only have to be stored for a short time before release to the sewage system, while liquid waste containing long-lived radionuclides may have to be stored for very long periods. Solid waste is sorted according to the same principles. Both solid and liquid waste can be collected in polyethylene containers shielded with lead or heavy concrete. The containers are often placed in metal drums to facilitate the further handling and transport to waste treatment plants when the containers are full.

Production Premises for the Aseptic Production of Radiopharmaceuticals

Aseptic production of radiopharmaceuticals will increase the requirements for the design and construction of the premises. Contained workstations and clean-room technology will be applied to a much

Figure 7.5. Manipulation of radioactive material in a shielded box unit via exterior tongs. (Photo courtesy of Institutt for Energiteknikk)

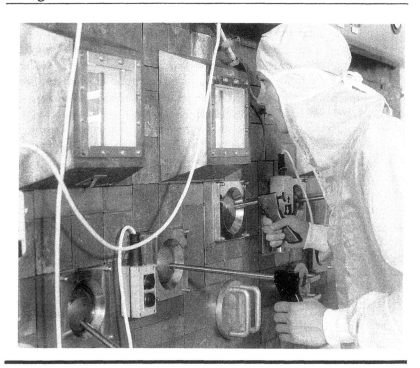

higher degree. The general requirements for the design of such premises are the same as for nonradioactive pharmaceuticals, including entry of staff and the introduction of materials through air locks. The main difference is found in the planning and design of the ventilation system. Laboratories for aseptic work normally have a positive pressure relative to the surrounding areas. But in laboratories for work with radioactivity, it is good practice to have a negative pressure to avoid the spread of radioactive material. In order to meet both pharmaceutical and radiation protection requirements, it is necessary to balance carefully the air pressures in the clean rooms, the air locks, and the surrounding areas. From a pharmaceutical point of view a negative pressure in the area designated for aseptic work can only be acceptable in special cases. There are various ways to solve these ventilation problems depending on the risk of airborne contamination. Most frequently, they are solved by using contained workstations with laminar airflow (LAF) units or sealed production boxes with negative pressure relative to the aseptic laboratory. The laboratory itself may then have positive pressure in relation to the surrounding premises. Warning systems must be installed to indicate failure in the filtered air supply to the laboratory; recording instruments should monitor the pressure difference between areas where this difference is of importance.

Sealed production boxes may be simple plastic boxes fitted with rubber gloves and hatches for the introduction of starting materials and production equipment. Such glove boxes are used mostly for simple labeling procedures; the amount of radioactivity that may be handled is limited. For large-scale manufacture special production boxes are designed. They are made of high quality stainless steel and are fitted with front and back plates made of plastic. A perforated stainless steel platform is fitted as a floor two to three cm above the bottom of the box. Production apparatus and equipment can easily be fitted to this elevated floor in the box. In the ceiling of the box a LAF unit is fitted together with appropriately designed lamps (Figure 7.6). Depending on the radiation characteristics of the radionuclide and the quantity of radioactivity to be handled, rubber gloves or handling tongs are mounted in the front plate of the box. The rubber gloves may contain lead or other shielding substances to reduce the radiation exposure to the fingers and the hands of the operator (Figure 7.7). In the

Figure 7.6. Box unit for aseptic production of radiopharmaceuticals. (Photo courtesy of Institutt for Energiteknikk)

Figure 7.7. Glove box for aseptic production of radiopharmaceuticals. (Photo courtesy of Institutt for Energiteknikk)

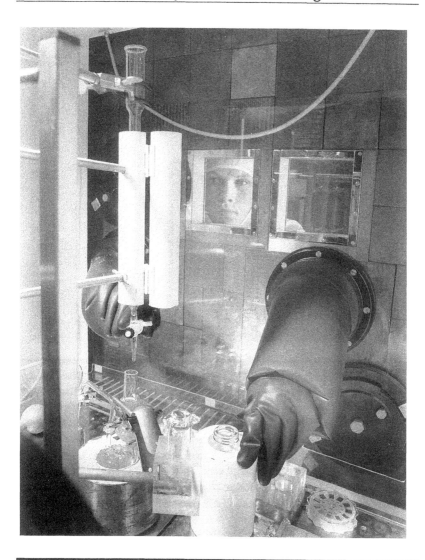

plastic back plate holes are cut out, which can be sealed tightly with specially constructed lids. These hatches are used for the introduction of starting materials and other production accessories (Figure 7.8). In practice the materials are introduced into a cleaned box and the ventilation system is started some time before the actual start of the production procedure. The air in the box is recirculated through high

Figure 7.8. Hatches for introduction of starting materials and equipment. (Photo courtesy of Institutt for Energiteknikk)

efficiency particulate air (HEPA) filters during the whole production until the product is dispensed and sealed either in a batch flask or in the final vial. Experience has shown that the continuous recirculation of air through the filter may lead to a considerable increase in the temperature in the box. This could be damaging for heat-sensitive products, such as labeled serum albumin and blood components. To avoid damage, a cooling loop may be placed on top of the box, where the air is cooled just before each passage through the LAF filter unit. In this way the temperature in the box is maintained at the same level as the room temperature (Figure 7.9).

The aseptic production boxes can also be shielded by walls of lead bricks. This is done when local shielding of production equipment within the box is not sufficient to reduce the radiation dose to the operator to a satisfactory level. Both rubber gloves and handling tongs can be incorporated in the protection wall. Special considerations must be made when designing the system for drains to be fitted to this type of production box.

Even in aseptic production laboratories it would be necessary to have sinks easily available for radioactive decontamination. An

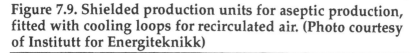

Figure 7.9. Shielded production units for aseptic production, fitted with cooling loops for recirculated air. (Photo courtesy of Institutt for Energiteknikk)

acceptable solution may be to place a sink with a heated water-lock just outside the door leading to the aseptic room.

Production of Lyophilized "Semifinished" Radiopharmaceuticals (Preparation Kits)

The extensive use of radionuclides with short half-lives has made it necessary to develop a range of preparation kits. These kits contain all the sterile ingredients needed to prepare the final radiopharmaceutical in the nuclear medicine department or in a centralized radiopharmacy. These products are, therefore, not radioactive until the radionuclide is added just prior to administration to the patient. However, these semi-manufactured products are, from the point of licensing, defined as radiopharmaceuticals, as they have no other application in medicine.

Most of these preparation kits have been developed for the labeling of various substances with ^{99m}Tc. Labeling is normally a single-step or two-step procedure, consisting of adding a solution of the radionuclide to a vial containing the ingredients needed for labeling. These ingredients include materials such as the substance or ligand to be labeled, a reducing agent to bring the radionuclide into a valency state with high reactivity, buffers for pH adjustment, and various stabilizers. The reason why most of these kits are supplied in a lyophilized form is to extend the shelflife for the product. Most of these products have to be produced aseptically, as they cannot be sterilized with other methods. (Sterilization by gamma irradiation has been tried for some of these products.) For the production of preparation kits, a conventional lyophilizer installed in a clean room can be used (Figure 7.10).

Stannous salts are often used as the reducing agent in these preparation kits. To ensure the required valency of the tin salt, it is important that all oxygen is removed from the vial during the lyophilization process. The vials are then filled with an inert gas, such as nitrogen, before the freeze-drying stoppers are closed completely. It is important to pass the gas through a membrane sterile filter to remove possible microbial contamination; in addition the gas may have to be dried.

Production of Radionuclide Generators

The essential part of the most common radionuclide generators is most often a simple chromatography column to which the mother radionuclide is absorbed on a suitable support material. The daughter radionuclide is a decay product of the mother nuclide. It is the daughter nuclide that is used in hospitals in combination with preparation kits to produce the final radiopharmaceutical. A typical example is the molybdenum/technetium generator system. In this system ^{99}Mo is fixed as molybdate to aluminium oxide in the column. The daughter nuclide ^{99m}Tc is eluted with sterile saline solution. After reconstitution of kits and the formation of various radiopharmaceuticals, this radionuclide is used in 90 percent of all nuclear medicine procedures.

The design of the generators and the accessories supplied for the elution of the daughter radionuclide varies from manufacturer to manufacturer. These systems can be very sophisticated in construction, making the generator easy and safe to use. But the complexity of the design may also create problems for maintaining the sterility and apyrogenicity of the eluate. The production hygiene aspects for radionuclide generators were not discussed much when the ^{99m}Tc generators were first introduced. Some experts claimed that these generators were self-sterilizing due to the large amount of ^{99}Mo that was fixed to the support material on the column. Other studies showed that the support material itself, most often aluminium oxide, acted as a very efficient filter bed that removed microbial contamination. Today no particular weight is put on these facts, but they are regarded as positive details in the total quality assurance (QA) program for the final product. Many producers autoclave the chromatography column after the

Figure 7.10. Lyophilizer installed in a clean room. (Photo courtesy of Institutt for Energiteknikk)

molybdate has been bound to the aluminium oxide. Other critical procedures during the production and assembly of the generators are now performed in designated areas fitted with LAF units with HEPA filters. Several manufacturers include a terminal membrane filter as an integral part of the generator design to ensure a sterile and apyrogenic eluate. As part of the development program, the generator system must also be validated to ensure that it gives a sterile eluate daily in routine use in a hospital environment. As an additional part of the QA program, sterility tests can also be performed on eluates from generators that have been returned from hospitals to producers.

Production of Ultra Short-Lived Radiopharmaceuticals

The installation of hospital cyclotrons and PET centers worldwide has lead to a great increase in the use of ultra short-lived radionuclides in nuclear medicine. These nuclides have a physical half-life from a few minutes to a few hours. The production of these radiopharmaceuticals will be the source of many new problems. The short half-life makes it impossible to introduce lengthy procedures in the routine quality assessment of the product. A strict QA program is, therefore, essential to secure the routine production of a high quality radiopharmaceutical. This type of hospital laboratory production has not been controlled until now and the products are not licensed to the same extent as other radiopharmaceuticals. Neither have pharmacopoeia monographs and international guidelines been fully established yet. As a supplement to the sale of the actual cyclotron, manufacturing companies do also offer to the PET laboratories special production units for the local manufacture of various compounds labeled with cyclotron-produced radionuclides. These units, so-called "black boxes," provide the production technology, starting materials, and equipment needed for production. But the use of such ready-to-use units may limit considerably the producer's knowledge of the properties and quality of the product. The production and QA programs for these ultra short-lived radiopharmaceuticals are presently under scrutiny by many official and unofficial groups; several reports and guidelines are to be expected in the future (Vera-Ruiz et al. 1990).

PRODUCTION EQUIPMENT AND PROCEDURES

Design of Equipment

Two principles are of utmost importance in the design of production equipment.

1. The equipment must be easy to repair after it has been installed in the production unit.

2. The equipment must have a simple construction and be easy to assemble, so a substitution can be done quickly when a total renovation of the equipment is necessary.

It is also important to keep in mind, when designing production equipment, that all sense of touch is lost when fingers are replaced by remote handling tongs.

The repair of production equipment for radiopharmaceuticals is more complicated than for nonradioactive drugs. Repair and maintenance work will have to wait until the radioactive contamination of the equipment has decayed to an acceptable level for handling by maintenance staff. To secure the continuous supply of important products, it may be necessary to construct two production lines in separate production boxes. The second unit is kept as a backup facility for production during breakdown and maintenance periods. The design of the equipment should be as simple as possible. It may, for example, be more advisable to use conventional pipettes made of glass for dispensing a radiopharmaceutical instead of installing the latest in advanced electronic dispensing equipment in the box. Changing a broken pipette is fast and easy, while the substitution of an automatic dispenser may require the removal of parts of the lead brick wall after the box has been left a period of time for radioactive decay. The introduction of new production technology for radiopharmaceuticals has, in fact, lead to unexpected problems, as the radiation effect on new components has not been considered. Some problems may be avoided by placing most of the electronic components on the outside of the production box. Glass is an important material in the construction of production equipment for radiopharmaceuticals. This material is also affected by radiation and will become discolored and brittle. When designing the equipment, the plans for necessary repairs in the future must be taken into consideration.

Sometimes it will be necessary to substitute not only parts but the whole of a production line. To facilitate this operation and thereby reduce time and radiation exposure, it is advantageous to build the whole production line on a stainless steel support frame, fitted with simple connections to electricity, water, and air supplies. The complete withdrawal of a production line from a box and the introduction of a new one can then be performed in a very short time (Figure 7.11).

The radiation protection of the personnel must always be an integral part in the design of the production equipment. Apart from the reduction of radiation doses obtained by shielding of the production box, the doses can be reduced considerably by using extra local shielding within the box. This can either be done by placing lead bricks within the production unit, or by constructing specially designed lead shields for critical parts of the production equipment. To obtain the most economical arrangement, the shielding should always be placed as close to the source of radiation as possible. The radiochemical purification of a labeled product is often performed on a chromatography

Figure 7.11. Entire production lane for a radiopharmaceutical fitted on a steel support frame (Photo courtesy of Institutt for Energiteknikk)

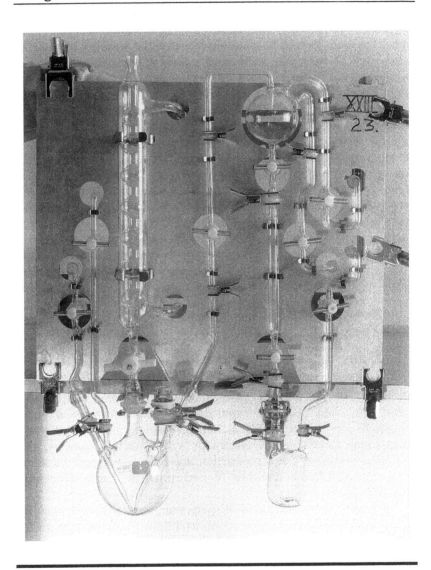

column. These can be shielded effectively by placing the column in a closefitting cylindrical lead shield. Hinges allow for the easy opening of the cylinder into two halves (Figures 7.12 and 7.13).

Figure 7.12. Wraparound lead shield for a chromatography column inside a unit for aseptic production. (Photo courtesy of Institutt for Energiteknikk)

Procedures

Procedures involving much direct handling of equipment with the hands and fingers should be avoided in radiopharmaceutical production. Advances in technology allow for the performance of many op-

Figure 7.13. Closeup view of wraparound lead shield. (Photo courtesy of Institutt for Energiteknikk)

erations by the use of control units placed outside the shielding of the production box. These operations include the transfer of liquids, distillation, or heating procedures. Membrane filtration is the process in the aseptic production of radiopharmaceuticals that still requires much direct handling by the fingers. Because of the small volumes produced, this filtration can be performed by removing the plunger from a sterile syringe fitted with a presterilized closed-membrane filter unit. The syringe is then placed in a syringe shield or a specially designed lead shield and placed on a stand. The product is transferred to the syringe, the plunger reintroduced, and pressed down, and the filtered product collected in a sterile vial.

For larger volumes or viscous solutions it may be better to incorporate a standard filter holder as part of the production line. The membrane filter has been sterilized in the holder by autoclaving before assembly in the line. In working with radioactive solutions, a vacuum system may be used to increase the speed of filtration. Filtration systems applying positive pressure are not used with radioactive products.

Sanitation and Maintenance

It is important to lay down detailed standard operating procedures (SOPs) for cleaning the premises and the equipment used in production. Personnel in charge of cleaning the premises must have special training that must include both clean-room aspects and radiation protection aspects. Only authorized personnel should be permitted to perform room sanitation. Personnel responsible for sanitation must also have training in how to monitor radioactive contamination of the premises and instructions for how to handle contaminated sanitation equipment.

Before any equipment or materials used during the manufacture process are removed from the production unit, a check for radioactive contamination must be performed. This can be done by simple hand-hold monitors with special probes that measure the various types of radiation. Limits must have been established for the contamination levels allowed on the items that are to be removed from the unit. If these limits of contamination are exceeded, further time for decay in the unit is necessary. After removal from the unit, the equipment is allowed to decay further in a special storage area, before it is cleaned and made ready for assembly again. For radiopharmaceuticals that are frequently manufactured on a routine schedule, it is necessary to have enough equipment for the assembly of several complete production lines. How often the various parts of the production line will be used depends on the amount of radioactive contamination and the decay of the radionuclide.

QUALITY ASSURANCE

General

The quality of a radiopharmaceutical must be assured by applying specially designed production and control systems so the final product is appropriate for the intended use. The QA procedures applied for nonradioactive pharmaceuticals are equally applicable to radiopharmaceuticals; but in addition, several tests relevant to the incorporation of radionuclides must be performed.

Documentation is an essential part of the QA program for a product. The documentation system should be such that the total history of a radiopharmaceutical batch can be traced. The guidelines for GMPs and GRPs state that such a system should include the following elements:

- *Specifications*, describing any product or material used in the preparation. This includes starting materials, packaging materials, and the final product.

- *Master formulas and preparation instructions.* These documents should be prepared for standard batch sizes.

- *Records*, providing the history of each starting material and batch of radiopharmaceutical

The recording of production data will make it necessary to bring batch documentation forms into the radioisotope laboratory. The paperwork should then be separated as far as possible from the handling of the radioactive material to avoid contamination of the documents.

Sterility and Pyrogen Testing

In the QA system applied for parenteral radiopharmaceuticals, several modifications of the normal procedures are needed for sterility and apyrogenicity testing. These procedures were often shown to be the factors that caused the highest radiation doses to the hands of personnel, as sampling and incubation make it necessary with repeated close contacts with radioactive sources. It is also a practical problem to have shielded incubators and not at least to find animal facilities that agreed to keep "radioactive" rabbits for the pyrogen test.

Even before the philosophy of QA was introduced, various pharmacopoeias allowed the release of radiopharmaceuticals before the test for microbial contamination had been completed. Most often, sterility testing would take more time than the physical half-life of the radionuclide incorporated in the product. At the end of the test, little or no product would be left for use. The general concept of regarding sterility testing as only a critical part of the QA system established by the manufacturer, is in line with the traditions within the field of radiopharmacy. The test for microbial contamination of radiopharmaceuticals is best carried out with filter methods. It is a great advantage to incubate only the filters instead of the radioactive solutions. The commercially available filter systems also reduce the need for close hand contact while performing the test. In this way both radiation doses to the fingers and the risk of microbial contamination from the operator to the sample are reduced. Even when such systems are applied, the test may have to be modified. For example, smaller volumes of the product than those required in the monograph are used to further reduce the radiation exposure.

The classical rabbit test for pyrogens was never a convenient test for parenteral radiopharmaceuticals. Apart from the practical problems of handling and housing radioactive animals for a prolonged period, the test itself had to be modified. The sample volumes of the products prescribed for the test were often not available, and extensive dilutions were also necessary to decrease the amount of radioactivity to be injected in each animal. The introduction of the limulus amoebocyte lysate (LAL) test for the detection of endotoxin-type pyrogens was, therefore, very welcome in radiopharmacy. The major advantages of the test are its rapid performance and the sensitivity. (Only small volumes of the product are needed.) The speed of the test makes it possible to perform the test on most radiopharmaceuticals before use. It was a great breakthrough when the European Pharmacopoeia gave the first approval for the use of the LAL test as a substitute for the rabbit test. This approval was first given only for radiopharmaceuticals, as the advantages of the new test were recognized to be particularly important for these products.

Visual Inspection of the Finished Product

As part of the QC all parenterals will be subject to an inspection for the possible content of particles. In most cases a normal visual control will not be allowed for radiopharmaceuticals, as radiation protection guidelines strongly discourage any direct eye contact with radioactive sources. However, an inspection for particles can be performed by using a mirror to avoid direct viewing. Another solution would be the use of a small television camera placed behind lead shielding, allowing personnel to assess the particle content on the screen.

For certain radiopharmaceuticals, such as suspensions and colloids, the control includes determination of particle number and size. This is preferably done using a microscope or electron microscope equipped with a recording camera. Membrane filtration can also be used for the determination of colloidal particle size. The product is then passed through a stack of polycarbonate filters with pores of defined sizes. After filtration the radioactivity remaining on each filter is measured. The activity level measured on the filters directly corresponds to the number of particles trapped, and the data can be used for the calculation of the particle size distribution of the product.

Validation and Control of Equipment and Procedures

The aseptic production of parenteral radiopharmaceuticals introduces new process elements that will have to be validated and for which control routines will have to be established. The same requirements applied for the approval and continuous survey of weighing, measuring, and recording equipment will have to be introduced for contamination and radiation monitors. Safety alarms will need to be connected to ventilation systems. Special emphasis must be given to the monitoring of critical equipment, such as dose calibrators, that are used to check the accuracy of the dispensing of patient doses.

REFERENCES

Nordic Council on Medicines. 1989. Radiopharmacy: Preparation and control of radiopharmaceuticals in hospitals. Uppsala, Sweden: NLN publication No 26.

Vera-Ruiz, H., C. S. Marcus, V. W. Pike, H. H. Coenen, J. S. Fowler, G. J. Meyer, P. H. Cox, W. Vaalburg, R. Cantineau, F. Helus, and R. M. Lambrecht. Report of an International Atomic Energy Agency's advisory group meeting on quality control of cyclotron-produced radiopharmaceuticals. *Nucl. Med. Biol.* 17 (5):445–456.

FURTHER READING

Radioisotope production and quality control. 1971. Technical Reports Series No 128. Vienna; International Atomic Energy Agency.

Saha, G. B. 1992. *Fundamentals of Nuclear Pharmacy* Heidelberg, Germany: Springer Verlag.

Sampson, C .B., ed. 1994. *Textbook of radiopharmacy*, 2nd enlarged ed. New York and London: Gordon and Beach Science Publishers.

8

Good Aseptic Practices: Education and Training of Personnel Involved in Aseptic Processing

Michael J. Akers

CHAPTER OBJECTIVES

Controversy continuously surrounds aseptic process manufacturing and control because of the lack of terminal sterilization and, consequently, the concern over the level of assurance of sterility for each lot or batch of product aseptically manufactured. Products prepared by aseptic manufacturing have historically been assumed to have a certain level of nonsterility or sterility assurance level (SAL). For years the U.S. Food and Drug Administration (FDA) has accepted SALs (as determined by aseptic process validation studies) for aseptically manufacturing sterile products of one in one thousand (10^{-3}) units of product; the FDA makes it clear that the manufacturer bears responsibility for any unit of product found to be nonsterile. Limitations certainly exist in aseptic manufacturing and testing for sterility of these products because of the need for human intervention in these operations. Advances in barrier technology are making progress in minimizing or eliminating direct human contact with sterile products during sterility testing and, in some cases, product manufacturing. Nevertheless, this technology is still in its infancy; even sterility testing using isolation chambers is not universally applied in all sterility testing laboratories. It will require several more years before this technology totally takes the place of human operators being needed in the manufacturing and testing of sterile products.

Because of human involvement with sterile products during aseptic manufacturing and control, great strides have been made to increase assurance of product sterility. Facility design, the use of laminar airflow (LAF) and high efficiency particulate air (HEPA)–filtration of air, sanitization and sterilization techniques for facilities and equipment, stricter environmental control programs, and aseptic process validation all have contributed to much greater control and sterility assurance of products prepared by aseptic manufacturing. However, one area vital to microbiological control and sterility assurance still needs significant improvement. That area is the education and training of people to work effectively and efficiently in clean room environments during aseptic product manufacturing and testing of sterile products produced.

The purpose of this chapter is to help the interested reader develop training and education courses and programs that thoroughly expose the employee not only to aseptic practices and techniques, but also to teach the basic principles behind aseptic practices; that is, basic microbiological knowledge, fundamentals of contamination control, and product characteristics that support microbial contaminants.

REVIEW OF THE LITERATURE

A few articles, but not many, have been published related to the training of employees in the pharmaceutical industry. Even fewer articles have been published that focus specifically on training and education of personnel working in clean room operations involving aseptic practices. One has to assume that every manufacturer of sterile products sustains an active training program on good aseptic practices, but very few have published information about these programs.

In 1974 the Parenteral Drug Association presented a panel discussion on the various aspects of educational training of on-line personnel. Avis (1975) presented reasons for and suggested a course outline for academic involvement in in-plant training courses. Table 8.1 gives the lecture outline Avis proposed using in these in-plant courses. Padovano (1975) discussed the training program utilized at that time by Warner-Lambert in their Puerto Rico facilities. Training involved orientation to the company's quality philosophies, health and hygiene, quality control (QC) practices, and periodic courses and programs on local procedures used in the laboratory and in the manufacturing areas. He concluded his presentation by listing the major reasons why quality training is so important (Table 8.2).

Although current Good Manufacturing Practices (cGMP) did not become official (backed by the power of the law) until September 29,

Table 8.1. Preliminary In-Plant Lecture Outline on Parenteral Dosage Forms*

Process Personnel

Importance of role of each person

The Patient

The need for safety, quality, purity, identity, and strength

Environmental Design and Control

- Construction features
- Housekeeping
- Surface disinfection
- Air distribution systems
- Laminar airflow personnel
- Traffic flow
- Evaluation of environmental control

The Container and Components

Types of materials—glass, rubber, plastic

Ingredients of Pharmaceutical Formulations

- Vehicles
- Solutes
- Stability evaluation

Processing of the Product

- Cleaning
- Preparation
- Filtration
- Filling
- Sealing
- Aseptic techniques
- Freeze-drying
- Packaging and labeling
- Sterilization procedures

Continued on next page

Continued from previous page

The Testing Program

- Sterility
- Pyrogen
- Clarity
- Leaker

Quality Assurance

- Role of operator toward quality production
- Compliance
- Polices and procedures
- Improve technical understanding

*Source: Avis, K. E. (1975). Reproduced with permission from the Parenteral Drug Association.

Table 8.2. Reasons for Quality Training Programs*

1. Quality training leads to quality minded employees who know their assignments.

2. Quality training leads to economical production.

3. Quality training leads to elimination of product rejections.

4. Quality training leads to increased productivity.

5. Quality training leads to success for everyone.

*Source: Padovano (1975). Reproduced with permission from the Parenteral Drug Association.

1978, the FDA began talking about cGMP requirements with respect to employee training in 1975. Haarmeyer (1976) presented a paper at the Parenteral Drug Association where he reinforced and emphasized the importance and expectations cGMP regulations make of employee training and that it is the responsibility of plant management to ensure that such training is accomplished. Note the following remarks Mr. Haarmeyer made:

I cannot recall how many times I have performed inspections of pharmaceutical manufacturing plants and have observed employees perform functions with total ignorance of Good Manufacturing Practices with obvious skeptical attitude towards them. I cannot recall how many times I have observed employees disregard recommended Good Manufacturing Practices Regulations procedures and then five minutes later remember to do it after the fact and because a Food and Drug Administration inspector was in the plant. I cannot recall how many times I have seen an employee take a short-cut step in his operation in order to lengthen his coffee break or give him more time to do something, without knowing that his short-cut could potentially maim or injure an innocent consumer.

These are extremely strong criticisms from an FDA spokesperson nearly 20 years ago. Could these comments be repeated today?

What is the right way to train employees in the pharmaceutical industry? Gallagher (1980) attempted to answer that question by emphasizing what is now referred to as "front-end analysis"; that is, first determining what the employee really needs with respect to training and identifying how training programs truly can fill those needs. According to Gallagher there are four major areas of training that should be developed:

1. Orientation for new employees

2. Job function training (detailed job instruction)

3. Ongoing training

4. Incumbent training

The FDA, through its cGMP regulations (21CFR 1978), has long held the belief that employee training is vitally important to the production of safe and effective drug products. Current views from the FDA regarding training were expressed in an article published by Levchuk (1991). Highlights of Levchuk's paper, representing the position of the FDA regarding employee training in the pharmaceutical industry, are presented in Table 8.3.

Luna (1993), in this author's opinion, has heretofore published the most comprehensive treatment of training for the aseptic operator. She does an outstanding job of presenting information and data on sources of contamination, training techniques, use of gowning, developing training programs, the role of management in training, and the criteria used to select clean room personnel. For example, her thorough discussion of the six personnel factors required to control microbiological

Table 8.3. Important Points to Consider in GMP Training Programs*

- Training is a dynamic process that should occur over an entire career.

- Training programs may exist, but most lack effectiveness.

- Training documentation does not verify quality.

- Systematic training plans for each job should be in place.
  — Performance objectives
  — Right methods for achieving these objectives
  — Assessment process to measure accomplishment of performance objectives

- Four main methods for measuring training effectiveness:
  — Testing
  — Evaluation of on-the-job error rates
  — Skill-related questions
  — Employee reports regarding their own assessment of their effectiveness

- Quality Assurance must do a better job of evaluating training effectiveness and ensure that training is current and of high quality.

- Training must be excellent, not to satisfy the FDA, but to ensure the manufacture of safe and effective drug products.

*Source: Levchuk (1991).

contamination (i.e., lack thereof) in the clean room are worth repeating in summary fashion here:

Effect of Bathing

Bathing will remove microorganisms, but will increase the number of particles emitted from the body. Amazingly, while the mechanical process of washing will remove bacteria from microcolonies accumulated on the skin, bacterial cells will be spread over the entire surface of the body, particularly near the perineum and on the face.

The washing process will remove the outer oily sebum layer of the skin, causing skin scales to dry, curl up, and peel off the body. This causes an increase in particulate dispersion rates immediately after showering. Within two hours after bathing the surface of the skin will resume its original pattern of microcolonies. Therefore, employees working in clean rooms should bathe at least two hours before they enter the clean room environment to minimize the extent of skin particulate shedding due to the bathing process.

Effect of Suntan

Suntanning dries the skin, causing it to flake and peel more easily. It is not surprising, therefore, that incidents of contamination occur more frequently in the summer months. Creams help to reduce skin shedding due to suntans, but do not entirely solve the problem. Employees working in clean rooms should be careful not to expose their skin to excessive sunlight.

Effect of Clothing

Friction between clothing and the skin will increase the rate of bacterial shedding from the skin. Up to 10 mg of skin particles may be deposited in a person's clothing during a 2-hour period. Hosiery also will increase skin dissemination as skin cells will pass through the pores of the stockings.

Good Personal Hygiene

Good personal hygiene includes the following:

- Bathing or showering routinely
- Washing the hair
- Trimming facial hair
- Cleaning the fingernails
- Wearing clean clothing and shoes

Clean room personnel must believe in and practice these basic steps in good hygienic practices.

Effect of Clean Room Garment

Clean room garments will largely, but not entirely, eliminate the dissemination of shedded skin cells into the clean room environment.

Particles from the skin and clothing will still find a way to escape the clean room garment and enter the environment. Garment areas vulnerable to particulate transport include seams; zippers; openings at the neck, wrists, and ankles; and even from the surface of the garment. Selection of clean room garments depends on the combination of the type of fabric and design of uniform that will balance the need to eliminate particulate and microbial contamination from the person while at the same time provide comfort to the wearer during usage. Luna (1993) considerably discusses the advantages and limitations of different types of clean room garments, including types of fabrics and different gowning designs. Five criteria must be met for a fabric to be used as a clean room garment:

1. Comfortable to wear

2. Function design to contain contaminating particles and bacteria shed from the skin and clothing of personnel

3. Possess antistatic characteristics

4. Ability to withstand sterilization

5. Durability

Effect of Traffic Movement and Control

Airborne contamination is directly related to the number of people working in a clean room and the types of activities they are doing. Therefore, the number of personnel working in a clean room should be kept to a minimum. Such personnel must be trained to work in a manner that minimizes air turbulence. This will be discussed in more detail below, but includes such aspects as avoiding the disruption of LAF, minimal talking and bodily movement, and avoidance of congregating within the same area where critical aseptic manipulations are performed.

Other literature relevant to this topic of personnel education and training will be referenced throughout the remainder of this chapter.

SCOPE OF THE PROBLEM

A contaminated parenteral product may cause one or more of the following problems or complications:

*   Deterioration of the product and the loss of potency

*   Pyrogenic reactions after administration to a patient

- Infection in the patient

- Colonization of microorganisms in the patient with the risk of secondary infection

Table 8.4 lists the most common types of general infections caused by contaminated pharmaceutical products (Ringertz and Ringertz 1982). While these examples stress the worst outcome of product contamination, it must be emphasized that any microorganism, pathogenic or nonpathogenic, found in a supposedly sterile pharmaceutical product is dangerous. Education and training in good aseptic practices, while pointing out the worst that can happen from a contaminated product administered to a human being, must emphasize that any type of contamination is wrong and indicative of poor practices.

Whyte et al. (1989) found that 12 of 19 (63 percent) small-volume injectable products not containing antimicrobial preservative agents supported the growth of various types of microorganisms. Even products containing antimicrobial preservative agents, 3 of 24 (13 percent) supported microbial growth. Biological products (blood, protein, and other natural products) usually support microbial growth quite readily because these products contain appropriate nutrient sources, have a neutral solution pH, and are essentially iso-osmotic. The famous report in *Lancet* (Phillips et al. 1972), although now relatively dated, still reminds us that at that time 56 percent of intravenous fluids containing additives prepared in a hospital pharmacy admixture service were contaminated.

This discussion points directly to the fact that sterile pharmaceutical products can and do, if conditions are right and lack of aseptic practices present, support the growth of microbiological contamination. It is imperative that operating personnel working in aseptic environments greatly appreciate and understand this fact.

Table 8.4. General Infections Caused by Contaminated Pharmaceutical Products*

- Septicemia caused by nonsterile infusion solutions
  Water contaminants such as *Pseudomonas species,*
  Enterobacter cloacae

- Bacteremia caused by intravenous infusion catheters
  Skin organisms such as *Staphylococcus aureus, group A Streptococci*

- Infections caused by contaminated subcutaneous injections

---

*Source: Ringertz and Ringertz (1982).

In the time period of 1965 through 1975, the FDA reported that 608 large-volume parenteral (LVP) recalls took place involving >43 million containers with the tragic result of 54 people being killed and 410 injured because of contaminated products (*Report of the Comptroller General* 1976). In 1970 nine people died and more than 400 cases of septicemia were reported as a result of improperly sterilized intravenous fluids (*Nosocomial bacteremias* 1971).

Between 1981 and 1991 the FDA reported that 40 human pharmaceutical products and 7 animal health products were recalled because of contamination. All of these products had been aseptically manufactured (*Federal Register* 1991). This was one of the justifications why the FDA proposed changes in GMPs to require that sterile products be terminally sterilized unless valid data demonstrate that the drug or drug product cannot withstand high temperatures or other conditions used in terminal sterilization.

The alarming rise in the incidence of microbial contamination of pharmaceutical preparations in the clinical setting was the subject of a review paper by Ringertz and Ringertz (1982). They cited over 80 published articles, letters, and other reports of infections caused by contaminated pharmaceutical products (parenteral, nonparenteral, and medical devices). More recent reviews of microbial contamination of pharmaceutical products have been published by Turco and King (1987, 383–384) and Bloomfield (1990).

The American Society of Hospital Pharmacists (now the American Society of Health-System Pharmacists) has published new guidelines that pharmacists must follow in the pharmacy in handling sterile dosage forms (*Draft Guidelines* 1992). These guidelines followed several unfortunate incidents where nonsterile products prepared in hospital pharmacies were administered to patients resulting in sepsis, ophthalmic contamination, and even death. These nonsterile products were a result of faulty practices employed by personnel who were inadequately trained in aseptic practices and techniques. A related document for the dispensing of sterile drug products for home use has been published (*Sterile Drug Products* 1995).

It seems ironic that people are required to produce and use sterile dosage forms, yet people are the primary source and the reason why these products become microbiologically contaminated! Although carelessness is a major reason why people working with sterile dosage forms contaminate these products, people usually try to be careful in their practices when handling these vulnerable products. The problem lies in the fact that people are not adequately trained to handle sterile dosage forms, and are not adequately educated on the basic principles of sterile products, such as their basic characteristics, how they are administered, why they are used, and why contamination is such a problem. Industrial training programs on aseptic techniques do a reasonably good job in improving a person's appreciation of how to minimize introducing microbial contamination into a product. Yet, many employees working in clean room areas in the pharmaceutical industry do not know *why* they must do what they do. They are not adequately educated about basic microbiological principles and how or why the particular products with which they are working are so susceptible to accidental contamination.

THE ROLE OF MANAGEMENT IN EDUCATION AND TRAINING

Management's role in education and training should not be a source of problems occurring in clean room operations, but, unfortunately, it is. Employees will admit that a major reason why attitudes are not more diligent toward excellence in aseptic practices is the perception that management is not involved and/or does not appreciate what is involved in good aseptic techniques and practices. There are at least six major responsibilities of supervisors in training people to know and follow good aseptic practices.

1. Supervisors themselves must recognize and fully appreciate the need to follow cGMPs and good aseptic practices.

2. Supervisors should hire people who are willing to accept and follow procedures assuring adherence to cGMPs and good aseptic practices.

3. Supervisors must effectively communicate and exemplify the importance of cGMPs and good aseptic practices without breeding ill feelings among employees.

4. Supervisors must work around the fact that good aseptic practices cannot be forced upon a person. The employee's attitude is so important and attitudes can be markedly influenced by the attitudes of supervision toward them and toward good aseptic practices.

5. Supervisors must support thorough and ongoing training programs in good aseptic techniques.

6. Supervisors should strive to be teachers and leaders in enabling their employees to want to learn and practice good aseptic techniques and to follow all aspects of good manufacturing practices.

Additional responsibilities and the impact management can have on the quality of training and conduct in clean room operations are listed in Table 8.5.

Table 8.5. Responsibilities and the Impact of Management on the Quality of Training and Conduct in Clean Room Operations

---

1. Impart belief that good aseptic practices are essential for the manufacture of sterile products possessing the GMP values of safety, identity, strength, purity, and quality.

2. Instill feelings of pride and confidence in clean room operators.

3. Stress the concept of teamwork.

4. Help people feel honored to be chosen for such a critical job function.

5. Keep employees informed continuously of what is going on internally and externally.
   - New or revised GMP regulations
   - Learning points from QA and/or FDA (or other regulatory) inspections
   - Advancements made from reading literature, attending outside meetings, seminars

6. Involve employees in goal setting, problem solving, decision making.

7. Actively listen and respond to employee feedback.

8. Creatively recognize conscientious and outstanding performance.

WHAT THE FDA LOOKS FOR WHEN OBSERVING ASEPTIC PRACTICES

The compilation of several literature reports authored by FDA inspectors find the following areas where the FDA inspector concentrates when he or she is conducting an inspection of an aseptic operation particularly focused on aseptic operator training (Tetzlaff 1987; Phillips 1989; Levchuk and Lord 1989; Avallone 1990). Phillips said that "The training of operators involved in aseptic processing operations is of significant interest to FDA investigators."

- Observation of operators during setup and actual filling
- Observation of the operator for bad habits such as gum chewing

- Observation of the gowning of the operator to see if hair is exposed

- Observation of improper hand washing and sanitization

- Determining the degree of supervision of aseptic operators

- Accuracy and thoroughness of the documentation of operator training on aseptic practices

- What are the means used by the manufacturer to evaluate the knowledge and practice of good aseptic practices? Are media fills conducted? How closely do they follow actual production conditions? What are the procedures for taking action when limits are met or exceeded? Are operator gowning surfaces sampled? Are operators periodically monitored by supervisory observation and are these documented? Are written tests administered and what are the criteria for passing or failing?

- What is the frequency of monitoring and evaluating operator practices of good aseptic techniques?

- What is the procedure used for retraining operators who fail microbial monitoring tests or who are observed to be practicing poor aseptic techniques in the clean room?

None of these concerns by the FDA when inspecting aseptic manufacturing operations should be a surprise or disputed by those responsible for these operations. Any quality education and training program should be designed and implemented to address these and other concerns. The motivation for education and training on aseptic practices should not be because the FDA expects such training. The motivation should be based on the desire of the company and its personnel to manufacture with the highest degree of assurance products that are sterile, pyrogen free, particulate free, and possess all the other high quality attributes of safety, identity, strength, and purity. Aspects of these education and training programs will now be discussed.

DIDACTIC EDUCATION

Requirements of Education and Training Personnel

Like any good teacher, an individual responsible for educating and training people to work with sterile products and processes must be committed to enabling the learner to learn. This cannot be a job given to an employee for whom no other position can be found. It is the special, high-performing employee with interest in teaching and follow-up who is the best choice for this type of assignment. Besides commitment to teaching and learning enablement, education and training personnel should have the following skills:

- Enthusiasm

- Responsiveness

- Demonstrable skill in performing good aseptic techniques

- Adequate knowledge of sterile production processing

- Good documentation and follow-up skills

One of the biggest complaints clean room operators express about their training program is the lack of experienced and qualified trainers. Unfortunately, in some cases, marginal or even poor performing employees are put into positions where they have responsibility for training others in good aseptic practices. This is absurdity and should never happen. The opposite ought to be practiced (i.e., the top performing operators ought to be positioned as trainers and instructors of others in the theory and practice of good aseptic practices and techniques).

Trainers should make good use of the available literature for supporting their instruction. People are more motivated when they see or can read specific examples or illustrations pointing out the importance of practicing certain behaviors. Literature citations throughout this chapter are good tools to use to supplement instruction on good aseptic practices. For example, the paper by Howorth (1988) showing the number of particles generated by different types of movement leaves a distinct impression on learners in this subject. Likewise, the showing of videotapes, such as those noted in Appendix B, provide pictures of what happens when inappropriate behavior occurs in the clean room. References from the clinical literature reporting on adverse reactions, even death, occurring from contaminated pharmaceutical products, help instruct and leave indelible impressions on clean room employees that what they do has life-saving impact.

Characteristics of Sterile Dosage Forms

Personnel working with sterile dosage forms must be thoroughly educated regarding the basic characteristics of these products with which they are handling. The big three unique characteristics of sterile dosage forms are as follows:

- Freedom from microbial contamination

- Freedom from pyrogenic contamination

- Freedom from particulate contamination

DeLuca (1983) wrote an article several years ago that addressed this need, whereby the approach to training personnel working with sterile dosage forms should be organized around the theme of micro-contamination control. This author has successfully used this approach in in-plant educational courses on good aseptic practices. Some of the learning objectives from teaching on each of the three characteristics of sterile dosage forms are summarized in Table 8.6. Many references (examples: Avis 1990; Akers 1994; Boylan and Fites 1990) are available to help education and training personnel prepare didactic material to educate clean room personnel on the basic characteristics of sterile products.

Table 8.6. Fundamental Knowledge in Aseptic Manufacturing Practices (focus on removal of and maintaining freedom from microbiological, pyrogenic, and particulate contamination)

Building and Facilities

- GMPs
- Design for cleaning and efficient movement

Equipment

- Cleaning
- Sterilization

Packaging

- Cleaning
- Sterilization
- Wrapping
- Sorption and leaching of rubber and plastic
- Glass particulates

Raw Materials and Compounding

- Bioburden

Environmental Control

- HEPA filters and laminar airflow
- Surface disinfectants
- Personal hygiene
- Personnel gowning
- Air and surface monitoring

Continued on next page

Continued from previous page

Water

- Different types and preparation
- Testing and monitoring

Sterilization

- Kinetics of microbial destruction
- Saturated steam under pressure
- Proper loading of steam sterilizers
- Dry heat and tunnel sterilizers
- Filtration
- Biological indicators

In-Process Quality Control

- Filter integrity testing
- Physical and chemical testing

Filling and Sealing

Lyophilization

- Basic principles and operation
- Sterilization of chamber
- Proper loading and unloading

Parenteral Quality Control Testing

- Sterility test
- Pyrogen test
- Endotoxin test
- Particulate matter test
- Package integrity test
- Antimicrobial preservative effectiveness test

Basics of Contamination

Our world is full of microbial life. Microbial contamination is everywhere, including the manufacturing environment and the materials used in product formulation and packaging (Underwood 1977).

The Atmosphere

Air is not a natural environment for microbial growth since it is usually too dry and absent of nutrients. However, there are several types of microorganisms that can tolerate the dry environment. They include spore-forming bacteria (*Bacillus sp., Clostridium sp.*), nonspore-forming bacteria (*Staphylococcus sp., Streptococcus sp., Cornebacter sp.*), molds (*Penicillin sp., Cladosporium sp., Aspergillus sp.*), and yeast (*Rhodotorula sp.*).

The degree of microbial contamination in the air depends on several factors. Dirt and particles in the air are contributed by machinery, people, and anything that causes air to move (e.g., ventilation, convection currents). Humidity actually decreases airborne organisms as they alight to moisture droplets and fall to the floor. Particles enter the air from people talking and sneezing; such particles will attract microorganisms. Handling of contaminated materials such as paper, chemicals, and packaging materials will introduce microbial contamination into the air.

Water

Water is used in sterile process manufacturing for the washing of equipment and packaging, process cooling, and compounding of the product. Microorganisms indigenous to freshwater include *Pseudomonas sp.* and other gram-negative bacteria. Bacteria introduced as a result of soil erosion, decaying plant matter, and heavy rain include *Bacillus subtilis*. Bacteria also are introduced by sewage contamination (e.g., *E. coli*). Bacteria from animals and plants usually die due to unfavorable living conditions in water.

Raw Materials

Many of the raw materials used in parenteral product formulations originate from animals and plants. These materials can be contaminated with pathogens (*E. coli, Salmonella*) from animals and organisms associated with plant cultivation (many gram-negative and gram-positive bacteria). Raw materials also can be contaminated with different types of molds and yeast. Plant-originated raw materials can also contain pathogens from animals via sewage. For these reasons raw materials used in parenteral formulations are mostly synthetic since they usually are free from all but incidental microbial contamination.

Packaging

Microbial flora of packaging material depends on its composition and storage. Glass, transported in cardboard boxes, usually contain mold spores (*Penicillin sp., Aspergillus sp.*) and bacteria such as *Bacillus sp.* Packaging materials that have smooth and impervious surfaces—polyethylene, polyvinyl chloride, metal foils—have low surface microbial growth. Cardboard and paper carry mold spores. Rubber closures can also have mold contamination.

Buildings

Most common flora of walls and ceilings are molds (*Cladosporium sp., Aspergillus sp.*). Nutrients come from plaster on which paint is applied. Contamination of floors occurs due to water and cracks. Edges and joints are culpable to contamination due to inadequate sealing.

Equipment

Contamination depends on the type of equipment, nutrients available, and environmental conditions (pH and temperature). The harder-to-clean areas such as screw threads, agitator blades, lubricated areas, valves, and pipe joints can be harbors for microbial contamination. Also brooms, mops, or anything used for cleaning can be contaminated and increase atmospheric contamination by raising dust and splashing water.

People

This area, of course, is the primary focus of this chapter and will be treated carefully throughout the remainder of the chapter. Indeed, people are the largest single source of contamination in parenteral products. From our skin originates many flora, but the most undesirable contaminant is *Staphylococcus aureus*. It is not removed upon washing the hands and face as it resides in the deep layers of the skin. Other bacteria, yeast, and fungi exist in the bodily areas that are moist, fatty, waxy, and wherever secretions occur. Bacteria other than natural skin flora are found on people as a result of poor personal hygiene. Bacteria from wounds can be numerous. Our nasal passages contain large numbers of *Staphylococcus aureus;* our throats are colonized by *Streptococci* as well as saprophytic *Streptococci* of the ciridian group lining our respiratory tract, including our mouths.

Data on how much particulate matter is generated by human beings are fascinating yet disturbing when this information is contrasted with the need to maintain Class 100 clean room conditions. Note some of these statistics (Luna 1993):

- Each adult loses approximately 6 to 14 grams of dead skin material every day.

- Each person loses a complete layer of skin about every four days, which is equivalent to a minimum shed rate of 10,000,000 particles per day!

- Ordinary walking movements emit approximately 10,000 particles per minute.

We human beings are generators of millions of particles daily, particles that can and do carry microbiological contamination. It is the responsibility of training personnel to ingrain this fact into the minds and attitudes of personnel working in aseptic operations. Once this has occurred, then, hopefully, actions and techniques taken will reflect care and concern on the part of the employee to do whatever is required to minimize or avoid breaches in aseptic practices.

The use of pictures, videotape, and other visual aids to teach aseptic techniques is highly recommended. Pictures such as that shown in Figure 8.1 can be used to help people identify errors in good aseptic practices and have some fun during this exercise. Staging good and bad aseptic practices and using video to demonstrate these practices also are good methods of didactic instruction. People love to participate in games. Games that advocate good aseptic practices help introduce aseptic operators to the importance of these techniques and will help reinforce good aseptic principles during "hands-on" training sessions.

Elimination of Sources of Contamination

Control and/or elimination of microbial contamination involves the following:

Engineering Design of Buildings and Facilities

Laminar airflow through HEPA filters is designed to provide an area with 99.97 percent sterile air. While HEPA filters do not sterilize objects, they are intended to maintain a sterile environment. The problem enters when people or objects interrupt the laminar flow pattern from HEPA filtration, as demonstrated in Figure 8.2. When laminar flow is interrupted, it is reestablished downstream within a distance equal to approximately three times the diameter of the object causing the interruption. The continuous flow rate (usually 90 feet per minute), considerably above the rate at which airborne particles will settle, sweeps suspended matter out of the LAF area and prevents microorganisms from being carried upstream.

Figure 8.1. Example of cartoon depicting errors in aseptic techniques during equipment setup. There are at least 14 errors. Can you find them? (1. Excess paper; 2. Tweezer on work surface; 3. Contact with open fill port; 4. Bare wrist; 5. Mask low on nose; 6. Goggles on forehead; 7. Gown unzipped; 8. Nonsterile supplies in aseptic area; 9. Arm resting on equipment; 10. Adjusting goggles; 11. Hood out of gown; 12. Second hood out of gown; 13. Rip in uniform; 14. Bare wrist)

Design of Equipment

As briefly discussed previously, equipment must be designed so that it can be cleaned and sanitized easily. Bacterial adhesion and colonization on stainless steel is well known although poorly understood (Vanhaecke and Van den Haesevelde 1991). Prevention of bacterial colonization is the most important step that is best achieved by physical (mechanical) removal of stainless steel surfaces followed by application of an effective disinfectant solution proven to reduce microorganisms to acceptable low limits.

Equipment must be properly designed so that during operation, should there be a need for minor repairs or adjustments, such interventions can occur without risk of product contamination. Change

Figure 8.2. Schematic of what happens when an object blocks the pathway of laminar airflow. (Courtesy of Mr. Steve Vogtman, Eli Lilly and Company)

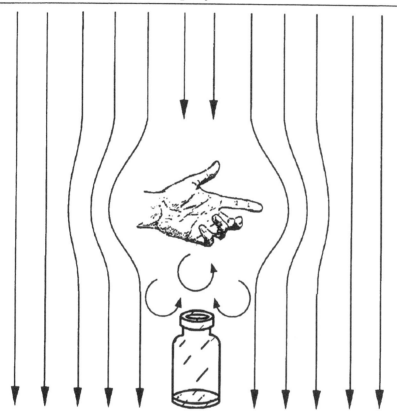

parts must be easily removed and/or replaced. Accessibility to frequently adjusted parts of a piece of equipment must be part of the initial design. For example, changing or adjusting filling nozzles or heads must be able to be performed without the need to shut down the filling operation and resanitize the entire piece of equipment.

Cleaning and Sterilization

Validated cleaning procedures and sterilization processes are essential to the control and elimination of microbial contamination. Cleaning must be carried out precisely by personnel according to procedures found to be optimal in removing microbial contamination and other sources of unclean material (dirt, chemicals, oils, etc.). Validation of cleaning of equipment has been the subject of significantly increased attention by the FDA over the past few years. Cleaning of surfaces, whether they are equipment, facility, packaging material, or any other surface that is part of aseptic operations, involves the use of reagents, including sanitization solutions, that are shown to remove, to a known limit of detection, any substance (microbial, chemical, etc.) known to be a contaminating agent. Cleaning of facilities usually is accomplished by manual methods, while cleaning of equipment depends on size. Larger equipment can utilize clean-in-place (CIP) technology, while smaller equipment is cleaned by industrial washing equipment. Once cleaning procedures are validated, there still must be reliance on personnel to strictly adhere to these procedures to assure the achievement of aseptic conditions.

Validated sterilization procedures are required to sterilize to a known SAL (usually 10^{-6}) all equipment, tubing, packaging materials, and anything that will come into contact with the sterile drug product. Sterilization usually is accomplished by heat (saturated steam under pressure or dry heat) using either stationary ovens (autoclaves or dry heat ovens) or steam-in-place (SIP) systems for large equipment.

The literature is filled with pertinent references to methods and validation approaches for cleaning and sterilization procedures.

In-Process and Final Process Quality Control

Environmental control, sterility, pyrogen, endotoxin, and particulate matter testing are all part of the QC process to assure control and elimination of bacterial contamination. The final four tests are primarily final product QC tests, while environmental monitoring involves tests performed during processing to evaluate the control of the microbiological and particulate quality in the processing area. Table 8.7 provides a summary of the types of airborne and surface environmental control tests commonly used in the pharmaceutical industry today.

Training

Personnel are key to the control of contamination. Three main aspects of personnel control of contamination are as follows:

1. Good personal hygiene

2. Proper use of gowning apparel

3. Good aseptic techniques

These will be discussed in more detail in the remaining sections of this chapter.

Table 8.7. Environmental Control Monitoring Methods

Airborne

Nonviable: Electronic particle counters

Viable: Settling plates
                Liquid impingers
                Membrane filtration
                Slit-to-agar (STA) samplers
                Reuter centrifugal air sampler (RCS)
                Sieve samplers

Surface

Viable: Rodac plates
                Swabs

Testing of Knowledge

Testing of employee knowledge on reasons why aseptic techniques are required, the appreciation of environmental control, and other aspects of aseptic manufacturing and control has not been a common practice in the pharmaceutical industry. However, in the era of total quality management (TQM), testing now is becoming more commonly practiced. One approach used will now be discussed.

Written testing can measure at least two areas of knowledge related to aseptic processing:

1.   Knowledge of elementary microbiology and principles underlying the control of contamination (**Knowledge**)

2.   Knowledge of aseptic procedures and practices (**Behavior**)

Examples of questions testing *knowledge* of microbiology and contamination control are found in Table 8.8. Likewise, examples of questions to test knowledge of aseptic practices (*behaviors*) are listed in Table 8.9.

Before any operator continues onward with their "hands-on" training (next section), he or she must pass a written test evaluating his or her knowledge in these two main areas. Management and human resources within a company must determine what score must be achieved for the operator to pass a written test. Scores ≥90 percent generally should be required. Any question missed by an individual

Table 8.8. Examples of Questions Used in Employee Testing of *Knowledge* Involved in Good Aseptic Practices

True or False

1. If a drug product is sterile, it is completely free from microbial contamination. (*T*)

2. Mold grows on the walls and ceilings of most homes and industrial buildings. (*T*)

3. Human hair falling from the body is a common source of product contamination. (*T*)

4. Skin cells are constantly dying and falling off the body while new cells are made. (*T*)

5. When gloves are sanitized, they are considered sterile. (*F*)

6. Jewelry does not contain microbial contaminants. (*F*)

7. If one coughs into a mask, particles of bacteria do not escape from the mask. (*F*)

8. Cotton clothing is not tightly woven; skin flakes and other particles can pass through. (*T*)

9. Sanitization achieves a sterile condition. (*F*)

10. Air coming through HEPA filters sterilizes equipment and product in the aseptic area. (*F*)

11. Opening more than one door to the critical area disrupts the unidirectional flow of air. (*T*)

12. The more a person moves, the more skin flakes are given off. (*T*)

13. Back rubbing causes a large number of skin flakes to fall from the body. (*T*)

14. Pyrogens are dead bacteria. (*T*)

15. Bacteria cannot be spread through the air; only by touch. (*F*)

16. Most particulate matter can be seen with the naked eye. (*F*)

17. Rodac tests measure nonviable particulate contamination. (*F*)

18. If bacteria are dead while accidentally mixed into a product that is injected, this is not harmful to the patient. (*F*)

Continued on next page

Continued from previous page

19. Wearing regular eyeglasses will protect the product from contamination just as well as goggles. (*F*)

20. Leaning over the production line does not disrupt the laminar airflow because the air comes down everywhere in the room. (*F*)

Multiple Choice

1. Unidirectional sanitizing means
   a. Wipe only in one direction.
   b. Wipe only with one hand.
   c. Wipe every square inch of surface. *Answer: a*

2. Which of these packaging materials is least likely to support microbial growth?
   a. Paper
   b. Cardboard
   c. Glass *Answer: c*

3. Which is better for operator gowning?
   a. A loosely knit fabric
   b. A tightly knit fabric
   c. It does not matter *Answer: b*

4. A Class 100 area means that
   a. The air is changed at least 100 times per hour.
   b. There are less than 100 particles per cubic foot.
   c. The area is 100 times as clean as the hallways. *Answer: b*

5. Which sanitizing agent kills all microbial life, including spores?
   a. LpH-SE
   b. Hydrogen peroxide
   c. No sanitizing agent kills all microbial life. *Answer: c*

6. The most likely place for bacteria to live on equipment is
   a. On parts that are directly in the laminar air flow where there is plenty of oxygen.
   b. In screw threads and around joints and valves.
   c. On flat, horizontal surfaces. *Answer: b*

Continued on next page

Continued from previous page

7. Particulates are particles that
    a. Can be seen with the naked eye.
    b. Can carry microorganisms.
    c. Come from the drug compound.   *Answer: b*

8. When laminar airflow is interrupted, it is reestablished downstream within a distance equal to
    a. Twice the diameter of the interfering object.
    b. Three times the diameter of the interfering object.
    c. Ten times the diameter of the interfering object. *Answer: b*

Table 8.9. Examples of Questions Used in Employee Testing of *Behavior* Involved in Good Aseptic Practices

True or False

1. Gloves must be sanitized each time before entering a critical area. (*T*)

2. Gum and candy are allowed in a critical area as long as they are kept within the mouth. (*F*)

3. A small amount of makeup may be worn in the aseptic area as long as a mask covers it. (*F*)

4. If goggles become fogged while working in critical areas, it is okay to move them up on the forehead for awhile. (*F*)

5. The sterile gown may be unzipped a little when taking a break from the critical area. (*F*)

6. Fallen containers in aseptic areas may be removed with your hands using careful techniques. (*F*)

7. It is acceptable to remove sterile tweezers from a critical area, then return them as long as you do not set them down. (*F*)

8. It is acceptable to sit down on ladders, but not on tables or countertops. (*F*)

9. Physical removal is as important as chemical destruction when you use a disinfectant wipe. (*T*)

10. Since a hood is used, it is acceptable to use hair spray. (*F*)

11. When moving from one production line to another, new gloves must be put on. (*T*)

Continued on next page

Continued from previous page

12. Operators with colds or other illnesses may work in aseptic areas as long as they appropriately cover their nose and mouth. (F)

13. Sanitizing solutions should be discarded if they are not used by the expiration date. (T)

14. The only thing in a critical area that should not be sanitized is the light switches. (F)

15. Daily bathing or showering is required for aseptic area operators. (T)

16. Sterile equipment such as forceps may be unwrapped in the critical adjacent area as long as they are quickly and carefully moved into the critical area. (F)

Multiple Choice

1. If you see a coworker gowning incorrectly, it is best to
   a. Report him or her to a supervisor.
   b. Ignore the issue.
   c. Politely point out the error.   *Answer: c*

2. When is talking allowed in critical areas during production?
   a. Talking is never allowed.
   b. Only when absolutely necessary.
   c. Any time.   *Answer: b*

3. If a rubber closure slips off a vial in a critical area,
   a. Carefully pick up the closure with gloved fingers and replace on the vial.
   b. Use sterile forceps to pick up the closure and replace on the vial.
   c. Discard the vial.   *Answer: c*

4. When gloved hands touch a piece of equipment on the filling line, you should
   a. Consider the gloves sterile because the equipment has been sanitized.
   b. Resanitize the gloves with the appropriate sanitizing solution or foam.
   c. Replace your gloves.   *Answer: b*

should be discussed with training personnel to assure that understanding of the question is known. Procedures should dictate action(s) to be taken should an individual fail to achieve a certain score on the test.

Testing generally is well received by aseptic processing operators, provided that the test questions are fair and straightforward (unambiguous, and given in a nonthreatening manner). Test results will provide valuable feedback on what the individual knows or, conversely, what he or she does not know or understand. Such results will stimulate further systematic training in areas of knowledge deficit. Tests help people understand reasons for specific behavioral rules. For example, understanding and correctly answering a question regarding the fact that sanitizing gloves is not sterilizing those gloves helps the aseptic operator appreciate those procedures that are taken to avoid touching exposed sterile surfaces with sanitized gloved hands.

Tests evaluating employee knowledge enables training personnel to focus subsequently on behaviors and attitudes, knowing that if inappropriate practices subsequently occur in the aseptic manufacturing area, the problem most likely is not due to lack of knowledge. Most operators are enthusiastic about taking tests because such testing constitutes a valuable review of important knowledge and procedures plus it stimulates them to think about their behavior on the job and about the reasons for the stringent rules they must follow.

"HANDS-ON" BEHAVIORAL TRAINING

Good aseptic technique can be defined as the appropriate use of our bodies, the hands in particular, to prevent microbial and particulate contamination when working with the product, the product container, and/or the two combined.

Personnel Requirements

Luna (1993) and others have elaborated on personnel requirements. Among the most important for proper selection of personnel to work in aseptic processing environments are:

- Good attitudes in general.
- Good personal hygiene.
- Good manual dexterity.
- Commitment to excellence.

- Attention to details.

- Willingness to endure personal inconvenience.

Gowning Procedures

Correct procedures for gowning go a long way toward assuring that the sterile product is protected from the person having some sort of contact with that product. One published study compared rates of contamination of a sterile product with employees dressed either in an ordinary protective coat or in clean room gowning (Aslund et al. 1977). Their results, not surprisingly, showed that the contamination rate increased 600-fold when people wore an ordinary protective coat rather than complete clean room garb.

Correctly putting on the various garments in a logical sequence and controlled manner is easier said than done. Table 8.10 displays some of the critical mistakes and/or problems encountered that must be addressed and avoided in order for employees to don gowning in a proper manner.

Training on appropriate gowning technique involves classroom instruction on the company's policies and procedures regarding clean room gowning, actual practice time where a trainer works with the employee to go through the gowning steps and the employee is given time to get used to gowning according to company procedure, and, finally, qualifying the employee to gown properly without contaminating any part of the gowning apparel.

Beer (1991) published an informative article on the use of video recording and microbial assessment in employee training on gowning techniques. Table 8.11 shows a reproduction of a table in Beer's article showing the correlation between microbiological recovery when a gown surface was sampled and event(s) recorded via videotape that might explain why the contamination occurred. The author obtained several conclusions on the effectiveness of videotaping gowning techniques on improving the training of clean room operators:

- Retrospective self-observation and critique

- Visual awareness of improper or subtle flaws in aseptic technique undetected during actual gowning

- Correlation of microbial counts on a gowning surface and improper technique recorded on videotape, thereby enabling the employee to realize the consequences involved with improper technique

- Recommendations to improve technique

Table 8.10. Critical Mistakes in Gowning Procedures

- Failure to follow established gowning procedures for the particular area (e.g., incorrect sequence of donning different parts of gowning items)

- Failure to scrub hands and fingernails thoroughly

- Using paper or fabric materials in the clean room

- Hair not completely covered

- Skin exposed between gloved hand and the uniform

- Reaching under parts of the garment with the gloved hand

- Dropping a part of the garment on the floor, picking it up, and using it

- Touching parts of the face with gloved hand; failing to sanitize the glove afterwards

- Face mask not appropriately covering the face

- Zippers not completely zipped; parts of gowning not completely tucked in

---

- Ascertaining the operator's ability to gown according to procedure

Broth Testing

Two types of tests are possible for broth testing. One involves a person hand-filling sterile containers with sterile broth, utilizing aseptic techniques throughout the filling and capping procedures. This is commonly referred to as the operator *broth test*. The other involves one or more individuals, whatever is the normal number of personnel involved in an aseptic filling process, using sterile broth as part of a media fill to validate the maintenance of sterile conditions during the aseptic manufacture of a simulated product. This is commonly called a *media fill* as part of the validation of an aseptic manufacturing process. Only the broth test will be discussed in this chapter.

The broth test is run in a Class 100 environment where the trainee must first be certified on donning sterile gowning correctly. A portable

Table 8.11. Correlation Between Microbial Contamination of Gown Surface and Improper Technique*

Operator	Gown Location	Microbial Monitoring (cfu)	(Organisms)	Video Replay
1	Zipper: Top	<40	*Staphylococcus sp.*	Gown collar zipper touched operator's neck
	Bottom	9	*Staphylococcus sp.*	Touched scrub suit
2	Zipper: Top	3	*Staphylococcus sp.* / *Micrococcus sp.*	Gown collar zipper was underneath head piece
	Boot gown interface	9	*Staphylococcus sp.*	Nonsterile booty touched sterile gowned leg
3	Glove cuffs (final pair): right	2	*Bacillus sp.* / *Micrococcus sp.*	Difficulty donning gloves
	left	5	*Bacillus sp.* / *Micrococcus sp.*	Difficulty donning gloves
	Gown chest	12	*Micrococcus sp.* / *Staphylococcus sp.*	Gown chest area touched scrub suit
4	Glove cuff (final pair): left	17	*Staphylococcus sp.*	Operator pulled over wrist with ungloved hand

*Source: Beer (1991). Reproduced with permission from the Parenteral Drug Association.

filling machine is set up beforehand, where the filling nozzle is connected via sterile tubing to a vessel containing sterile culture media (usually sterile trypticase soy broth). The trainee manually fills 250 to 300 vials per setting with sterile culture media, aseptically closes the vials with sterile rubber closures, then the vials are incubated to determine if any have become contaminated. This exercise is repeated two more times usually on separate days such that each operator must fill up to 900 vials without contaminating any of them. Also, after each of the three broth test runs, the trainee's fingers and chest are sampled with a Rodac (*replicate organism detection and counting*; see next section for more detail) plate to determine whether contamination existed during the broth test. Should any failures occur, either with the appearance of contaminated vials or the Rodac plates show contamination, the trainee must repeat the entire broth test procedure after remedial training. Usually the trainer observes some or all of the broth test and will be able to point out technique errors during the test.

The broth test is a reasonably valid determinant of operator aseptic technique because it requires the person to handle sterile vials, fill these vials aseptically, and aseptically close these vials with a sterile rubber closure. These steps offer several opportunities for breach of technique. However, simply passing the broth test does not guarantee that the person is highly and totally trained in good aseptic practices. This author has witnessed instances where the trainee performed poor technique, even touch contaminating the opening of a vial or the surface of a closure in contact with sterile media, yet the vial did not reveal any evidence of contamination. Such breaches of technique must be pointed out by the trainer to the operator trainee. The trainee must be taught that any touching of a sterile surface renders that surface nonsterile. Sometimes this is difficult for people to appreciate since (1) they feel their gloved hands are sterile and (2) they do not see contamination resulting from vials of broth media that have been touch contaminated. However, theoretical positions must be taken; that is, whatever a gloved hand touches, that surface must be considered nonsterile, and that broth media does not always support microbial growth particularly with low level contamination as what probably results when inadvertent touch contamination occurs during a broth test.

Despite the limitations of operator broth testing, it is still the best test available in today's state-of-the-art practices for ascertaining a person's skill and behavior in performance of good aseptic techniques and practices. More sensitive methodology for measuring and certifying good aseptic practice is needed, but, to this author's knowledge, not yet available or acceptable.

Monitoring of Practices by Microbiological Techniques

A variety of techniques are available for ongoing monitoring of aseptic practices. One of the most common is the Rodac plate test. A Rodac plate is a sterile 60-mm petri dish containing a semisolid medium nutrient agar (usually trypticase soy agar). The agar is steam sterilized and while still molten is poured into the dish until a convex surface appears at the rim of the dish. It is this convex surface that is capable of making contact with any surface onto which the agar is pressed upon. Thus, Rodac plates are used for monitoring surfaces of equipment, facilities, and, in this discussion, bodily parts of personnel working in the aseptic environment. Normally, following time spent working in the aseptic area and prior to exiting the area to degown, the operator's index finger is gently rubbed over the surface of the Rodac plate. The plate cover is replaced onto the plate and the plate is incubated to determine if any microbial growth occurs. If so, then this indicates that the operator's finger was contaminated during at least part of the time while working in the aseptic environment. This must be considered not only in the decision to release the sterile product, but also in ascertaining the need for additional training of the operator in good aseptic practices.

In a 1992 survey of current practices in the validation of aseptic processing, personnel monitoring was discussed in some detail. Highlights of this survey are summarized in Table 8.12. It is obvious that personnel involved in aseptic processing are routinely monitored during media runs and the results used to assess future training needs.

Observation and Feedback

Most human beings do not like to be observed while performing their jobs. However, good training programs in good aseptic practices must involve some routine observation of techniques. Such observation can be done among the operators critiquing each other and/or done by training personnel on a periodic basis. The use of videotaping, while very controversial, can be a highly effective tool for observing techniques, then reviewing the videotape with the operator and pointing out practices and techniques needing improvement. Ongoing and active observation and feedback practices indicate a dynamic and healthy aseptic practice training program. Of course, a major key in the continued success of observation and feedback, as is also the case with other parts of the total training program, is the style and professional manner in which the observer shares feedback with the operator. Praise for a job well done should be freely shared as well as

Table 8.12. Current Practices in Personnel Monitoring*

86 percent of all firms monitor personnel involved in media runs for surface microbial load.

Surface Monitored	% Who Monitor Personnel at Different Times During Media Run			
	Upon Entering	During Run	After Run	Not Tested
Hands	42	52	60	2
Arms	24	24	45	29
Face Masks	9	7	14	74
Chest	27	22	44	32
Other Gown				
Surfaces	14	14	24	60
Foot Covering	9	2	9	85

Only 10 percent of firms routinely videotape media fill operations.

*Source: Current practices in the validation of aseptic processing—1992. 1993. Technical Report No. 17. *J. Paren. Sci. Tech.* 47:Suppl.

constructive criticism and coaching when poor or wrong practices and techniques are performed.

In another way of approaching feedback, an interesting survey was conducted with clean room operators regarding their concerns and problems working everyday in Class 100 environments. Their feedback is found in Table 8.13. It is imperative that training personnel and production management seriously consider such feedback information and take steps to resolve these issues. This enables aseptic operators to feel important because their input was heeded.

SUMMARY

It is hoped that this chapter has provided essential information with pertinent references for those involved in personnel training to use as

Table 8.13. Feedback from Aseptic Process Operators on Concerns Working in Clean Room Environments

- Personnel required to do too many tasks at the same time. People start to rush and neglect proper aseptic technique.

- Insufficient space within laminar airflow units; work area can be too confined.

- Frequent and repeated breakdowns of equipment.

- Training and conduct sometimes different among sites and shifts.

- Room temperature sometimes either too hot or too cold.

- Eye goggles become foggy, retarding ability to see and work well.

a resource in planning, organizing, implementing, and/or reviewing their education and training programs for employees involved in aseptic operations.

ACKNOWLEDGMENTS

The author expresses his grateful appreciation to Mr. Greg Wheatley and Ms. Billie Jo Keith, both of Eli Lilly and Company, and to Dr. Daniel J. Mueller, Indiana University, for their significant contributions in providing materials for parts of the text and several of the tables, figures, and appendices.

REFERENCES

Akers, M. J. 1994. *Parenteral quality control: Sterility, pyrogen, particulate matter, and package integrity testing,* 2nd ed. New York: Marcel Dekker, Inc.

Aslund, B., O. T. Olson, and E. Sandell. 1977. Aseptic work under various hygienic conditions. *Acta Pharm. Suec.* 14:517–524.

Avallone, H. 1990. Current regulatory issues regarding sterile products. *J. Paren. Sci. Tech.* 44 (4):228–230.

Avis, K. E. 1990. Parenteral preparations. In *Remington's pharmaceutical sciences,* 18th ed., edited by A. R. Gennaro. Easton, PA: Mack Publishing Co.

Avis, K. E. 1975. Training on-line personnel with an educator's input. *Bull. Paren. Drug Assoc.* 29 (4):205–209.

Beer, C. L. 1991. Gowning training: The use of video recording together with microbial assessment. *J. Paren. Sci. Tech.* 45:128–131.

Bloomfield, S. F. 1990. Microbial contamination: Spoilage and hazard. In *Guide to microbiological control in pharmaceuticals,* edited by S. Denyer, and R. Baird. West Sussex, UK: Ellis Horwood Ltd.

Boylan, J. C., and A. L. Fites. 1990. Parenteral products. In *Modern pharmaceutics,* 2nd ed., edited by G. S. Banker, and C. T. Rhodes. New York: Marcel Dekker, Inc.

Code of Federal Regulations 21. 1978. Human and veterinary drugs: Current good manufacturing practice in manufacture, processing, packaging, or holding. *Federal Register* 43FR190:45014–45089.

DeLuca, P. P. 1983. Microcontamination control: A summary of an approach to training. *J. Paren. Sci. Tech.* 37 (6):218–224.

Draft guidelines on quality assurance for pharmacy-prepared sterile products. 1992. *Amer. J. Hosp. Pharm.* 49:407–417.

Federal Register. 11 October 1991; 56 (198):51354.

Gallagher, D. B. 1980. Training pharmaceutical employees: What's the right way? *Pharm. Tech.* 4:39–43.

Haarmeyer, R. J. 1976. In-house training requirements of current good manufacturing practices. *Bull. Paren. Drug Assoc.* 30 (5):255–260.

Howorth, H. 1988. Movement of airflow, peripheral entrainment, and dispersion of contaminants. *J. Paren. Sci. Tech.* 42 (1):14–19.

Levchuk, J. W. 1991. Training for GMPs. *J. Paren. Sci. Tech.* 45 (6):270–275.

Levchuk J. W., and A. G. Lord. 1989. Personnel issues in aseptic processing. *Biopharm.* September:34–40.

Luna, C. J. 1993. Personnel: The key factor in clean room operations. In *Pharmaceutical dosage forms: Parenteral medications,* vol 2, 2nd ed., revised and expanded, edited by K. E. Avis, H. A. Lieberman, and L. Lachman. New York: Marcel Dekker, Inc.

Nosocomial bacteremias associated with intravenous fluid therapy. 1971. *USA Morbid. Mortal. Wkly. Rep.,* Special Suppl., 6 March.

Padovano, R. A. 1975. New employee training and quality assurance. *Bull. Paren. Drug Assoc.* 29 (4):210–212.

Phillips, I., S. Eykyn, and M. Laker. 1972. Outbreak of hospital infection caused by contaminated autoclaved fluids. *Lancet* i:1258.

Phillips, J. X. 1989. FDA Inspections—What does the investigator look for during inspections of parenteral manufacturers. *Pharm. Eng.* 9:35–38.

Report of the comptroller general of the United States: Recalls of large-volume parenterals, MSD-76-67. 12 March 1976.

Ringertz, O., and S. Ringertz. 1982. The clinical significance of microbial contamination in pharmaceutical and allied products. In *Advances in pharmaceutical sciences*, vol 5, edited by H. S. Bean, A. H. Beckett, and J. E. Carless. London: Academic Press.

United States Pharmacopeia. 1995. *Sterile drug products for home use.* United States Pharmacopeia 23/The National Formulary 18, pp. 1963–1975. Rockville, MD: United States Pharmacopeia Convention, Inc.

Tetzlaff, R. F. 1987. FDA regulatory inspections of aseptic processing facilities. In *Aseptic pharmaceutical manufacturing*, edited by W. P. Olson, and M. J. Groves. Buffalo Grove, IL: Interpharm Press Inc.

Turco, S., and R. E. King. 1987. *Sterile dosage forms*, 3rd ed. Philadelphia: Lea & Febiger.

Underwood, E. 1977. Ecology of microorganisms as it affects the pharmaceutical industry. In *Pharmaceutical microbiology*, edited by W. B. Hugo, and A. D. Russell. Oxford: Blackwell Scientific Publications.

Vanhaecke, E., and K. Van den Haesevelde. 1991. Bacterial contamination of stainless steel equipment. In *Sterile pharmaceutical manufacturing*, vol 2, edited by M. J. Groves, W. P. Olson, and M. H. Anisfeld. Buffalo Grove, IL: Interpharm Press, Inc.

Whyte, W., L. Niven, and N. D. S. Bell. 1989. Microbial growth in small-volume parenterals. *J. Paren. Sci. Tech.* 43 (5):208–212.

APPENDIX A

Example of a Training Program for Personnel Working in Aseptic Environments

1. Prerequisites

- Safety
- Overview of GMPs
- Good documentation practices
- Garment practices
- Good hygiene practices

2. General Aseptic Practices

- Laminar flow hoods
- General aseptic procedures
- Understanding the aseptic environment (see Appendix B for more details on this course)
- Aseptic connections
- Introducing equipment through air locks
- Preparation of areas for aseptic processing
- Use of goggles
- Sanitization of aseptic areas
- Environmental monitoring of aseptic areas
- Recent internal audit findings
- Review of FDA inspections

3. Specific Aseptic Practices

- Introduction to aseptic gowning
- Aseptic gowning procedure
- Aseptic operator monitoring
- Gowning certification
- Broth test procedure
- Broth test certification

APPENDIX B

Outline of Course on Understanding the Aseptic Environment

Purpose:

- To increase knowledge of basic microbiology
- To increase awareness of contamination
- To understand how the employee can help control contamination through proper behavior and technique

Part I: Basic Microbiology

- Definitions (parenteral, sterile, clean, microorganisms)
- Viewing and discussion of videotape *Basic Microbiology**
- Origin of microbial contamination, including human contamination
- Growth requirements for microorganisms

Part II: Basic Contamination Control

- Viewing and discussion of videotape Basic Contamination Control*
- Definition of contamination
- Sources of contamination
- Elimination of contamination
- Laminar airflow

Part III: Behavior in the Clean Room

- Viewing and discussion of videotape Behavior in the Cleanroom*
- Good aseptic techniques: "Dos and Don'ts" (see Appendix C)
  1. Dress
  2. Work
  3. Act
  4. Move

---

*Videotapes produced by **Micron Video International, Inc.**, 222 North 6th Street, P.O. Box 486, Chambersburg, PA 17201.

APPENDIX C

Dos and Don'ts of Good Aseptic Practices

Dress Correctly

- Proper gowning *at all times.*

Work Correctly

- Proper cleaning and sanitization of room and all work surfaces.

- Never store sanitizing solution in a critical area.

- Sanitize your gloved hands *each time* before you enter a critical area.

- No paper except sterile bioshield paper is allowed in a critical area.

- Pens, calculators, and so on must never be placed inside a critical area.

- Keep laminar hood doors closed as much as possible. When you must enter a critical area, make sure that all other doors are closed. Failure to do this will disrupt the laminar airflow significantly.

- During setup bring sterile equipment as close to the critical area as possible before transferring the equipment.

- All wrapped sterile equipment must be unwrapped in the critical area. Utmost care must be taken to protect the sterility of this equipment.

- Never touch any "contact part" with your gloved hands.

- Nothing should be placed between the product or contact part and the laminar airflow hood.

- If you must lean or reach over sterile product or containers (unfortunately necessary at times), you must discard that product or container.

- Never use anything to remove a particulate from a contact part.

- Once unwrapped sterile forceps, sterile tweezers, sterile hemostats must remain in the critical area.

- Never pick up anything off the floor while product is being produced unless it has been determined that a safety hazard exists. If you must pick something up, you must change your gloves. Discard the pickup.

- Never go from one critical area to another without at least changing your gloves.

Act Correctly

- Never have anything in your mouth while in any aseptic area.

- Never touch exposed skin and avoid touching others' clothing as well as your own. Absolutely no back rubbing.

- Sitting on tables, ladders, waste receptacles, and so on is prohibited. Only approved chairs suitable for aseptic areas can be used.

- Never put your feet on anything that could come into contact with your hands or gown.

- Talking should be held to an absolute minimum.

- Never open your sterile gown unless you are gowning or de-gowning.

Move Correctly

- Avoid unnecessary motions in any aseptic area. When you move, your movements should be slow and deliberate.

9

Predictive Sterility Assurance for Aseptic Processing*

Colin S. Sinclair
Alan Tallentire

Terminal sterilization and aseptic processing constitute the two principal methods of manufacture of sterile pharmaceutical products. Of these two methods, aseptic processing has become associated more often with sterility failure. Inevitably, this has led aseptic processing to be placed under considerable scrutiny from Regulatory Authorities, none more so than the Food and Drug Administration. Authorities have increasingly expressed the view that the sterility assurance of products processed by aseptic methods must be improved in order to reduce the risk of public harm (Barr 1993).

Conventional methods to estimate sterility assurance for products prepared by aseptic processing are based upon 'medium-fill' studies. Typically in such a study, sterile microbial growth medium is processed through the entire manufacturing and filling operation, followed by incubation of the final product and subsequent examination for bacterial growth (Sharp 1987). Routinely, at least 3,000 units are filled to detect a contamination rate of not more than one contaminated unit per thousand units filled. The resultant estimate of contamination level of 10^{-3} compares unfavorably with a sterility assurance level (SAL) of 10^{-6} targeted for products processed by terminal

*Sinclair, C. S., and A. Tallentire. 1993. Predictive sterility assurance for aseptic processing. In *Sterilization of medical products*, Vol. VI, edited by R. F. Morrissey, pp. 97–114. Montreal: Polyscience Publication. Reprinted with permission.

sterilization. However, it is often claimed that advanced aseptic processing technologies which eliminate or minimize human intervention can achieve contamination rates considerably lower than the usual target rate of 1 in 1,000. To test the validity of such claims, a new approach to estimating sterility assurance for aseptic processing is essential.

Blow/Fill/Seal processing is a fully automated technology that is designed to be operated remotely when contained within a controlled environment. Thus it is an ideal test system for examination of the relationship between the extent of product contamination by airborne microorganisms during medium-fill studies and the level of airborne microorganisms in the environment. Moreover, noting experimentally how changes in operating conditions influence the level of product contamination may well provide a better understanding of the manner in which processing factors impact the microbiological quality of products. Given this understanding, prediction of sterility assurance for aseptic processing becomes a realizable goal.

This chapter describes a series of experiments (designated Study 2) that is a follow-up to earlier published work describing Study 1 (Bradley et al. 1991). Together, these two studies are aimed at defining the exact relationship between fraction of product contaminated and level of microbial challenge over an extensive range of challenge concentrations for Blow/Fill/Seal processing operating under a particular set of conditions. Study 2 was designed to examine whether the mode of air shower operation, variation in production cycle time, and 'interventions' may produce discernible effects on the fraction of product contaminated.

MICROBIOLOGICAL CHALLENGE STUDIES— DESIGN CONSIDERATIONS

The following are certain specific problems to be considered in designing an airborne microbiological challenge test for aseptic processing.

Nature and Production of the Microbial Challenge

In view of the diversity of microorganisms present in the environment and the form in which they occur, it can be justifiably argued that more than one type or species of organism should be employed in challenge studies. In choosing any organism, however, due regard must be given to non-pathogenicity, ease of production and recognition, viability, average cell size, and size distribution. Furthermore, an airborne challenge may consist of the microorganisms contained within liquid droplets, located on solid particles, or as a dispersion of discrete cells in air. For a given species of microorganism, a dispersion comprising discrete cells is made up of the particles of the smallest possible dimension for that species and is very likely the most rigorous challenge that can be devised. Logically, such a dispersion is the challenge of choice. The production of a dispersion of this type can be achieved by aerosolization of a liquid suspension of cells into droplets of sufficiently small size that, on expansion into a gaseous fluid, instantaneous evaporation of the liquid droplet occurs, leaving the discrete cells suspended in the gas. Dispersions so formed are referred to as 'dry'. Nebulizers restricting droplet size and thus giving rise to dry dispersions have been described (May 1973).

Sampling of the Microbial Challenge

In order to estimate the concentration of the microbial dispersion used to challenge the process, there is a requirement to detect accurately, with a high degree of precision, microorganisms present in a known volume of dispersion. Quantitative detection is best achieved by making use of the property that any one viable microorganism, recovered from the dispersion, produces a visible colony when incubated on solidified growth medium. In practice, the collecting device must be able to recover microorganisms from air dispersions containing a wide range of concentrations. The characteristics, specifications, and performance of various devices for collecting microorganisms have been detailed (Public Health Monograph 1964). An air sampler, based upon filtration, has proven particularly useful for sampling microbial challenges over a wide range of challenge concentrations.

Containment of the Microbial Challenge

In order to present the microbial challenge to the Blow/Fill/Seal machine in a controlled and reproducible fashion, it is necessary for the air-dispersed microorganisms to be contained within defined limits. However, all contained dispersions of airborne particles undergo decay primarily due to the influence of gravity, i.e., particles fall out (Dimmick 1969). A form of decay known as 'stirred-settling' occurs if sufficient air turbulence is created so that, even though the dispersion is decaying under the influence of gravity, a uniform distribution or particles generally exists throughout the contained volume. Under these conditions, the rate of decay for a given particle mass is independent of particle concentration but inversely related to the height of the containment volume (Dimmick 1969). To maintain dispersion concentration under stirred-settling conditions, it is then necessary to balance the loss of particles from the dispersion through decay with the controlled introduction of dispersed particles into the containment volume.

MICROBIOLOGICAL CHALLENGE STUDIES— DEVELOPMENTAL WORK

Laboratory-Based Studies

In order to develop the basic techniques to generate controlled microbial challenges, laboratory-based studies have been carried out to investigate the production and containment of air-dispersed microorganisms.

As shown in Figure 9.1, a containment vessel with a volume of 0.13 m^3 was constructed from sheet aluminum with the specified dimensions. Four access ports were located in the chamber walls. Two of these ports were used in charging the chamber with dispersion; the inlet port was connected directly to a modified Collison nebulizer (May 1973), and the outlet port exhausted to atmosphere via a filter to allow equilibration of pressure within the chamber during nebulization and sampling. The two remaining ports were used for sampling purposes. An 'upward drift' electrically-driven fan (~300 r.p.m.) was positioned centrally on the base of the chamber.

Figure 9.2 shows the behavior of air-dispersed spores of *Bacillus subtilis* var. *niger* contained under stirred-settling conditions within this laboratory vessel. Dispersions at concentrations ranging from 10^1 to 2 $\times$ 10^6 spores dm^{-3} were generated by a 30-minute aerosolization of spore suspensions at different concentrations ranging from 1 $\times$ 10^4 to 7 $\times$ 10^8 spores cm^{-3}. Following the initial 30-minute charging of the vessel, the spore dispersions were held under stirred-settling conditions, without further nebulization, and sampled at 60-minute intervals over a 300-minute period. It is clear from this Figure that spore dispersions undergo decay at a constant rate over the 300-minute period (half-life of 188 minutes) with essentially the same results over a 10^5-fold change in spore concentration.

The behavior of spore dispersions held within the laboratory vessel under stirred-settling conditions during the continuous aerosolization of spore suspensions at different concentrations ranging from 1 $\times$ 10^4 to 7 $\times$ 10^8 spores cm^{-3} is shown in Figure 9.3. It is evident from this Figure that, for each of the four different dispersions, the concentration of air-dispersed spores attains a plateau level after approximately 60 minutes of aerosolization, and that this level is maintained throughout the subsequent period of nebulizer operation.

Figure 9.1. Schematic representation of the laboratory containment vessel.

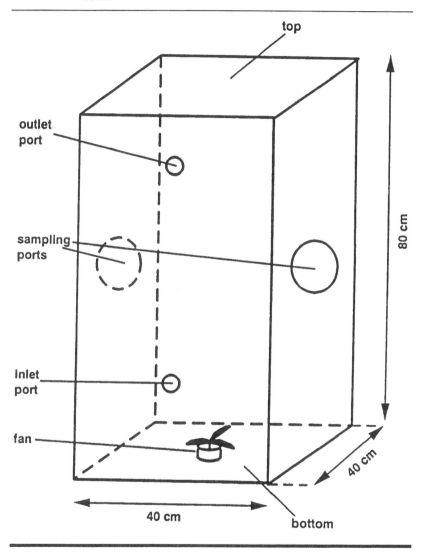

Furthermore, the spore concentration at the plateau level is directly related to the spore concentration of nebulized suspension. These findings demonstrate the potential of employing containment conditions of stirred-settling, coupled with continuous introduction of dispersion to the contained volume, to generate microbial dispersions of fixed and controlled concentration.

Figure 9.2. Stirred-settling decay of spore dispersions held within the laboratory vessel.

(○)   concentration of aerosolized suspension 1×10^4 spores cm^{-3}
(△)   concentration of aerosolized suspension 9×10^5 spores cm^{-3}
(□)   concentration of aerosolized suspension 4×10^7 spores cm^{-3}
(●)   concentration of aerosolized suspension 7×10^8 spores cm^{-3}

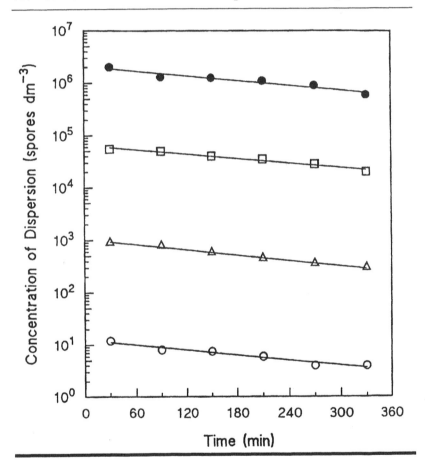

Scale-Up Studies

To develop the above laboratory-based methods in order to allow generation and maintenance of controlled airborne microbial challenges over prolonged time periods in the environment surrounding a Blow/Fill/Seal machine, it has been necessary to perform initially a number of scale-up experiments. This work was conducted using a

Figure 9.3. Spore concentration of dispersions within the laboratory vessel during continuous aerosolization of suspensions at different spore concentrations.

(O)   concentration of aerosolized suspension 1×10^4 spores cm^{-3}
(△)   concentration of aerosolized suspension 9×10^5 spores cm^{-3}
(□)   concentration of aerosolized suspension 4×10^7 spores cm^{-3}
(●)   concentration of aerosolized suspension 7×10^8 spores cm^{-3}

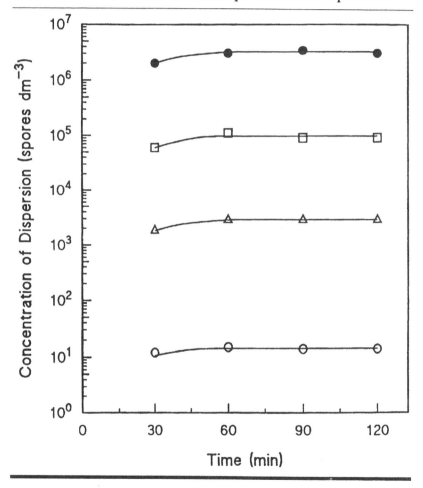

'mobile cabin' modified to serve as a containment room (Figure 9.4); the volume of the cabin was 24.6 m^3, representing about a 190-fold increase in volume over the laboratory vessel. A sampling port was located on each of the two end walls of the cabin, Port 1 at a height of

Figure 9.4. Schematic representation of modified mobile cabin.

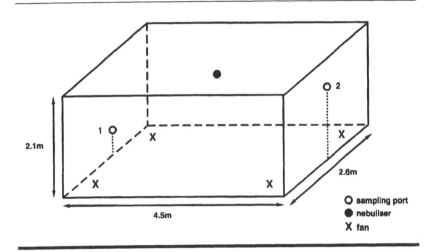

0.5 meter and Port 2 on the opposite wall at a height of 1.5 meters. A multi-jet climbing column nebulizer (Bradley et al. 1991) was positioned centrally within the cabin at a height of 1 meter; a supply line to the nebulizer ensured that it could be recharged with spore suspension during operation. Four electrically-driven 'upward-drift' fans were placed at floor level and located on diagonal axes 1 meter from each corner.

Figure 9.5 shows the behavior of a given spore dispersion held within the mobile cabin as a function of time; this particular dispersion was generated by continuous aerosolization of a 3×10^3 spores cm^{-3} suspension. It can be seen from this Figure that, over an extended experimental period (0.5 to 9.25 hours), spore concentration as estimated at the two distant locations of differing heights, falls around a mean level of 6.5×10^2 spores m^{-3}. The constancy of spore concentration within the room throughout the 8.75-hour experimental period provides strong evidence of the feasibility of producing controlled microbial challenges within the relatively large volume needed to enclose a Blow/Fill/Seal machine.

MATERIAL AND METHODS

Study Set-Up

A previous communication (Bradley et al. 1991) gave a detailed description of the experimental set-up used to generate and maintain

Figure 9.5. Estimates of dispersion concentration within the mobile cabin generated by continuous aerosolization of a suspension 3×10^3 spores cm^{-3}. Solid points represent estimates made at Port 1 while open points represent those made at Port 2.

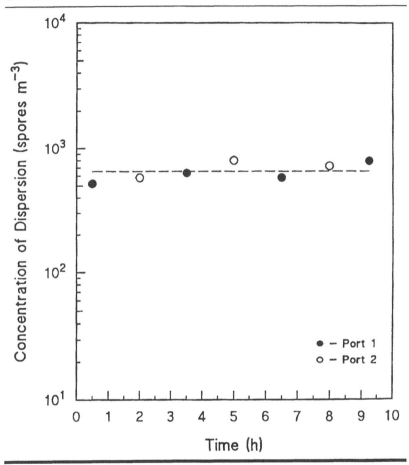

over specified time periods controlled airborne microbial challenges in the enclosed space surrounding an operating Blow/Fill/Seal machine. The test method, basic test design and Blow/Fill/Seal machinery used in the present work were identical to those previously described.

The containment room (67.5 m^3) was sited at Weiler Engineering, Elk Grove, Arlington Heights, Chicago, a location where no sterile pharmaceutical production is carried out. Blow/Fill/Seal machine, ALP 624-017, was tooled to the same specification and mold configuration as those of ALP 624-015, the machine used in Study 1. The configuration provided 24 molded ampoules, each with a 2 cm^3 fill volume. The operation of the machine was identical to that used in Study 1, namely, production of 24 ampoules (1 cycle) every 12 seconds to give an overall production of 120 ampoules min^{-1}.

The air shower unit of machine ALP 624-017, developed to provide 'local protection' of filling mandrels, was of the same basic design as previously described (Bradley et al. 1991). Prior to commencement of the experimental work, the HEPA filter unit within the air shower assembly was DOP tested and rated as 99.99% efficient. By controlling the speed of the fan located in the unit, the velocity of air emerging from the air shower outlet (a slot running the length of the base of the shower enclosure) could be varied between 1.6 and 3.7 m s^{-1}; at a given fan speed, the velocity of emerging air was found to be constant over the entire length of the slot.

For each challenge test, the Blow/Fill/Seal machine was set-up according to the machine manufacturer's protocol for 'medium-fill validation'. The fill medium used was heat-sterilized Tryptone Soya Broth, tested to meet the minimum USP fertility level. The medium was delivered to the point-of-fill after passing through two in-line liquid filters (nominal pore size 0.2 μm) that were tested for integrity at the end of each day's experimentation.

Microbial Challenges

Air-dispersed spores of *B. subtilis* var. *niger* (NCIMB 8056) comprised the microbial challenge. Aerosolization of an aqueous spore suspension at a predetermined concentration gave spores dispersed throughout the air of the containment room at a given challenge concentration over the range 3×10^2 to 3×10^6 viable spores m^{-3}. The test duration, during which the challenge spore concentration was maintained at a nominal level, varied according to concentration, the lower the concentration, the longer the duration. A longer test duration was required at a low spore challenge concentration to allow detection of the frequency of ampoule contamination. In practice, test durations ranged from 1 to 10 hours. The concentration of spores in the air of the containment room was monitored intermittently during each challenge test by collecting spores present in sample volumes of air, drawn isokinetically from the room via three sampling ports (Bradley et al. 1991).

Test Design

Irrespective of test conditions, the following activities were carried out throughout the duration of the test:

1. continuous operation of the Blow/Fill/Seal machine employing medium-fill

2. continuous aerosolization of appropriate spore suspension

3. periodic sampling of the containment room air

In practice, a minimum of four sampling operations, comprising replicate sampling at the three access ports, was carried out for each challenge test. As previously described (Bradley et al. 1991), each port was positioned at a selected location on one of three walls of the containment room, i.e., east (E), south (S), and west (W). For a given challenge test, the spore challenge concentration is the mean value of spore concentration derived from individual estimates of concentration made at the different sampling locations during the test time.

Immediately after production, all ampoules were incubated at 30–35°C for 14 days so that contamination of ampoules could be assessed by appearance of visible growth. To allow measurement of fraction of product contaminated, expressed in terms of the ratio of number of contaminated ampoules to total number of ampoules produced, each individual ampoule was identified relative to time of production and filling location.

RESULTS

Stability and Uniformity of the Spore Challenge

One challenge concentration generated by nebulization of a spore suspension containing approximately 6×10^7 spores cm^{-3} has been chosen to illustrate the general findings when spore challenges were generated within the contained environment housing an operating Blow/Fill/Seal machine.

Figure 9.6 is a plot of the estimates of the concentration of spores in the containment room air sampled at the three different locations (E, S, and W) against time covering the entire challenge period (530 minutes). Aerosolization of the spore suspension commenced at t = 0 minutes and was maintained throughout the entire challenge period. Estimates of spore concentration were made at the three different sampling locations at t = 30 minutes, and at regular intervals (~ 30 minutes) throughout the challenge period. It is evident from Figure 9.6 that, at each sampling occasion, the three estimates of spore concentration fall within a 2–3 fold range; this relatively narrow range indicates active dispersal of spores throughout the air within the containment room to give rise to a reasonably homogeneous distribution. It can also be seen from this Figure that the spore concentration of

Figure 9.6. Estimates of dispersion concentration made at the three sampling ports (east [E], west [W], south [S]) of the containment room during aerosolization of a suspension of 6×10^7 spores cm^{-3}.

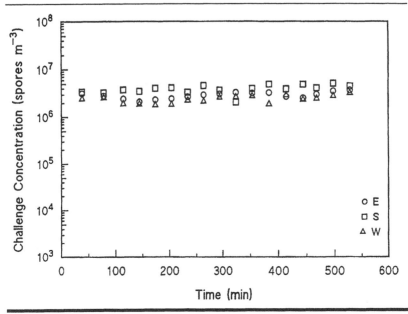

the dispersion was maintained at a fixed level (around 3×10^6 spores m^{-3}) over the challenge time extending from 30 to 530 minutes. As in the earlier study (Bradley et al. 1991), the first 30 minutes of aerosolization were excluded from the challenge period as this time interval was utilized to establish the concentration of dispersed spores within the containment room.

Relationship Between Fraction of Product Contaminated and Microbial Challenge Level

Figure 9.7 is a plot, on logarithmic scales, of fraction of contaminated ampoules against spore challenge concentration for the Blow/Fill/Seal machine operating without the air shower functioning. The four closed points represent data generated in Study 2, whereas the three open points represent previously reported data generated in Study 1 (Bradley et al. 1991) using the identical machine specification and operating conditions. It is immediately evident that both sets of datum points fall around the same curve (the dashed line is the extrapolation of the curve to lower levels of product contamination). Datum points from both

Figure 9.7. Fraction of contaminated ampoules as a function of spore challenge concentration for the blow/fill/seal machine operating without the air shower operating. Dashed line is an extrapolate of the curve.

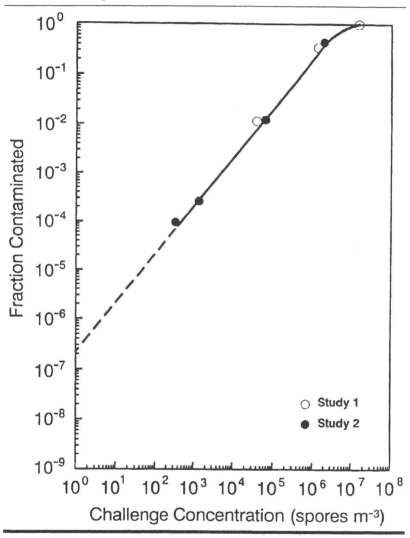

studies comprehensively define the form of the relationship between fraction of product contaminated and level of spore challenge concentration over a 50,000-fold range of spore concentration. Moreover, the datum points show that the linear portion of the curve defining the direct relationship is experimentally demonstrable, and holds over an approximately 7,000-fold change in challenge concentration.

Operation of Air Shower

Figure 9.8 gives data generated for the Blow/Fill/Seal machinery operating with the air shower at maximum setting (open points represent data generated in Study 1 and solid points those generated in Study 2). The uppermost curve and its extrapolate depicted in Figure 9.8 represent the common behavior defined by data generated in both studies with the air shower not operating (taken from Figure 9.7). These plots reveal that, for each study, there is a distinct relationship between fraction of product contaminated and spore challenge concentration with the air shower operating maximally. Furthermore, the two curves derived with the air shower operating at maximum setting and the common curve derived with the air shower off are, in effect, parallel. However, the two curves generated with the air shower operating maximally are shifted downwards from that seen without the air shower operation, the magnitude of this shift being approximately 9- and 70-fold for machines employed in Study 1 and Study 2, respectively.

Variation in Cycle Time

Production using Blow/Fill/Seal technology is an incremental process. The production cycle comprises container molding, filling, sealing, and discharging of product, with transfer of molded container(s) from the molding station to the filling station being achieved via a mold carriage which, at the end of the cycle, returns to the molding station. In the present study, adjustment in cycle time was achieved through changing the period of time taken for the mold carriage to move from the molding station to the filling station; the filling, sealing, and return of mold carriage elements of the cycle were held constant.

At cycle times of 12 and 14 seconds, determinations were made of the fraction of ampoules contaminated for the Blow/Fill/Seal machine (Study 2) with air shower both off and operating maximally. The upper solid line of Figure 9.9 is the curve, originally depicted in Figure 9.7, relating the fraction of product contaminated and level of spore challenge concentration for a cycle time of 12 seconds without the air shower operating, and the lower solid line (taken from Figure 9.8) is the curve for the same cycle time derived in Study 2 with the air shower at maximum operation; these two lines are given for comparative purposes. The two datum points represent the findings obtained with a 14-second cycle time, the closed point indicates the air shower off and open point indicates the air shower on at maximum. As observed, each datum point falls above the corresponding curve generated for a cycle time of 12 seconds; the magnitude of the displacement

Figure 9.8. The relationships between fraction of contaminated ampoules and spore challenge concentration for the blow/fill/seal machine operating with the air shower at maximum setting for Study 1 and Study 2. Dashed lines are extrapolates of the two curves; the uppermost curve and its extrapolate are the common behavior seen with air shower off (taken from Figure 9.7).

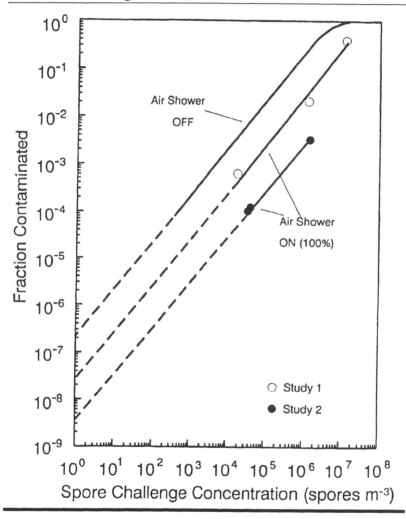

is small for air shower off but is approximately 20-fold greater for air shower operating maximally.

Figure 9.9. Datum points generated for the blow/fill/seal machine functioning with a 14-second cycle time. The closed point indicates the air shower off and open point indicates the air shower on; the upper solid line is the curve for a cycle time of 12 seconds with air shower off (taken from Figure 9.7) and the lower solid line is the corresponding curve for 12-second cycle time with air shower operated maximally (taken from Figure 9.8).

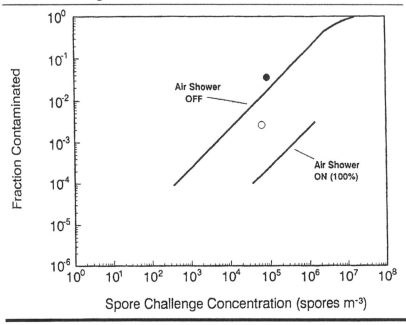

Simulated Interventions

At a given spore challenge concentration (nominally 3×10^6 spores m^{-3}), two gross interventions were simulated through switching the air shower off from maximum setting (in so doing, compromising local protection of the filling mandrels) and then, after a fixed time period (5 or 15 minutes) switching the air shower back on to maximum. Figure 9.10 is a plot of the fraction of contaminated ampoules observed in each successive five production cycles (~1-minute production) throughout the challenge period (330 minutes). The two sets of dashed lines delineate the boundaries of the two intervention periods of 5- and 15-minute duration, designated X and Y, respectively, and values in parentheses give the overall fraction of contaminated ampoules

Figure 9.10. Fraction of ampoules contaminated for incre-
ments of five production cycles throughout the experimental
period covering the two simulated interventions (X and Y).
Intervention is defined as the time during which air shower
was switched off from maximum setting. The dashed lines
delineate the boundaries of the two interventions of 5- (X)
and 15-minute (Y) duration.

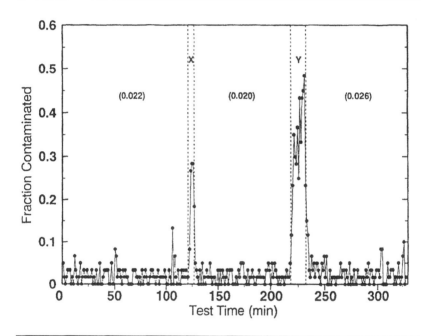

during the experimental periods before intervention X, between interventions X and Y, and after intervention Y.

It can be seen from Figure 9.10 that, prior to the first intervention
X, the fraction of ampoules contaminated, at production intervals of
five cycles, fall within the range 0 to 0.13, the overall fraction of ampoules contaminated throughout this first period (test time 0 to
121 minutes) was 0.022. At the start of intervention X (test time
122 minutes) switching off the air shower was seen to have an immediate impact on the fraction of ampoules contaminated, the level increased to a maximum of 0.28 (around a 13-fold increase above the
mean contaminated fraction that existed prior to intervention X). At
the end of intervention X (test time 127 minutes) restarting operation

of the air shower was seen to bring about an immediate fall in the fraction of contaminated ampoules; within one 5-cycle increment of production (corresponding to 1 minute), the fraction fell to a level within the pre-intervention range. For the period following intervention X up to the start of intervention Y (test time 128 to 217 minutes), the overall fraction of ampoules contaminated was 0.020. This level did not differ significantly ($p = 0.05$) from the level of 0.022 recorded prior to simulated intervention X, demonstrating that machine performance, in terms of fraction of product contaminated, returns fully to pre-intervention level.

As indicated in Figure 9.10, a similar behavior for intervention Y is seen with a duration of 15 minutes (test time 218 to 233 minutes). Again, switching off the air shower (start of intervention Y) is observed to bring about an immediate increase in the frequency of ampoules contaminated; in this instance, the fraction contaminated attains a maximum level of 0.48 (around a 22-fold increase above the original pre-intervention level). On restarting the air shower (end of intervention Y), the fraction of ampoules contaminated is again observed to fall rapidly to pre-intervention levels. Following the simulated intervention of 15 minutes, however, the minimum time required to achieve pre-intervention contamination levels is approximately 3 minutes. For the test period following intervention Y (test time 234 to 330 minutes) the overall level of fraction of ampoules contaminated was 0.028. Again, this level did not differ significantly ($p = 0.05$) from that recorded prior to the two simulated interventions, indicating full recovery of machine performance.

DISCUSSION

The experimental approach underlying the present fundamental investigation of automated aseptic processing has been to establish, over extended time periods, controlled challenges of air dispersed spores distributed throughout the environment within which the Blow/Fill/Seal machine operated. To set a spore challenge concentration within the operating environment, it was necessary to balance the rate of production and the rate of loss of air-dispersed spores. This balance was achieved by continuous aerosolization of spore suspension at a predetermined rate and by regulating dispersion decay via stirred-settling conditions in the containment volume around a machine. Under these experimental conditions, controlled challenges of air-dispersed spores were generated over a wide range of spore concentrations through aerosolization of spore suspensions at different spore concentrations. In practice, it has been possible to generate controlled challenges at spore concentrations extending over a 50,000 fold range for periods of time ranging up to 10 hours.

The results of Study 2, in which an operating Blow/Fill/Seal machine was challenged with air-dispersed spores, again unequivocally demonstrated that the quality of the microbiological environment surrounding the machine impacts upon the fraction of product contaminated. In general, for the Blow/Fill/Seal machine operating under a fixed set of machine conditions, there is a regular and definable relationship between the fraction of product contaminated and the level of airborne microorganisms. For a machine operating with the air shower off, the constancy of the behavior between Study 1 and Study 2 provides strong evidence that, under controlled conditions, Blow/Fill/Seal machine performance with respect to product contamination is highly consistent and reproducible. Overall, when the air shower is off, the relationship between product contamination and the spore challenge concentration has been comprehensively defined over a 50,000-fold range in spore challenge concentration. Moreover, the linear portion of the relationship is amenable to extrapolation providing a means for predicting air quality under which the frequency of product contamination is low and acceptable (see dashed line on Figure 9.7). For example, for the particular mode of machine operation employed here, a level of one organism per cubic meter of air is predicted to provide a rate of product contamination of 2.3×10^{-7} (i.e., one contaminated ampoule in $\sim 4.3 \times 10^6$ ampoules produced).

Study 1 had previously shown that operation of an air shower around the filling mandrels reduced the probability of product contamination (Bradley et al. 1991). In the present study, local protection was again seen to reduce the level of product contamination. For Study 2, however, maximal operation of the air shower was observed to bring about a 70-fold reduction in the level of product contamination for a given challenge concentration as represented by the common curve derived with the air shower off as opposed to approximately a 9-fold reduction for air shower operation in Study 1. The difference between the two studies in the extent of protection afforded by the air shower suggests that, for a given machine type, the configuration and arrangement of the air shower is critical to the efficiency of local protection. Nonetheless, the different curves derived for air shower operation in the two studies appeared regular and amenable to extrapolation (lower two dashed lines on Figure 9.8). They provided, at an average challenge concentration of one organism per cubic meter of air, predicted contamination rates of around 2.6×10^{-8} and 3.5×10^{-9} for air shower operation in Study 1 and Study 2, respectively. It is worth noting that none of these predicted contamination rates for an average challenge of 1 spore per cubic meter of air, with or without air shower operation, could be assessed by any practical medium-fill study. Furthermore, the above findings provide clear evidence that microbial product contamination occurs during the filling element of the Blow/Fill/Seal process.

The observed increase in rate of product contamination consequent upon an increase in cycle time (both for air shower off and on at maximum) is indicative that the length of time taken by the mold carriage to move from the molding station (Parison head) to the filling station is a critical determinant of product contamination. Clearly, present data are limited and caution must be exercised in making definitive interpretations. Nonetheless, the behavior is in keeping with a second mode of product contamination occurring before the filling process (i.e., during container molding and/or transportation). It is also interesting to note the impact of changes in cycle time on the efficiency of local protection afforded by operation of an air shower around the filling mandrels. For the same air shower operating at maximum setting, a 70-fold reduction in product contamination was recorded at a cycle time of 12 seconds as opposed to about a 10-fold reduction at a 14-second cycle time. This behavior could be explained by hypothesizing that the element of product contamination which occurs prior to filling increases with increasing cycle time and, in so doing, reduces the apparent protection of the air shower operating around the filling mandrels.

Results of simulating gross interventions that compromise local protection of an air shower, achieved through switching the air shower off, have revealed that such interventions immediately impact upon the rate of product contamination. Resumption of operation of the air shower was also shown to impact immediately upon the level of product contamination. The rigor of the intervention, as controlled by the length of time when the air shower was not operated, also influenced the level of product contamination. Thus, for an intervention of 5-minutes duration, the fraction of ampoules contaminated was observed to achieve a maximum of 0.28, whereas, for the intervention of 15-minutes duration, a maximum of 0.48 was recorded. Furthermore, the minimum recovery time required to achieve pre-intervention levels was less than 1 minute following a 5-minute intervention as opposed to around 3 minutes following a 15-minute intervention. Clearly, both the establishment and the destruction of local protection afforded by the air shower are time functions. This behavior is in keeping with local protection, provided by filtered air emerging from the air shower, being achieved in part through establishing a compartment of 'clean' air within which critical Blow/Fill/Seal operations are conducted. However, further experimental work is required to provide conclusive proof for the existence of such a compartment. The constancy of the rate of product contamination for the production periods prior to, between, and after the two simulated interventions is also worthy of note. It provides further evidence that, under controlled conditions, Blow/Fill/Seal machine performance, in terms of microbiological quality of product, is highly reproducible and predictable.

It is essential to recognize that the findings described above, and their interpretation, apply only to the particular test machinery operated under the specified conditions. Nonetheless, they demonstrate the potential of employing 'medium-fill' under conditions of controlled microbial challenge to rationalize the performance of Blow/Fill/Seal technology in aseptic processing. The findings reported here have shown that:

a. The fraction of product contaminated is determined by the microbiological quality of the Blow/Fill/Seal machine environment.

b. Under fixed operating conditions, the relationship between the fraction of product contaminated and the level of airborne microorganisms is regular, highly consistent, and reproducible.

c. The protective shower of air around the filling mandrels reduces the frequency of product contamination; the efficiency of local protection is dependent upon air shower design and machine operating conditions.

d. The time necessary for the mold carriage to move from the molding station to the filling station is a critical determinant of the rate of product contamination.

e. 'Interventions' that compromise local protection afforded by the air shower have an immediate and detrimental impact upon the frequency of product contamination; however, the observed effects are fully reversible.

Our work also serves to demonstrate that responses to controlled microbial challenges can provide an effective approach to estimating rates of product contamination for advanced aseptic processing. Furthermore, they allow prediction of operating conditions, including machine operating conditions and environmental quality, under which an acceptably low frequency of product contamination is attained.

ACKNOWLEDGEMENTS

The authors wish to thank the following companies for facilities and/or support in undertaking controlled microbial challenges of Blow/Fill/Seal machinery:

Automatic Liquid Packaging Inc., Illinois, U.S.A.

Fisons Pharmaceutical Division, Cheshire, England

Astra Pharmaceutical Production AB, Södertälje, Sweden

Invaluable technical assistance was provided by Mrs. Rita Taylor which is greatly appreciated.

REFERENCES

Barr, D. B. 1993. Aseptic processing: Proposed regulation. *J. Paren. Sci. Tech.* 47 (2):57–59.

Bradley, A., S. P. Probert, C. S. Sinclair, and A. Tallentire. 1991. Airborne microbial challenges of blow/fill/seal equipment: A Case study. *J. Paren. Sci. Tech.* 45 (4):187–192.

Dimmick, R. L. 1969. Stirred-settling aerosols and stirred-settling aerosol chambers. In *An introduction to experimental aerobiology*, edited by R. L. Dimmick, and A. B. Ackers. New York: John Wiley and Sons.

May, K. R. 1973. The Collison nebulizer: Description, performance and application. *J. Aerosol Sci.* 4:235–243.

Public Health Monograph No. 60. 1964. *Sampling microbiological aerosols.* Washington, DC: U.S. Department of Health, Education and Welfare.

Sharp, J. R. 1987. Manufacture of sterile pharmaceutical products using "blow-fill-seal" technology. *Pharm J.* 239:106–108.

10

Aseptic Processing of Biopharmaceuticals

Nelson M. Lugo

Genetically engineered biologicals are the significant products that drive the race of biotechnology research and development to the forefront of medicinal therapy. The discovery and application of these biopharmaceuticals has progressed if not exceeded traditional pharmacologic formulations. Biopharmaceuticals differ from traditional pharmaceutical indications in that they act in the same manner as the compound that occurs naturally in the body. The challenge in mimicking cellular biochemical processes comes in identifying the etiological agent of interest, expressing it as a stable substance, and purifying it to an acceptable level of biological activity. Since peptides and proteins comprise many of the biological indications in use or under development, the challenge comes in the need for processing these molecules efficiently through the barrage of established bioprocessing methodologies. In addition, many, if not most, of these biologicals need to be processed and analyzed under regulatory guidelines for current Good Manufacturing Practices (cGMPs) and analyzed using Good Laboratory Practices (GLPs). These established regulatory controls are necessary to ensure process consistency and final product safety.

The focus of this chapter addresses the aseptic processing of biopharmaceuticals derived from recombinant technology. Humulin® (Human Insulin, Lilly, Indianapolis, IN/Genentech, So. San Francisco, CA) was one of the first biopharmaceuticals approved (October 1982) by the Food and Drug Administration (FDA) for the treatment of diabetes; it is produced by recombinant means. Well over a hundred similar biopharmaceutical products await regulatory approval each year, with hundreds of others still in the research and development (R&D) stage. Prior to approval, these products must have processes that are defined, characterized, and validated—the level of each increasing as the regulatory submission process nears FDA approval and commercial production. Here we will introduce the reader to the various components necessary for the aseptic processing of biopharmaceuticals. A typical process flowchart for the isolation and purification of biopharmaceuticals derived from prokaryotic cells (Figure 10.1) includes the identification and establishment of the gene to encode the native protein, isolation of the gene, insertion into an appropriate expression vector, transformation, and culturing in a suitable growth medium. The transformed cells are characterized and isolated for storage into a Master Cell Bank (MCB). The MCB is further expanded into a second tier or Manufacturing Working Cell Bank (MWCB) to maintain the requirements of manufacturing for the duration of the product's life time. Cells from the MWCB are cultured, harvested, and disrupted to release the protein of interest. The protein is immobilized, purified, formulated, and filled for final disposition or stored as an intermediate for further processing. Processes utilizing prokaryotic (bacterial) and eukaryotic (mammalian) cells will be discussed.

CELL BANK

Generation and Characterization

In order to identify and characterize the structural gene that encodes the native protein, the gene must first be isolated or synthesized. Since most of the human therapeutic proteins are quite specific, it would be impractical to systematically derive the gene from a body cell that has between 30,000 and 100,000 functional human genes, totaling 3 billion DNA base pairs. A simpler method is to locate a body cell that produces the native protein naturally and in sufficient quantities. For example, genetic material needed for insulin production would most likely be found in pancreatic tissue. Once the gene is identified, it is exposed to a restrictive enzyme that recognizes a specific base sequence on a DNA molecule and causes the DNA molecule to split or cleave at this site (Stwertka and Stwertka 1989). The restrictive enzyme EcoR1 cleaves between the bases adenine and guanine for the sequence CTTAAG. The complementary sequence, GAATTC, is cleaved between guanine and adenine. The cleavage process results in "sticky

Figure 10.1. Process flowchart for the isolation and purification of a biopharmaceutical derived from prokaryotic cells.

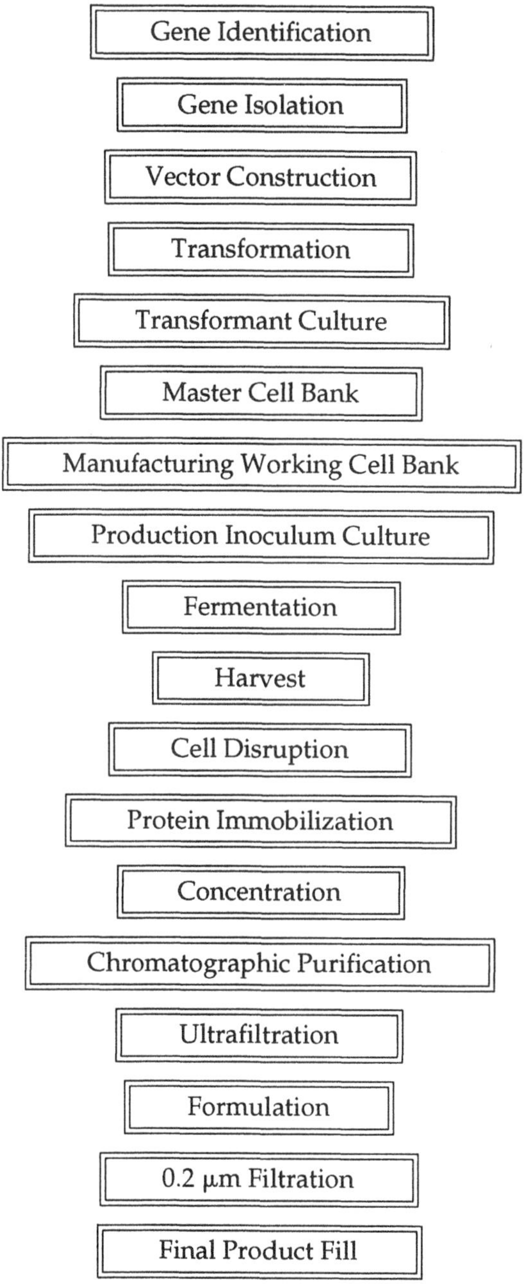

Gene Identification

Gene Isolation

Vector Construction

Transformation

Transformant Culture

Master Cell Bank

Manufacturing Working Cell Bank

Production Inoculum Culture

Fermentation

Harvest

Cell Disruption

Protein Immobilization

Concentration

Chromatographic Purification

Ultrafiltration

Formulation

0.2 μm Filtration

Final Product Fill

ends" remaining at the ends of the DNA fragment (Figure 10.2). In order to express the gene, an appropriate expression vector or carrier is identified to receive the gene and insert it into the host cell. A bacterial plasmid, used as the vector, is exposed to the same restrictive enzyme, causing cleavage of the identical base sequences. A "ligase" is then used to "glue" the complimentary base sequences resulting in the formation a "hybrid" (Figure 10.3). Transformation occurs when the "hybrid" is introduced into a solution of host cells for the generation of identical cells (clones) containing the modified plasmids.

Mammalian cell culture suspensions, which are used for the production of specific monoclonal antibodies, are generated by the fusion

---

Figure 10.2. Cleavage of DNA base pair.

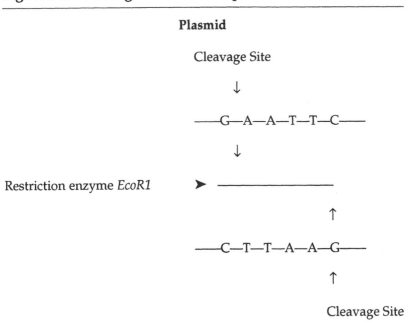

Plasmid

Cleavage Site

↓

——G—A—A—T—T—C——

↓

Restriction enzyme *EcoR1* ➤ ————————————

↑

——C—T—T—A—A—G——

↑

Cleavage Site

Base pair sequence is cleaved at a specific site in the plasmid.

——G   A—A—T—T—C—

Sticky End   ←→   Sticky End

——C—T—T—A—A   G——

The plasmid opens, leaving two "sticky ends."

Figure 10.3. Recombinant plasmid.

DNA Sequence Fragment in the presence of Ligase

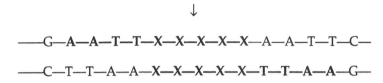

of eukaryotic cells with tumor cells. Here immunized animal anti-body-producing cells (mouse spleen) are fused with myeloma cells to form hybrids called "hybridoma" cells. The hybridoma cells generated are then cultured either in vitro or in vivo for the production of the desired monoclonal antibody (Figure 10.4). The in vivo production of these monoclonal antibodies, in the form of ascites fluid, generated from perhaps thousands of mice for a single production run, is slowly being replaced by in vitro batch and continuous cell culture suspensions (Birch et al. 1985). The major advantages are lower per unit cost of producing larger quantities, reproducibility, low level or absence of extraneous antibody, and the reduced risk of contamination due to adventitious agents from mice or rats.

Production Cell Lines

The majority of production cell lines are specifically selected for their genetic uniformity and stability. Most importantly, the selection is based on their ability to generate the highest expression levels of the desirable component with the lowest levels of associated contaminants. Prokaryotic-producer cells have certain inherent secretion characteristics that are particular to the strain and aid in the selection of the organism as a host. The gram-positive bacteria *Bacillus subtilis* has a single surface membrane barrier and can normally secrete several enzymes at significant levels. To generate the protein of choice, the N-terminal signal is fused with the corresponding genes and expressed from an appropriate promoter. Earlier, the use of gram-positive organisms for secretion of foreign proteins was problematic with poorly characterized and many times unstable plasmid vectors. In addition, the release of several proteases reduced protein yield significantly (Holland et al, 1986). Given these initial obstacles human alpha-interferon and human proinsulin are foreign proteins that have been successfully produced in *B. subtilis* on the lab scale. *B. subtilis* is not typically used to generate product on the manufacturing scale.

Figure 10.4. Production of monoclonal antibodies from hybridoma cells.

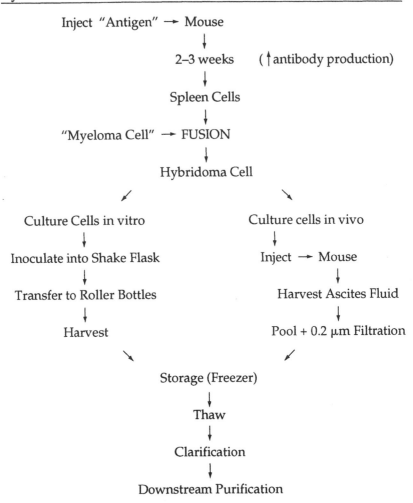

Escherichia coli (*E. coli*) is one of the most widely utilized gram-negative organisms for the production of foreign proteins. Specifically *E. coli* K12 has been used extensively in the field due to its well-defined genetic makeup and the considerable knowledge attained from its use over the last decade. It has a well-characterized genetic and physiological history. Exposed to the proper growth conditions, it can multiply by a factor of two in less than one hour. Although laden with endotoxin, due to its lipopolysaccharide outer membrane, toxicity levels in the product medium are reduced effectively via the downstream purification process. Its ability to be transformed with a variety of comids, plasmids, and bacteriophages allows the production of an impressive assortment of foreign proteins. Human growth hormone, human beta-interferon, and human tumor necrosis factor (hTNF) are but a few examples of the biopharmaceutical proteins produced using this organism.

Yeast cells such as *Saccharomyces cerevisiae* have also been used to produce foreign proteins of biopharmaceutical interest such as human epidermal growth factor (EGF) and hepatitis B surface antigen. Yeasts are prokaryotic organisms that lack the lipopolysaccharide outer membrane associated with endotoxins found in *E. coli* preparations. This has made them especially useful in industrial fermentation processes for the production of foreign proteins used in the food and healthcare industry. Yeast cultures, in general, have well-established genetic characterization and fermentation conditions. *S. cerevisiae* cultures, for example, are generated quite easily and are economical to maintain. Its population doubles in approximately 90 minutes when grown in a rich medium that contains a glucose carbon source. In addition, it grows on minimal medium and uses a diversity of nonglucose sources of carbon (Marino 1989).

Eukaryotic mammalian cell systems utilizing murine hybridoma cells, used for the production of monoclonal antibodies, have been instrumental in the field of diagnostics for the detection of assorted diseases. Hailed as "magic bullets," the monoclonal antibodies produced have also been employed to recognize and locate tumors and associated abnormal tissues sites within the body (Klausner 1986). Therapeutically, they are introduced as a naked antibody to specifically bind to the site and kill tumor cells. Although this approach does have some limitations, a more conventional approach is to conjugate the antibody to cytotoxic agents and steer them to a specific treatment site. Orthoclone OKT®3 is a murine monoclonal antibody manufactured and approved for the prevention of organ transplantation rejection. Mammalian cells must be used for the production of complex biologically active proteins, such as tissue plasminogen activator (TPA) and erythropoietin. Certain processing events such as gamma-carboxylation and glycosylation can be performed only in mammalian cells (Tung et al. 1988).

There are essentially four cell lines that are used as host cells for heterologous expression of foreign proteins. Murine hybridoma Sp2/0-Ag14 and murine myeloma NSO are non-Ig secreting cells used for the expression of antibodies. Baby hamster kidney (BHK) and Chinese hamster ovary (CHO) cells are favored for the expression of other proteins (Hoffman 1992).

Industrial microbial cultures can be obtained from established culture collection centers. These centers provide a means to permanently preserve cultures of interest for the industrial and research communities. There are well over 100 such centers located throughout the world. A partial listing of some of these centers is shown in Table 10.1.

Storage

Prokaryotic and eukaryotic cell systems need to be preserved in an adequate manner to ensure continued proliferation of the host cell line. The stability of the strain is critical at all phases of the product development process. Once a organism has been altered to selectively express the foreign protein, its continued existence must be maintained to provide a supply of equivalent cells for use over the lifetime of the product. This will be covered more closely in the following section regarding regulatory considerations.

There is no single method available to properly maintain, preserve, and store biological microorganisms or animal cell cultures. Each species of microorganism has a particular storage condition that will maintain its stability and viability. *E. coli* can be easily preserved

Table 10.1. Cell Culture Collection and Preservation Centers

AFRC—Agriculture and Food Research Council, London, UK

ATCC—American Type Culture Collection, Rockville, MD

BBL—Baltimore Biological Laboratory, Baltimore, MD

CCF—Culture Collection of Fungi, Department of Botany, Charles University, Prague, Czechoslovakia

DSM—Deutsche Sammlung von Mikrooganismen, D-3400 Gottingen, West Germany

FDA—Food and Drug Administration, Washington DC

IFO—Institute for Fermentation, Osaka, Japan

NCTC—National Collection of Type Cultures, Central Public Health Laboratory, London, UK

WHO—World Health Organization, Geneva, Switzerland

as a suspension of viable cells in 50 percent glycerol/water mixtures. At -85°C the suspension can remain viable for several years (Kelly 1992). Subtle differences within strains of the same species may require unique preservation techniques and conditions. The method of choice for the preservation and maintenance of microorganisms include subculturing, drying, freeze-drying, freezing, or some modification of those methodologies.

Traditionally, subcultures were inoculated in a suitable medium and incubated at an appropriate temperature. The cultures were stored and are reinoculated at intervals to ensure fresh cultures. This method allows the cultures to be easily resuscitated and is applicable to a wide range of microorganisms. This technique is rarely used in the field of biopharmaceutical manufacturing. It is highly labor intensive and there is a risk of contamination and a loss of viability.

Drying involves the removal of water and the prevention of rehydration. There are several drying methods available. Drying is most widely used for particular organisms such as yeast and some bacteria. Contamination is less likely than for subculture methods.

Freeze-drying is the removal of water through sublimation of frozen aliquots. The organism is suspended in an appropriate medium, frozen, and exposed to a vacuum under controlled conditions. This method is applicable to most organisms including some viruses. It is not suitable for plant or animal cells. The disadvantages of this method include the high cost of capital equipment, changes in population, and that some species may experience genetic changes and loss of plasmids.

In freezing cultures the organisms can be classified according to the freezing temperatures used. The most common storage range is from -20°C and -196°C. One method involves immobilizing the organism on specially prepared glass beads that are subsequently stored at -70°C. Each bead carries enough material for one subculture. This method is fast and easy to perform, but has a high cost of maintenance. One of the most widely used freezing methods is storage in liquid nitrogen or vapor phase nitrogen. Although this method does have some drawbacks, such as high equipment cost and the potential for population loss (depending on the organism), it allows for the successful preservation and storage of viruses, bacteriophages, and prokaryotic and eukaryotic cell lines.

Regulatory Considerations

It is imperative that a cell bank system be generated, from the cell line of choice, to ensure a supply of equivalent cells for use over the duration of the product's lifetime. The *Points to Consider in the*

Characterization of Cell Lines Used to Produce Biologicals (Zoon 1993) recommends that the MCB be identified, well characterized, and tested extensively for cell homogeneity and adventitious agent contamination. This ensures the establishment and stability of all significant cell components prior to banking. The Center for Biologics Evaluation and Research (CBER) defines the MCB as a collection of cells of uniform composition derived from a single tissue or cell. Additional information on the recommended control procedures, identification, and characterization of fermentation processes can be found in the *Guideline for Submitting Supporting Documentation in Drug Applications for the Manufacture of Drug Substances*, Feb. 1987 (CBER, Rockville, MD).

The MWCB, as defined by CBER, is derived from one or more ampoules of the MCB and is expanded by serial subculture up to a passage number selected by the manufacturer. The number of generations cultured should be kept to a minimum.

FERMENTATION

Process Fermentation

The early Egyptians were one of the first people known to have utilized the process of fermentation, in the act of brewing beer. The chronology of the field of process fermentation includes the first large-scale batch production breweries of the 1700s, through the war-driven 1940s era with the manufacture of the first commercial antibiotics. Genetic manipulation of strains was accomplished in the early 1960s. But it was not until the 1970s that foreign genes were introduced into a host cell for the production of viable biopharmaceutical protein compounds. Cell fusion technology followed with the generation of mammalian cell culture systems, which borrowed much of the technology already in place from bacterial and fungi fermentation processes.

There are various fermentation processes that are currently utilized commercially: those that generate cells or biomass as an end product, ones that generate metabolites (e.g., vitamins, nucleotides, and ethanol) and enzymes (e.g., protease, amylase, and glucose oxidase), and those used for the transformation of a compound as is in the conversion of ethanol to acetic acid in the manufacture of vinegar (Stanbury and Whitaker 1984).

Growth Characteristics and Kinetics

The microbial growth kinetics of fermentation differs for each type of system utilized. Batch culture processing is a relatively simple, "closed culture" system that utilizes a specific amount of sterilized nutrient media to which an inoculum is added and allowed to incubate. There is no further addition of nutrients added, aside from the introduction of oxygen and agents to control foam and pH. The rate of cell growth is measured over time over five phases (Figure 10.5). The initial or "lag" phase is the period in which no cell growth is measured. Here the cells adapt to the new substrate environment with alteration of some cell metabolic systems to the surrounding conditions of the new substrate. This phase of the fermentation process is analyzed closely during the development stage to minimize and reduce its duration and allow cell growth to occur soon after inoculation of the vessel.

The exponential or "log" phase is the period in which the cells grow at their maximum rate. This phase of growth can be expressed by the following equation:

$$dx \div dt = \mu x \qquad [1]$$

where x = microbial biomass concentration (g/ℓ), t = time in hours, and μ = specific growth rate.

Equation [1] can be used to express the rate of increase in cell number by substituting cell density N (1/liter), for the biomass concentration, as follows:

$$dN \div dt = \mu N \qquad [2]$$

where N = cell density, t = time in hours, and μ = specific growth rate.

Figure 10.5. Typical growth phases of bacteria in batch culture systems.

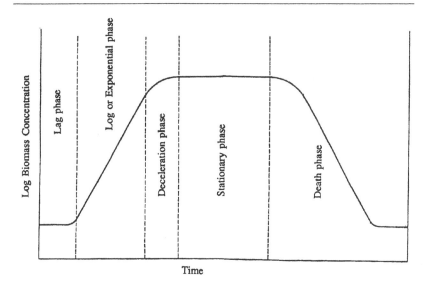

The doubling time, or the time it takes for one cell to become two, can be integrated with μ by the equation

$$\mu = (1 \div x)\,(dx \div dt) = d(\ln x) \div dt = \ln 2 \div t_d \qquad [3]$$

Taking the natural logarithm (ln 2), transforms the equation to

$$\mu = 0.69 \div t_d \qquad [4]$$

Overall μ values vary due to the specific organism's metabolism, the available nutrients, and the supply of oxygen. Examples of maximum specific growth rates (μ_m) for microorganisms are listed on Table 10.2.

Since μ is a function of other parameters, such as the residual growth-limiting substrate, the relationship can be represented in the (Monod 1942) equation

$$\mu = \mu_m\,(S \div K_s + S) \qquad [5]$$

where S = the concentration of the limiting substrate and K_s = substrate specific constant or the substrate concentration at which $\mu = 0.5\ \mu_m$ (50 percent of the maximum specific growth rate is achieved).

Figure 10.6 shows the relationship between the specific growth rate and the limiting substrate concentration. The limiting substrate concentration therefore coincides with the "deceleration phase" of the growth curve. When the growth rate has declined to zero the culture system is at its "stationary phase." The rate of biomass generation increases slowly or remains constant. There is a formation of metabolites not generated during the growth phase. The final stage of the growth curve is the "death phase" where the cells begin die at an exponential rate. All substrate has been depleted reducing the number of viable cells.

Table 10.2. Maximum Specific Growth Rates for Microorganisms

Microorganism	(μ_m)	(t_d)
E. coli	2.1	0.33
S. cerevisiae	0.45	1.5
P. chrysogenum	0.28	2.5
A. niger	0.2	3.46

Figure 10.6. μ vs. limiting substrate concentration.

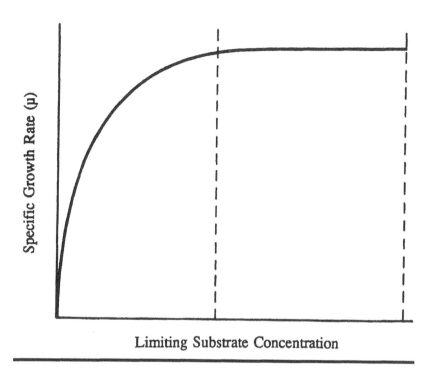

Limiting Substrate Concentration

Media Considerations

Microorganisms used for the synthesis of biopharmaceuticals require an array of simple and complex chemical nutrients for growth and propagation. Carbon, one of the most important chemicals in all life-forms, is required for the generation of carbohydrates and proteins. Glucose and sucrose are readily used as sources of carbon, but not as a sole supply in media preparations. Starch, malt extract, cellulose, whey, methanol, and ethanol (ETOH) are metabolized by microorganisms as an alternative carbon source or as a cosubstrate. Oxygen, although not required for all cells, plays a pivotal role in microbial metabolism. Adequate gas exchange is critical to the operation of fermentation systems. Although a complete understanding of the utilization of oxygen is beyond the scope of this work, further study is recommended to fully comprehend the principles of gas exchange and mass transfer and its influence on fermentation and bioreactor function. Nitrogen, required for the formation of proteins, amino acids,

DNA, and RNA, is derived and utilized from components such as ammonium salts and urea. Some complex nitrogen sources include corn steep liquor, peanut meal, soybean meal, caseamino acids, and peptones. Minerals such as magnesium, phosphorus, sulfur, calcium, iron, and copper are critical to many media preparations. The concentration and selectivity of each component can be crucial to the metabolic system targeted and the resulting yield of the product of interest. Testing for trace elements in media formulations can help define the associated linear relationships between metal concentration and product yield. Other media considerations are vitamins and growth factors. Vitamins are employed by microorganisms to create cofactors such as NAD and NADP. Growth factors allow the cell to expand and produce without having to expend energy synthesizing needed amino acids and pyrimidines. Water, one of the key components of all life-forms, makes up approximately 90 percent of cells (most cells). Its purity and use is critical to the balance of the substrate matrix. Variance in pH and levels of contaminates may effect batch-to-batch substrate consistency.

A typical growth medium used for the production of a recombinant protein from *E. coli* may contain g/ℓ quantities amino acids, yeast extract, phosphate salts, potassium salts, ammonium salts, and glycerol at a specific percent v/v (Sherwood et al. 1991). Those components that can be sterilized together are normally combined in a separate container and transferred into the fermenter for sterilization prior to the introduction of the inoculum from the shake flask. The remaining components are presterilized and aseptically added to the sterile fermenter.

Eukaryotic or mammalian cell cultures (hybridomas) require a special extracellular medium that imitates the cell's in vivo conditions. For optimal cell growth the use of complex, chemically defined media containing 5–20 percent serum is commonly used. Serum components provide binding proteins, albumin, transferrin, hormones, growth factors, vitamins, minerals, lipids, and protease inhibitors (Mauer 1992). Although serum has been found to be essential to the growth and proliferation of many cell systems, it has some fundamental disadvantages, including batch-to-batch variability, cytotoxic effects, and risk of contamination due to adventitious agents such as viruses and mycoplasma. The design of low-serum, serum-free, and total protein-free substrates involves the supplementation of hormones and growth factors normally found in serum with component concentration and condition adjustments for the specific cell line (Fike et al. 1991).

Equipment Specification

The equipment necessary for the growth and production of viable organisms encompasses the use of laboratory-scale apparatus through large-scale fermenters or bioreactors. A typical scaleup of the fermentation process for prokaryotic cells involves reviving the organism from the stored MWCB ampoules. The organism is inoculated into agar plates, then transferred into the shake flask until sufficient cell density is achieved to allow R&D studies to be performed or for continued scaleup processing. The shake flask inoculum is then transferred into a small-scale fermenter for the generation of pilot-scale quantities of the protein of interest or processed (expressed) further and used as the seed stock for inoculation of large, production fermenters (Figure 10.7).

To prepare the inoculum, critical transfer steps must be performed under extreme aseptic conditions in order for the host cell to survive and multiply in a homogeneous state. The lack of adequate facilities and/or poor laboratory procedures can lead to the contamination of the process mediums and the eventual loss of the batch. The risk of contamination is eminent the moment the MWCB is reconstituted or thawed for the initial starter inoculum. Prior to initiating a culture for production, it is important to verify baseline levels of potential contaminants through the use of environmental monitoring techniques, such as particle counters and Rodac (replicate organism detection and counting) plates. Contact plates of the laminar flow hood and the microbial process room are also needed for the identification and concentration of background flora. Sanitization of the work area is accomplished by wiping down exposed sections (floors, walls, ceiling, hood, and chair) with commercially available disinfectants. Aseptically filtered (0.2 μm) ETOH can be used prior to and during all inoculum transfer operations. Depending on the host and vector, the host strain can be streaked onto selectivity agar plates for pretesting of growth and expression. Once incubated the germinated colonies are transferred into the shake flask containing the specific substrate. Shake flasks are placed into shaker incubators until a minimum concentration of organisms for seeding the larger production fermenter is obtained. This concentration varies for each species and may range from 1.0 percent to 10.0 percent, depending on the inoculum cell density, for bacterial expression systems. The shake flask inoculum is then aseptically introduced into the production fermenter, usually through a pre-sterilized septum. The fermentation can progress as a batch, fed-batch, continuous, or recycled process operation (Figure 10.8).

Mammalian cell cultures are scaled similarly, but utilize specific culture labware, depending on the particular attachment properties of the cell line (Figure 10.9). Anchorage-dependent cells, such as fibroblast, form a monolayer of cells when attached to a surface. Anchorage-independent cells (lymphocytes) are unattached and propagate as a suspension. Some of the more common fermentation systems in use include roller bottle, suspension, and hollow fiber (Figure 10.10).

Figure 10.7. Bacterial process fermentation scaleup.

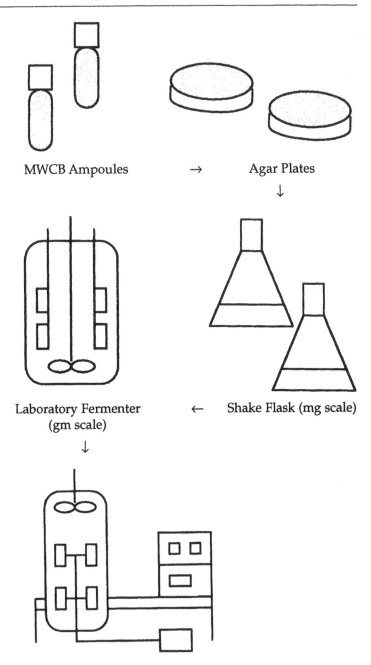

MWCB Ampoules → Agar Plates

Laboratory Fermenter (gm scale) ← Shake Flask (mg scale)

Manufacturing (Production) Fermenter (multiple gm scale)

Figure 10.8. Typical prokaryotic fermentation processes and equipment requirements.

Fermentation	Operation and Specifications	Process Configuration
Batch	Inoculum added into a "closed environment." No nutrients added throughout cycle. Fermentation is short and simple.	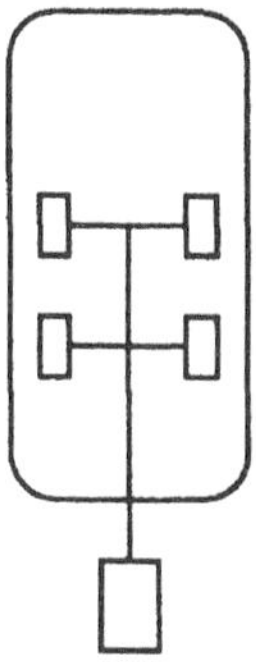
Fed-Batch	Nutrients added during cycle. Metabolism parameters monitored. Makeup substrate must be sterile and added aseptically to avoid contamination.	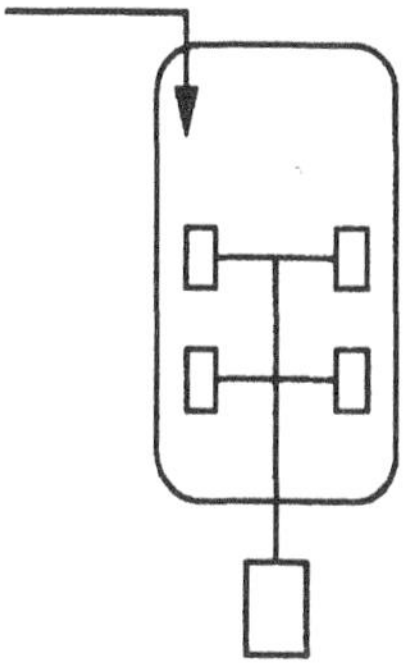
Continuous	Nutrients added during cycle. Metabolized substrate removed at an equal rate to the feed stream. Maintenance of "steady state" conditions critical.	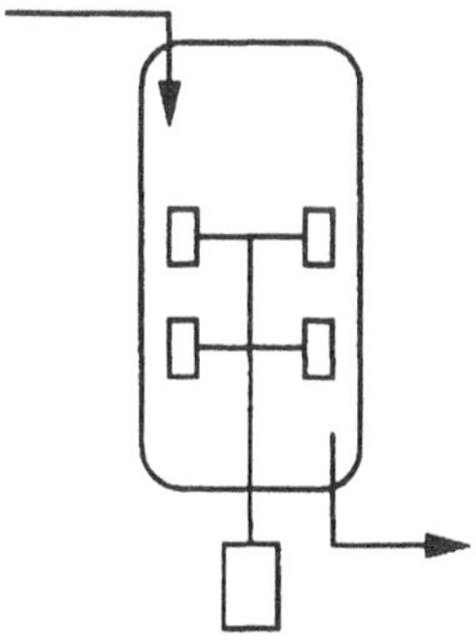

Given the variety of fermentation systems available, the most widely used fermenter for either bacterial or animal cell culture expressions is the stirred tank.

Figure 10.9. Mammalian cell culture process fermentation scaleup.

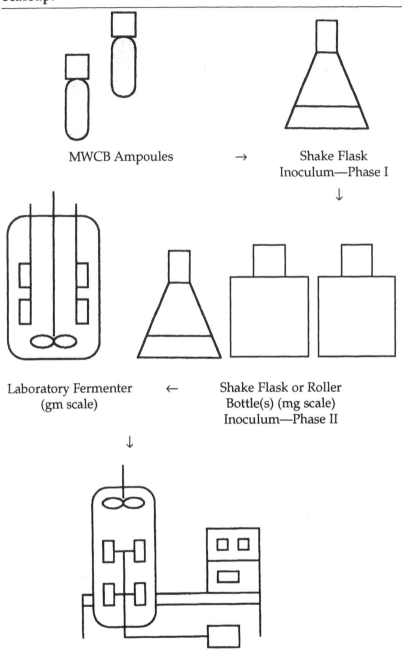

MWCB Ampoules → Shake Flask
Inoculum—Phase I

↓

Laboratory Fermenter ← Shake Flask or Roller
(gm scale) Bottle(s) (mg scale)
Inoculum—Phase II

↓

Manufacturing (Production) Fermenter (multiple gm scale)

Figure 10.10. Typical eukaryotic fermentation processes and equipment requirements.

Fermentation	Operation and Specifications	Process Configuration
Roller Bottle	Glass or polystyrene bottles are used. A monolayer of cells is generated. Bottles are harvested, washed, and re-feed. Process is labor intensive and has a high risk of contamination.	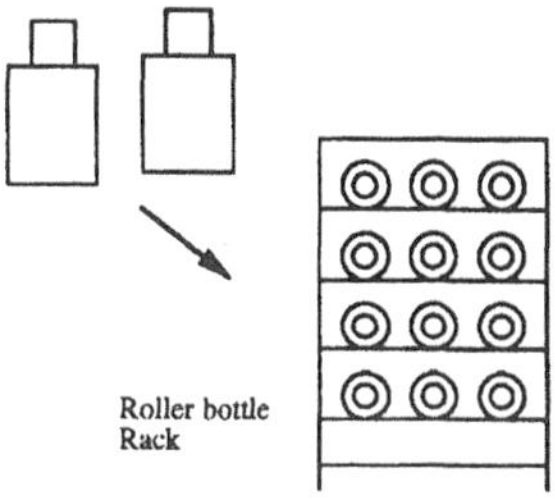
Suspension	Stirred tank or airlift fermentation is used. Batch, fed-batch, or continuous systems are used.	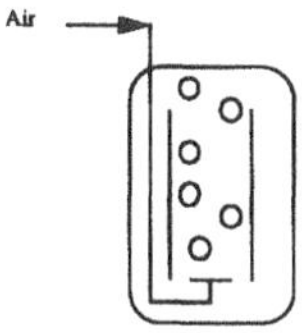
Hollow Fiber	Cells adhere to fiber matrix. Adequate oxygen transfer is required. Overgrowth of cells obstructs pores. Cartridge size may limit productivity.	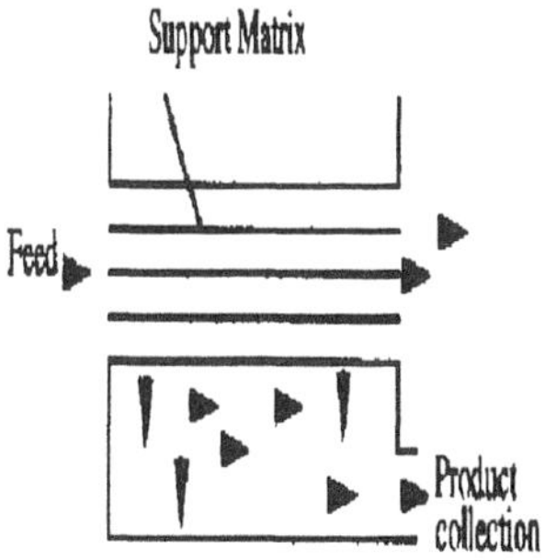
Microcarrier	Cells are grown on porous or nonporous microcarrier beads. System can be continuously perfused with fresh medium thus increasing longevity. May be costly at larger scale.	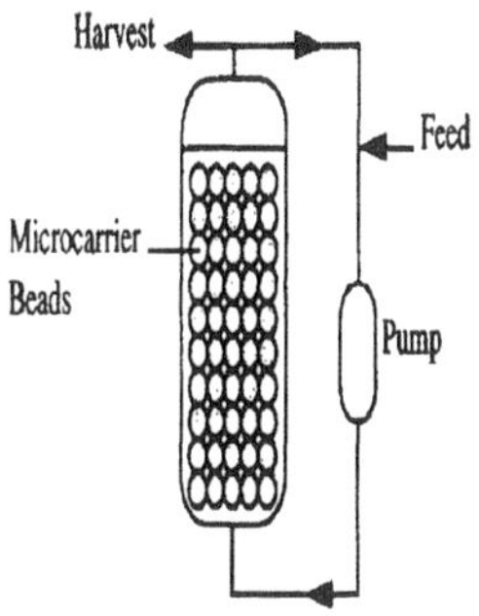

Sterilization of Fermenters and Fermentation Equipment

Heat sterilization is the most common method used for destroying trace viable organisms from fermentation and process equipment surfaces. It ensures that the host organism of interest is the only one present as it is transferred from one stage of the fermentation process to the other. Other methods of sterilization utilize chemical and radiation mechanisms, both of which may be hazardous to the host cell and toxic to personnel and to the environment. Heat, wet or dry, under pressure, has the ability to denature cell proteins irreversibly. Moist heat (at 60–70°C) destroys proteinaceous and enzymatic systems within the organism. Dry heat (at 150–250°C) attacks oxidative processes that require higher temperatures and longer exposure times. Bacterial spores are the most resistant cell forms known. Bacterial spores have been found on remnants found in archeological digs hundreds of years old. When inoculated into a suitable substrate, these spores have been able to germinate into full viable colonies. Wet heat or saturated steam under pressure is used most often as the means of destroying bacterial cells and associated spores. Moist heat is used to sterilize stainless steel instruments, containers, and aqueous solutions. Dry heat is reserved for nonaqueous solutions and glassware. Other methods of sterilization utilize chemical and radiation mechanisms, both of which may be hazardous to the host cell and toxic to personnel and to the environment.

Sterilization of Lab Equipment and Fermentation Vessels

There are a number of procedures and associated equipment used for the sterilization of lab equipment and fermentation vessels. Most general lab glassware (and plasticware), instruments, hoses, and fittings are cleaned, wrapped, and autoclaved on a steam porous cycle for a minimum of 30 minutes at 120°C. To ensure that the equipment being sterilized comes in contact with the steam, a vacuum is pulled in the chamber of the autoclave prior to steam injection. The entire cycle is qualified using time, temperature, and time of exposure. Verification of load sterilization is accomplished using biological indicators (BIs) that contain spores of *Bacillus stearothermophilus*. The BI is inoculated into sterile media and is compared to a nonautoclaved BI (control) after culturing for a specific time period. Elimination or inactivation of a high concentration of these organisms constitutes an effective sterilization process.

Small volume (lab scale) fermenters can be effectively sterilized in an autoclave. The systems are cleaned, prepped, and sterilized under a steam porous cycle. Setup and additions to these fermenters are then carried out under aseptic conditions using presterilized instruments and fittings. Production scale fermenters are normally sterilized-in-place (SIP). After the unit is cleaned (CIP), a pressure hold is performed on the tank to ensure integrity of the system prior to use. Prepared culture media is added to the vessel and the entire volume is heated to within normal autoclave conditions (120°C, 15 psi) and held for a preestablished time. The feed inoculum is then added through sterilized septums attached to the vessel. Inactivation of a failed run can also be accomplished by increasing and holding the process temperature to sterilization conditions. The inactivated cells can then be disposed of safely.

Fermenter Process Control and Automation

Control and optimization of the fermentation process requires the use of special sensors and monitors that detect and analyze the metabolic state of the bioreactor. Some of the parameters that are monitored on large-scale systems are as follows:

- Temperature
- pH
- Pressure
- Dissolved O_2 (DO_2)
- CO_2 in solution
- Foam
- Feed rates
- Airflow
- Impeller speed
- Volume or weight of fermenter contents

A typical fermentation system is shown in Figure 10.11.

Prokaryotic and eukaryotic cells are sensitive to temperature fluctuations and require a narrow range in the circulating substrate for optimal growth. There are a variety of utility systems used for controlling the temperature in a fermenter. Large-scale fermenters commonly use direct steam inject (sterile), heated water, heat transfer coils, and processed chilled water. Small systems use a water/glycol jacket and/or electric coils. Temperature control is achieved by installing dual resistance temperature detectors (RTDs) in the fermenter vessel. One RTD is placed inside of a thermoweld extending into the vessel

Figure 10.11. Process controlled fermentation system.

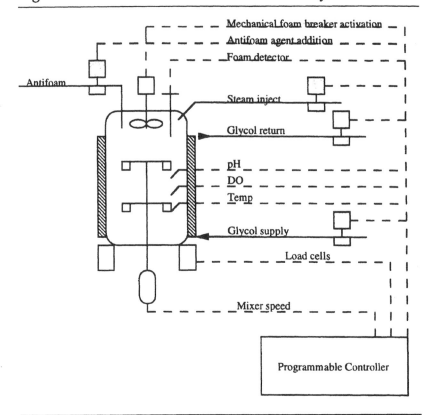

interior. The other is placed in the jacket of the vessel to monitor the temperature of the controlling fluids. A feedback loop attached to a controller monitors both RTDs and regulates the temperature to within a tenth of a degree on some systems. Steam-sterilizable probes are recommended for the monitoring of pH in the fermenter. Combined electrodes that have built-in temperature compensators are highly recommended due to pH probe sensitivity to temperature fluctuations.

Dissolved oxygen probes are composed of an anion electrode and a cation electrode that are bathed in an electrolyte solution and are covered with a glass or stainless steel sheath. Separation of the fermenter substrate and the internal probe is maintained by a gas permeable membrane attached to the tip of the probe.

Dissolved CO_2 (off gas) measurements can be done outside the fermenter or with the newer in-place DCO_2 (Dissolved CO_2) probes. The probes contain a bicarbonate solution that encapsulates a CO_2 permeable membrane.

Antifoam mechanisms reduce the extent of foam generated from agitation aeration. Mechanical foam breakers remove foam by disrupting the generated foam above the liquid level. On stirred-tank fermenters an additional impeller is attached to the upper section of the head plate and is controlled by a conductivity probe that senses generated foam. Automatic chemical antifoam addition systems also have a conductivity probe extending from the top of the reactor into the vessel interior. If foam comes in contact with the probe, a signal is sent to a controller, activating the addition of antifoam agent until the tip of the probe is clear of foam. Sensitivity of the probe must be adjusted to prevent the addition of excess agent. Too much antifoam can cause problems with the oxygen transfer rate.

Viscosity measurements are normally performed off-line and indicate the flow characteristics of the process fluid and the potential shear effects of the impeller on the surrounding medium. Traditional tube, cone, and plate viscometers can be adapted for measuring most process mediums. Most viscometers have inherit drawbacks and limitations due to the solids content in most fermentation broths.

A data acquisition and control system (DACS) provides consistent monitoring of most fermentation parameters (Omstead 1990). The system adjusts running parameters more accurately and effectively than can be performed by system operators. The systems acquire an analog current/voltage that represents a value for a specific parameter. Once acquired the system analyzes the information and implements control algorithms that compares the parameter set point to its current status. A feedback loop control allows the system to then send out a correction signal to a control unit adjusting the parameter to within the set point.

PURIFICATION

Upstream Processing and Clarification

Once the host cell has generated the protein of choice, it must be isolated from the cell remnants and purified to meet full product specifications. The recovery mechanism is dependent on the host cell (prokaryotes or eukaryotes) and whether the expression was intracellular or extracellular. If intracellular, the cells are disrupted or lysed, phase separated clarified, then purified using chromatographic procedures. If extracellular, the medium is clarified and either precipitated or exposed to specific chromatographic resins to isolate the protein.

Further chromatographic steps are performed for increased purity levels. These will be discussed in the next section.

The techniques available to initially remove the cells from the surrounding medium are limited. Filtration, centrifugation, sedimentation, and flocculation are the methodologies most widely used. The method of choice is dependent on cell processing history and development data.

Filtration systems can be classified into several categories. Nominal filters remove a percentage of particles within a certain range of the filter's pore size rating or larger. Absolute filters remove all particles larger than the filter's largest pores. Most of the filters in these categories are associated with static filtration elements.

Clear, cell-free filtrates can be obtained using charge-modified depth filters. Hybridoma cultures with starting densities of 1.6×10^6 cells/ml have been clarified using CUNO® Zeta Plus depth filters. These filters are composed of inorganic filter aids and positively charged-modified cellulose. Some grades have stronger positive zeta potentials. This zeta potential allows for the adsorption of particulates that would normally be too small to be retained by the filter through mechanical straining. The filters come in a variety of sizes and configurations from flat sheet stock to ready to use capsules and cartridges. An example of a disk cartridge depth filter is shown in Figure 10.12. Figure 10.13 shows an equipment configuration for the clarification of mammalian cells using a depth filtration cartridge and 0.2 μm membrane filter.

In cross-flow filtration the culture solution is passed across a filter membrane element. Filtrate passes through the membrane and the culture cells are swept along and retained in the process stream (retentate) (Figure 10.14). These devices come as flat cassette modules, which can be stacked for increasing the surface area necessary for optimal performance and recovery, as hollow-fiber bundles or as a spiral-wound cartridge. Spiral-wound cartridges have similar membrane properties as the cassette modules, but are wound in a tubular fashion around a hollow core (permeate stream) (Figure 10.15). Some initial development work should be performed to determine the optimal system to use, based on host cell size, culture density, volume of batch, pump requirements, holdup volume of the system, automation needed, and overall cleanability. Figure 10.16 shows an automated cross-flow filtration system used in the recovery and clarification of prokaryotic cells.

The centrifugation of cells and cell debris is governed by Stokes's Law, which is represented by the following expression:

$$v = \frac{d^2 w^2 r (P_s - P_L)}{18\,\mu} \qquad [6]$$

Figure 10.12. Depth filtration cartridge element. (*Reprinted with permission from CUNO® Incorporated.*)

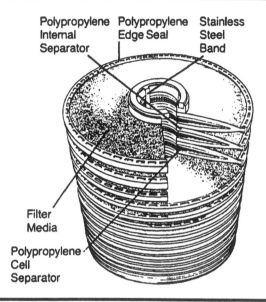

Figure 10.13. Clarification of mammalian cells using depth filtration elements.

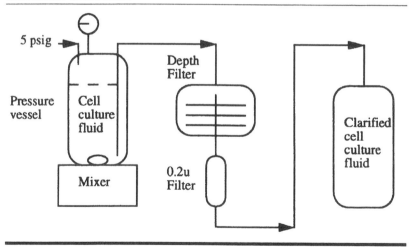

where v = sedimentation rate (cm/s), d = diameter of particle (cm), w = angular velocity (rads/s), r = particle distance from axis rotation (cm), P_s = particle density (g/cm^3), P_L = liquid density (g/cm^3), and μ = liquid viscosity (g/cm s).

Figure 10.14. Cross-flow filtration system.

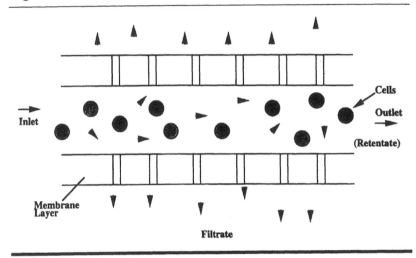

Figure 10.15. Spiral-wound cross-flow filtration device.

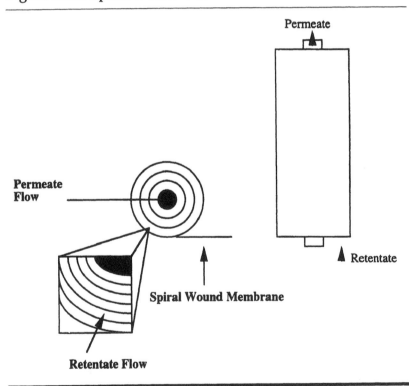

Figure 10.16. Large-scale automated cross-flow system. (*Reprinted by permission from Sartorius Corporation.*)

Separation of cells and cell remnants is not as easy as starting a balanced basket centrifuge and returning 20 minutes later to find a clear layer above the packed cells. Consideration must be given to the percent solids content of the feed, the size of the particles to be separated, and the viscosity of the feed fluid. Equipment considerations include the size of the unit, its hold volume, the radius of the bowl, and its angular velocity. Centrifuges are divided into two broad types, centrifugal filters and sedimentation centrifuges. They can be further divided by the means of which they feed the process stream and discharge the separated solids and liquid phases (Ambler 1988). Figure 10.17 gives a cross-sectional representation of three types of centrifuge configurations.

Tubular bowls have a high centrifugal force ("g force") ($\approx 16,000$), a small rotation radius, mid-scale solids capacity, and the ability to

Figure 10.17. Types of centrifuges.

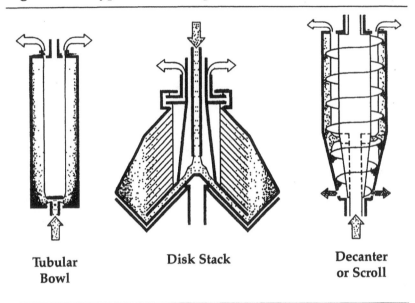

Tubular   Disk Stack   Decanter
Bowl   or Scroll

remove particles in the range of 0.01 μm to 100 μm. Disk stack and multichamber centrifuges have a low centrifugal force (g) ($\approx$6,000), a large rotation radius (due to large bowl diameters) a large solids capacity, and particle size removal in the range of 0.01 μm to 100 μm. Decanter or scroll centrifuges have very low centrifugal forces (g) ($\approx$3,000), a mid-scale rotation radius, a very large solids capacity (due to the constant discharge of solids), and particle size removal in the range of 1 μm to 3000 μm. There are a few other variations of these same units. Aside from the tubular bowl model, that have a set bowl capacity, most units discharge solids during the run cycle, eliminating cleaning the unit at intervals. Most of the newer centrifuges have the option to be set up for clean-in-place (CIP), eliminating the need to disassemble the system after each batch or run.

Flocculation of the cell/medium mixture increases the sedimentation rate of the particles. A flocculation agent, such as PEG (polyethylene glycol) or polyethyleneimine, is added to the mixture. These agents increase the diameter of the particles, thereby increasing the sedimentation rate. Heat and adjustments in pH are also used to enhance flocculation. The ability of the cells, and most importantly the protein of interest, to withstand the addition of sedimentation enhancers should be developed along with the overall purification scheme prior to large-scale implementation.

Flotation of the cells involves aerating the mixture with gas, entrapping the cells to a foam layer and mechanically removing the cells from the top of the fermenter. The size of the gas bubble governs the separation. A fine bubble formation (foam) is created from dissolved gases and enhanced by promotion of insoluble long-chain fatty acids or amines.

After harvesting and separating, prokaryotic cells, yeast and some mammalian cultures are disrupted to release the protein molecule. Some cells produce small nodules or inclusion bodies that contain the protein within each segregated node. Cell walls of microorganisms can be disrupted by mechanical methods (shear and pressure sources) and nonmechanical methods (biological and chemical sources). Mechanical methods include mills of various types, ultrasound, freeze/thaw, presses, and homogenizers. Nonmechanical methods (such as biological enzymes) utilize lysozymes that break down the cell wall. Chemical lysis is achieved by using detergents, salts, and surfactants. The method of choice depends on the type of cell and the relative ruggedness of the cell wall. Eukaryotic cells are more sensitive to physical forces than are prokaryotic yeast and bacterial cells. Gram-positive bacteria are more difficult to lyse then are gram-negative bacteria.

A common method used for the liberation of proteins in the production of biopharmaceuticals is high pressure lysis or homogenization. Impingement of a rapidly moving bacterial feed stream against a hole in a valve seat causes the cells to lyse as they move around a hardened rod in the valve assembly (Figure 10.18). Throttling of the valve against the seat increases the pressure generated. Higher pressures lyse more cells, causing an increase in the release of cell contents (increasing product recovery). Pressures up to 20,000 psig are typical, depending on the homogenizer used and the cell to be lysed.

Chromatographic Isolation and Purification

It is virtually impossible to find a biopharmaceutical protein that is derived by recombinant means that is not exposed to some form of chromatographic medium. Most if not all need to be processed to some level of specified purity. The main focus in "downstream processing" is to isolate and concentrate the protein molecule in as few steps as possible while minimizing the levels of variants and contaminants. Dunhill and his associates discovered that precipitation was found to be the most favored method to follow homogenization of the host cell in the initial purification phase (Bonnerjea et al. 1986). The latter stages include chromatographic support media. Precipitation relies on the solubility of the protein through the manipulation of temperature, pH, and ionic strength in the surrounding medium.

Figure 10.18. High-pressure homogenizer outlet valves.

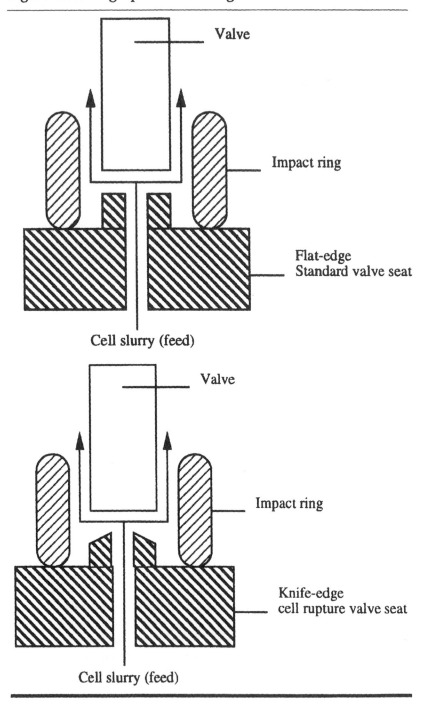

Chromatographic processes resolve low and high molecular weight species in a heterogenous pool consisting of the main molecule and closely related forms. Resolution of the parent protein from cell contaminants (HCP, DNA, adventitious agents) is also accomplished. Selectivity is accomplished by the use of a myriad of chromatographic mediums and resins. Traditionally, the sequence of chromatographic resins has been ion exchange followed by affinity, with the final step being gel filtration, (although this final step is slowly being replaced with other methodologies such as ultrafiltration).

Separations with ion-exchange resins involve the reversible adsorption of charged ions in a solution to a fixed ion-exchange group of the opposite charge. Ion-exchanger types can be either anion exchanger (having positively charged groups on the resin) or cation exchanger (having negatively charged groups on the resin).

Affinity chromatography purification separates specific solute molecules using a ligand immobilized to a support medium. Ideally suited to be used in the earlier stages of the process, it is costly and would foul on exposure to the heavy mix of proteins and associated contaminants found in lysates and early upstream product solutions.

Gel filtration media are composed of cross-linked polymer gels, such as agarose, that separate according to the differences in size of the molecules in solution. The molecules are eluted in order of decreasing molecular size. Small molecules can diffuse into the gel matrix, slowing their progression through the column; whereas larger molecules are force to travel around the matrix, causing them to elute first.

Equipment for Aseptic Processing

Chromatographic Columns and Specifications

There is a wide assortment of chromatographic columns available for lab scale development of biopharmaceutical products. Unfortunately the availability of off the shelf columns for processing large quantities is not as extensive. Chromatography columns used in manufacturing can range in size from 3 cm in diameter to over a meter in diameter. The associated bed heights, depending on the resin used, its capacity and the level and type of separation or resolution needed, can be as small as 5 cm diameter column to as tall as 100 cm. Certain applications require the manufacture to have the vendor build a custom column to specific height, width, and pressure rating requirements. Some standard column configurations used in pilot and process scale chromatography are shown in Figure 10.19. These columns are specifically designed for aseptic processing applications. The necessary requirements include product compatible (nonreactive) contact surfaces such as glass, acrylic or stainless steel. Glass and acrylic columns are used many in low pressure chromatographic applications. Stainless steel columns are used for analytical and preparative high pressure liquid chromatography (HPLC) applications. In addition sanitary flow adapter configurations and inlet and outlet connection ports are critical. To ensure that the columns are operated and maintained under aseptic conditions the manufacture may recommend unpacking the resin on a scheduled basis and sanitize using chemical agents such as ETOH, sodium hydroxide, or sodium hypochlorite (an oxidizing agent). Engineered columns may be designed for routine clean-in-place (CIP) where the entire system is configured for cleaning and/or autoclaving. For processes requiring sterile column operations in-situ sterilization has been accomplished using formulated peracetic acid solution (1,500 ppm peracetic acid in 0.5 M sodium acetate, pH 5) (Jungbauer 1994).

Figure 10.19. Column configurations for pilot and preparative chromatography.

Low to medium pressure sanitary column (Reprinted with permission from Amicon®, a GRACE company)

Continued on next page.

Figure 10.19 continued. Preparative scale HPLC Stainless steel column (Reprinted with permission from EM Separations Technology)

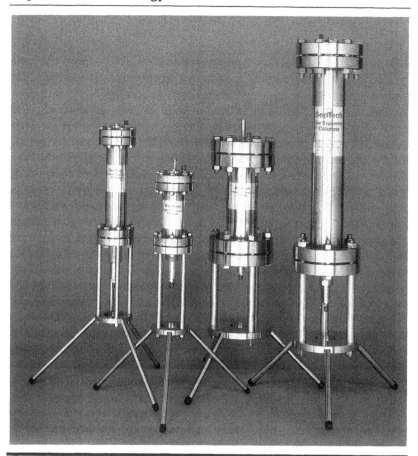

   Since most chromatography processes are carried out under aseptic conditions it is necessary to reduce and minimize the exposure of the column and medium to adventitious agents. The choice of regeneration and cleaning agents depends on its compatabilty to the medium used and the stage of processing. Most chromatography medium suppliers have "recommended" guidelines for cleaning, regenerating, and sanitizing. A careful analysis of the product's components will allow the end user to determine the best CIP reagent and method to use. Most soluble and degradated/precipitated proteins can be removed from chromatography medium using sodium hydroxide, sodium chloride, acetic acid or water. Lipids on the other hand may require ETOH, isopropanol, nonionic detergents or acetonitrile for removal and decontamination. The removal of endotoxin and the inactivation of some viruses can be accomplished using varying concentrations of sodium hydroxide.

Pumps

Pumps come in various sizes and configurations. Those currently used in the field include tubing (peristaltic), positive displacement (PD), centrifugal, and diaphragm. There are addition pumps and pump configurations available, with some designed for very specific applications. For example, magnetically driven pumps are available with sealed rotor pump heads that are removed and autoclaved for aseptic filling operations. Some of the factors that need to be taken into consideration when choosing a pump include the product (viscosity), scale of process operation, flow rate, and pressure. Shear and temperature sensitivity of the biological fluid should also be consider. For instance, many proteins do not hold up well to the extreme shear forces generated from constant speed centrifugal pumps. These pumps are better suited for the transfer of water and buffer solutions. Many of the rotary lobe (positive displacement) pumps used in the pharmaceutical and biopharmaceutical field have been transplanted from the food and dairy industry. The 3A sanitary standards on these pumps indicate their ability to be disassembled and cleaned effectively. In addition, all wetted parts are constructed of sanitary compatible materials such as 316 SS. Most of the pump manufacturers now have options for CIP and SIP for pumps used in the biopharmaceutical field. Tubing pumps have been used most effectively for small scale and pilot scale operations due to the flexibility of pump and tubing sizes available. Although the newer tubing materials are more durable and cleanable and they can be discarded after each run to ensure no lot-to-lot or product-to-product contamination.

Ultrafiltration Systems

Biological process intermediates can be concentrated and/or buffer exchanged consistently at scale with sanitary design ultrafiltration systems. The systems can be "off the shelf" models with specific capacities and flow rates or they can be custom designed for individual needs. The minimum requirements for a pilot system should include the appropriate membrane holder, a vary-drive sanitary pump, retentate flow rate, retentate pressure (in and out), sanitary piping, valves, and fittings. Custom designed systems may include a vary-drive sanitary pump (with sanitary flush seals), pump RPM indicator, sanitary retentate and permeate flow rate meter/indicator (ℓ/min) (non-turbine type), retentate pressure (in and out), sanitary piping, valves, and fittings. If a process vessel is part of the system a form of level indication system is recommended.

Chromatographic Systems

Chromatographic systems or "skids" come in a variety of sizes and configurations. Most of these systems are custom made for each client. There basic function is to allow the operators to systematically run a chromatographic process without having to set up the necessary equipment each time. The general design includes a pump, UV detector, conductivity and pH sensor, a recorder to trend one or all aspects of the process, flow and pressure transmitters. The systems are normally designed with SS process piping configured with sanitary inlet, outlet, and internal ports and valves. Classifications range from low pressure, high pressure, aqueous, solvent (explosion proof), sanitary, or nonsanitary. Many units are specified with a few of these features as standard items. Table 10.3 outlines a comparison of chromatography systems for preparative use.

CIP Systems

Clean-in-place (CIP) systems are usually designed specifically for a particular application of process system. The concept of CIP originated from the cleaning specifications developed for the food and dairy industry and was published in the *3-A Accepted Practices for Permanently Installed Sanitary Products—Pipelines and Cleaning Systems.* Some general guidelines needed for designing a particular CIP unit include the type of tank or piping to be cleaned, its size, and the number of tanks and line circuits. This also includes any fermentation tank configurations and cleaning requirements. Consideration should also be given to the product residue remaining in each tank or line circuit. The removal of protein material from the walls of tanks and piping may require specific cleaning patterns and cycles that would not necessarily be used for the removal of salts from dedicated buffer tanks and systems. The physical system can be designed as an in-place remote unit that is located in a dedicated part of the facility. The CIP solutions are pumped through dedicated CIP piping to the required areas, and are either recirculated back to the CIP unit or allowed to flush through once at the cleaning site (Adams and Agarwal 1989). This type of configuration is preferred in facilities that have a fairly large dedicated tank and piping system. Portable CIP units are used to perform the same functions as the remote units but can be

Table 10.3. Comparison of chromatographic systems.

	Company		
	A	B	C
Pump			
Design	Piston	DBL Diaphragm	Piston
Mixing	Low Pressure	High Pressure	Low Pressure
Flow rate (ml/min)	50–500	100–2000	30–250
Gradient capable	No	Yes	Yes
Max. operating pressure (psig)	350	1500	150
Automation			
Computer (optional)	No	486/66	486/50
PLC	Yes	Yes	No
Recycle	Yes	Yes	Yes
Validatable	No	Yes	Yes
Added Features			
	Fraction Valves	Fraction Valves	Sanitary design
		Explosion proof	
		Sanitary design	
Price			
	$$	$$$$	$$$

transported from vessel to vessel or from area to area as needed. They are not as sophisticated as their larger counterparts but can be operated efficiently in multiuse areas. These units have all of the necessary components and equipment needed to clean tanks and piping. The disadvantages with these systems include space requirements (for moving the unit around), noise level during operations, setup and assembly, and transporting between sites. Both large and portable units are controlled by a PLC configured for each application and/or cycle. Figure 10.20 shows a portable CIP system configured for cleaning a fermentation vessel. Effective cleaning also requires the design of an appropriate cleaning cycle that includes the design configuration for the spray heads (for complete internal vessel coverage), cycle time (prerinse/wash/postrinse), temperature, and type and concentration of detergent used. These parameters are normally validated to ensure consistent cleaning of the equipment and to minimize residual contaminants from lot to lot and from product to product.

Figure 10.20. Portable CIP system cleaning configuration.

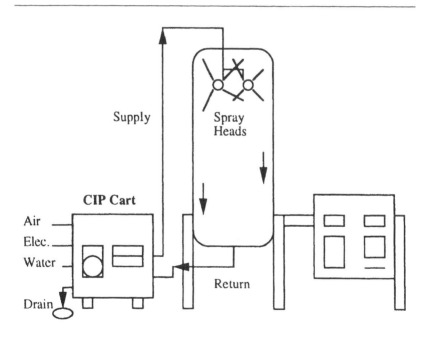

Quality Control Testing

Quality control (QC) testing of biopharmaceuticals requires the identification, characterization, purity, and potency of the expressed and isolated protein as well as identification of associated contaminants. There are a multitude of assays associated with determining the profile necessary to ensure a biopharmaceutical host organism has been transformed and expressed properly. A detailed description of all the characterization assays available is beyond the scope of this chapter. Further readings on this subject can be found in the references (USP 1990; Garnick et al. 1991; FDA 1985; ASTM 1989). Table 10.4 illustrates some of the more common assays used during the initial stages of R&D and throughout the life of the product of interest.

Formulation of Bulk Biopharmaceuticals

Proteins vary greatly in their number of different chemical and physical properties. This is the reason that one is able to purify proteins from a mixture of thousands of proteins. These properties also contribute to the characterization aspects of the molecule under a variety conditions. Maintenance of a stable formulation is necessary to ensure the stability of the product over the product's full shelf life. Handling of the specific protein of interest is as critical from the initial generation (isolation) to the final fill container awaiting administration to a patient. There are a number of factors that affect the stability of biologicals. Degradation mechanisms and denatured products must first be understood in order to appreciate their impact on overall stability. Some degradation mechanisms include deamination, fragmentation, aggregation, oxidation and denaturation. These have been illustrated in closer detail in Figure 10.21. Deamination occurs by the lost of α-amino groups on asparagine and glutamine that results in the formation of aspartic acid and glutamic acid, respectively. Hydrolysis of amide bonds, which link amino acids together, cleaves the protein into fragments (fragmentation). Aggregation can arise from the hydrophobic interaction between protein molecules. This can occur through covalent or noncovalent bonding. Oxidation can be caused by various pathways resulting in the production of free radicals such as the oxidation of methionine to its sulfoxide derivative. Alteration of the parent molecule structure results in denaturation. Denaturation may or may not be reversible.

Stability monitoring assays ensure that degradation products are detected and quantified. A clear understanding of the degradation products allows the manufacture to control process parameters closely and focus on problem areas. Once identified, modifications to the

Table 10.4. General Assays for Biopharmaceuticals

Test Parameter	Analysis	Determination
Identity	Amino acid analysis	Amino acid composition
	Degradation sequencing	N-terminus; C-terminus
	HPLC	
	Mass spectroscopy	Molecular weight
	Peptide mapping	Elucidation of amino acid sequence
	SDS–PAGE	Characterization by molecular size
Potency	Biological activity	
	Concentration	
Purity	Adventitious agents	Bacteria, virus, mycoplasma
	DNA	Host nucleic acid
	Circular dichroism	Three-dimensional structure
	Endotoxin	Level of pyrogenic activity
	Host cell protein (HCP)	Host cell impurities (antigen)
	HPLC	Identification and isolation of single species
	SDS–PAGE	Determine oxidized, denatured, and dimer forms

process can be implemented prior to initiation of a long term stability program for the product.

Analytical methods to detect degradation products can involve standardized characterization test procedures or specially developed techniques specific to the protein molecule of interest. Test methodologies available for the detection of some of the aforementioned

Figure 10.21. Mechanisms of protein degradation.

Deamidation $H_2N-Asn-COOH \Rightarrow H_2N-Asp-COOH + NH_2$

Fragmentation $H_2N-Lys\ Ala-COOH \Rightarrow H_2N-Lys + Ala-COOH$

Aggregation $n\ (H_2N-COOH) \Rightarrow (H_2N-COOH)n$

Oxidation $H_2N-Met-COOH \Rightarrow \overset{O}{\overset{\uparrow}{Met}}-COOH$

Denaturation

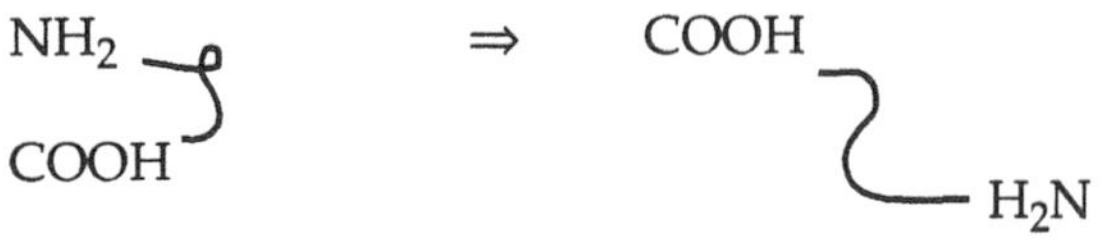

degradation products vary in their ability to identify and quantify changes to the parent molecule. The combination of several assays with overlapping sensitivity ranges allows formulators to evaluate the mechanisms responsible qualitatively and to some degree quantatively. Below are some degradation mechanisms and the associated test methods utilized for their detection and identification:

Degradation State	Test Methods*
Deamidation	RP–HPLC, IEF, IEC, tryptic mapping
Fragmentation	SDS–PAGE, HPSEC, AAS, tryptic mapping
Aggregation	HPSEC, SDS–PAGE
Oxidation	Amino acid analysis, tryptic mapping, HPLC, RP–HPLC

Monitoring for stability of recombinant derived products begins at the plasmid generation phase and extends through all aspects of the isolation, purification, and final fill container stage. Although we've identified some important aspects of protein degradation and modes for detection, the formulator's role is to address the prevention of

* Note: Assay Abbreviations: AAS—amino acid sequencing; IEF—Isoelectric focusing; IEC—Ion-exchange chromatography; HPSEC—High performance size-exclusion chromatography; HPLC—High performance liquid chromatography; RP–HPLC—Reverse phase high performance liquid chromatography.

denaturation of a product and to reasonably predict the inherent shelf life of the molecule under various internal and external conditions. Understanding degradation mechanisms is one aspect of the formulator's role in the overall stability program. Minimizing, or at best, preventing these mechanisms from occurring is another. Environmental conditions are monitored to determine areas of concern throughout the process. The exposure of the product to heat, cold, agitation, aeration, and shear are investigated. During the process development stage the effects of buffer conditions are monitored and augmented as needed to ensure consistent product purity and recovery yields. The need to formulate and effectively stabilize proteins as therapeutic biopharmaceuticals has given rise to a field far more different than the traditional chemical compound formulations. Experimental models to study denaturation and evaluate stabilizers have been established. Wang (1988) did an extensive search of the available literature and outlined the use of excipients to stabilize parenteral formulations of proteins and peptides. Albumin, from various sources, is used in over 50 drug formulations as a primary stabilizer. Amino acids have also been used effectively as chelating agents, aggregate formation inhibitors, heat denaturation stabilizers, and reduce surface adsorption agents. Other noted stabilizers include surfactants (e.g., polysorbate), metals (calcium), polyols (glycerol), reducing agents (thiol compounds) and sugars.

Process Automation

Biopharmaceutical processing, in general, is inherently slow involving any where from a dozen steps to complete (e.g., manual pH determination and adjustment), to several hundred different manipulations (e.g., initiating and maintaining a fermentation run) in the course of batch run. In addition, process conditions need to be monitored, recorded and corrected, if necessary at regular intervals. In many cases reaction to process conditions need to be performed accurately and in a timely manner. Process automation allows operations to proceed in a continuous, organized, and predetermined mode. One of the prerequisites to cGMP compliance is to ensure that the process is reliable and reproducible. Automatic batch processing can accomplish both of these conditions. An added benefit is the potential to reduce overall processing cost. Downstream purification processes are ideally suited for automated batch mode configuration. Chromatographic separations and ultrafiltration (concentration/diafiltration) are two unit operations which have straight forward applicability for process control (Ransohoff 1990).

Filling and Finishing of Biopharmaceuticals

The majority of recombinant derived products are filled as small volume parenterals (SVPs). Small-volume parenterals are packaged in volumes up to 100 ml. Most of the containers used in this category range in size from 1 ml to 50 ml. Product formulations range from injectable aqueous solutions to topical or ophthalmic applications.

Prior to identifying the type and size of container to be used the manufacture investigates and agrees on the final product formulation specifications. This will include concentration, how the product will be administered and the proposed dosage levels. Some formulations may require special handling conditions (storage at 4°C) or additional processing (lyophilization). The type and size of container to be used may also be driven by the formulation specifications. Product vs. container compatibility should be reviewed carefully to ensure that once the product is placed into the container it will still contain and deliver the specified volume and retain its original concentration.

Lot sizes for the majority of biopharmaceutical firms in phase I and phase II clinical trials are relatively small compared to those for traditional or established pharmaceutical companies. Smaller lot sizes entail additional monitoring and handling considerations. Large companies have dedicated staff who specifically monitor and control automated vial and stopper washing operations. That staff may also participate in the filling operations that take place in Class 100 areas under Class 100 hoods. Handling of the final product and container/closures is minimized. Small companies normally have a well-trained staff that perform not only the purification duties but the final fill operations as well. Operators working in aseptic areas have been cited as being the primary factor for introducing and dispersing microbial and nonviable particulate contaminants into controlled areas. To minimize potential contamination of the final product, it is critical that operators be well trained and have specific SOPs and documented procedures. Media or mock fills should be performed at scheduled intervals to ensure filling consistency. Environmental conditions are crucial to any sterile filling operation. The filling area is the most clean and environmentally controlled region in the entire facility. The room air balance pressure is normally provided by a separate HVAC system and should be positive to all areas. This prevents contamination from anywhere in the facility from finding its way into the fill room. A careful review of general disinfectants used is necessary. Alternating agents prevents microbial strain resistance. Contact plating of filling rooms before and during operations is coordinated and performed by QC personnel. They should also monitor the staff working at filtration/fill stations by taking contact plates of operator hands and arms. For small filling operations the best guidelines to follow include minimizing the number of staff needed for the operation, preventing air turbulence during the procedure (talking, standing, walking), keeping personnel with infectious diseases out of aseptic areas, minimizing handling of product and containers, training personnel in aseptic procedures, and automating procedures where possible (Levchuk and Lord 1989).

SUMMARY

The field of biopharmaceutical processing is expanding and changing as new biological products are discovered and brought to market. Multistep biopharmaceutical processes need to be monitored and modified, if possible, to optimize consistency, improve yield, and reduce overall cost. To accomplish this, manufacturers need to look at their operations critically and identify those areas that maximize growth and streamline processing operations. Excellent advances have been made in the field of aseptic processing in the last 7 to 10 years. New methodologies in the fields of genetic engineering, microbial cell expression, fermentation, purification, and formulation have allowed manufacturers to advance through the regulatory maze with better defined processes and with greater confidence. Equipment vendors are now involved early on in the development of manufacturing procedures and have responded with state-of-the-art equipment and instrumentation. This work has focused on some of the historical processing methods and their influence on current manufacturing procedures. A foundation was generated on the origin of biopharmaceutical products and processes utilizing either prokaryotic or eukaryotic expression systems. It is hoped that the procedures and equipment described in this chapter will serve to clarify and enhance the understanding of aseptic biopharmaceutical processing.

REFERENCES

Adams, D. G., and D. Agarwal. 1989. Clean-in-place system design. *BioPharm*. June: 48–54.

Ambler C. M. 1988. In *Handbook of separations techniques for chemical engineers*, 2nd ed., edited by H. B. Crawford, and S. Thompson. New York: McGraw-Hill, Inc.

ASTM. 1989. Standard guide for determination of purity, impurities, and contaminants in biological drug products. In *Annual book of*

ASTM standards, 11.04 (designation E 1298-89): 898–900. Philadelphia: American Society for Testing and Materials.

Birch, J. R., P. W. Thompson, K. Lambert, and R. Boraston. 1985. *Large scale mammalian cell culture*. Orlando: Academic Press.

Bonnerjea, J., S. Oh, M. Hoare, and P. Dunhill. 1986. Protein purification: The right step at the right time. *Biotechnology* 4 (Nov):954–958.

FDA. 1985. *Points to consider in the production and testing of new drugs and biologicals produced by recombinant DNA technology* (draft). Bethesda, MD: Food and Drug Administration, Center for Drugs and Biologics, Office of Biologics Research and Review.

Fike, R., J. Pfohl, D. Epstein, D. Jayme, and S. Weiss. 1991. Hybridoma growth and monoclonal antibody production in protein-free hybridoma medium. *BioPharm*. March: 26–29.

Garnick, R. L., M. J. Ross, and R. A. Baffi. 1991. Characterization of proteins from recombinant DNA manufacture: Drug biotechnology regulation. New York: Marcel Dekker, Inc.

Hofmann, F. 1992. Scale-up and production of mammalian cell expressed biopharmaceuticals: GMP production of recombinant proteins and monoclonal antibodies. In *Proceedings of the 1992 Technical Program, Interphex-USA*. New York.

Holland, I. B., N. Mackman, and J. Nicaud. 1986. Secretion of proteins from bacteria, *Biotechnology* 4 (May): 427-431.

Jungbauer, A., H. P. Lettner, L. Guerrier, and E. Boschetti. 1994. Chemical sanitization in process chromatography, part 2: In situ treatment of packed columns and long-term stability of resins. *BioPharm*. July/August: 37–42.

Kelly, W. S. 1992. Scale-up and production of first generation therapeutic proteins. Paper presented at the 1992 Technical Program of Interphex-USA, 1992. New York.

Kirsop, B. E., and A. Doyle, eds. 1991. *Maintenance of microorganisms and cultured cells*, 2nd ed. London: Academic Press Ltd.

Klausner, A. 1986. Taking aim at cancer with monoclonal antibodies. *Biotechnology* 4 (March): 185–194.

Levchuk, J. W., and A. G. Lord. 1989. Personnel issues in aseptic processing. *BioPharm*. Sept.: 34–40.

Marino, M. H. 1989. Expression systems for heterologous protein production, *BioPharm*. July/August: 18–33.

Maurer, H. R. 1992. Towards serum-free, chemically defined media for mammalian cell culture. In *Animal cell culture: A practical approach,* 2nd ed., edited by R. I. Freshney. Oxford: Oxford University Press.

Monod, J. 1942. Recherches sur la croissance des cultures bacteriennes. Paris: Herrman and Cie.

Omstead, D. R. 1990. *Data acquisition and control systems: Computer control of fermentation processes.* Boca Raton, FL: CRC Press, Inc.

Ransohoff, T. C., M. K. Murphy, and H. L. Levine. 1990. Automation of Biopharmaceutical purification process. *BioPharm.* March: 20–26.

Sherwood, R., F. Plank, J. Baker, and T. Atkinson. 1991. Large-scale production of proteins from recombinant DNA in E. coli. *Soc. Appl. Bacteriol. Tech. Ser.* 28: 315–335.

Stanbury, P. F., and A. Whitaker. 1984. *Principles of fermentation technology.* Oxford: Pergamon Press.

Stwertka, E., and A. Stwertka. 1989. *Genetic engineering,* revised ed. New York: Impact.

USP. 1990. The United States Pharmacopeia, 22nd Rev. Rockville, MD: United States Pharmacopeial Convention.

Tung, A. S., J. V. Sample, T. A. Brown, N. G. Ray, E. G. Hayman, and P. W. Runstadler, Jr. 1988. TPA production through mass culturing of Chinese hamster ovary cells. *BioPharm.* Feb.: 50–55.

Waterson, R. M. 1991. Novel depth filtration technologies: Strategies for process development. *Pharm. Eng.* 10:22–29.

Zoon, K. C. 1993. *Points to consider in the characterization of cell lines used to produce biologicals.* Rockville, MD: Food and Drug Administration, Center for Biologics Evaluation and Research.

Wang, J., and M. A. Hanson. 1988. Parenteral formulations of proteins and peptides: Stability and stabilizers. *J. Paren. Sci. Tech.* Tech. Report No. 10: 42.2S.

11

Lyophilization

Edward Trappler

THEORY AND HISTORY

Lyophilization is a process that is growing in its application as a method of production for preserving a wide variety of healthcare products—vaccines, therapeutic agents, and novel products developed from advances in biotechnology. This unique method of preservation is a process of removing water by first freezing and then freeze-drying to produce a stable product. The principal purpose of applying this process is for long-term preservation of materials that are unstable in the presence of water.

In the circumstance where a pharmaceutical preparation is not stable as an aqueous solution, the product presentation may take the form of a dry powder. When the combination of instability in an aqueous environment and sensitivity to heat exists, and there is incompatibility in alternate solvent systems, then the preferred preparation is as a lyophilized product.

Since the academic interest of the 1800s, through the application to processing of heat-labile and moisture-sensitive antibiotics in the mid 1900s, freezing and freeze-drying of materials has progressed to be the method of choice for the newest diagnostic and therapeutic products—the fruits of biotechnology. The development of this technology of low temperature vacuum drying, slow to progress as a fully developed science, has been predominantly empirical.

291

Process Overview

Although principally considered a drying process, lyophilization consists of multiple processes. Although distinctly different, each part of the process has a common influence on finished product qualities, along with subsequent steps in the process. Simply considering the name given to the process of low temperature vacuum drying, the steps consist of first freezing the material, then drying the product to a suitable level of moisture where acceptable stability of the dried product is achieved. The principal advantage to this method of drying is the retention of the initial characteristics of the material preceding the drying operation: The dried product maintains the qualities that existed in the frozen state. In essence, the process provides the advantages of frozen storage without the need for such low temperature conditions.

The process of freeze-drying is a specialized application of vacuum technology. The process occurs through the manipulation of environmental conditions of subambient temperatures and subatmospheric pressures. These conditions are created and need to be maintained by the equipment. Therefore, the equipment must be able to evacuate and maintain a special environmental chamber at low pressures. In addition, a heat transfer system must be able to first remove heat from the product via the shelves to completely solidify the solution during freezing. This same system then supplies controlled quantities of heat to the shelves for maintaining an acceptable drying rate at a safe product temperature for sublimation and desorption.

In the processing of lyophilized materials, the product, most often a solution, is loaded into a freeze dryer and placed upon the shelves. The shelves are then cooled to a temperature where all the material is solidified. A condensing unit, used to collect the water that is removed from the product, is then chilled and the air within the system is evacuated. At this point sublimation begins. In order to achieve acceptable rates for sublimation, the shelf temperature is raised, maintaining the product temperature to below where the material would melt and lose the structure that is established during freezing. When the product temperature reaches the shelf temperature, indicating that all the ice has been sublimed, the shelf temperature is again adjusted to complete the removal of residual water to an acceptable level for the long-term stability of the product.

Freezing, the first step in the process, is necessary to complete the subsequent step, drying. In addition, how the material is frozen has an influence on the drying processes to follow, along with the final product characteristics such as efficacy and stability. Freezing often consists of chilling the material to a temperature where complete solidification occurs. Such complete phase transitions from a liquid to a solid often occur at very low temperatures, well below the theoretical freezing-point depression of the solution (DeLuca and Lachman 1965).

During the freezing step the objective is to "fix" the product in the preferred physically and chemically stable form. In addition, freezing also prepares the product for subsequent freeze-drying. The freezing occurs in two principal steps. First, when the temperature of a dilute aqueous solution is lowered to some temperature below 0°C, ice forms. During the proliferation of ice crystals through the volume of the solution, the remaining solutes increase in concentration. At some point, with continued cooling, the remaining material solidifies to either a crystalline material or an amorphous material, and sometimes to varying degrees of both (Gatlin and DeLuca 1980).

Assuming that the material is successfully solidified at the completion of the freezing step, the majority of the solvent component of the formulation, most often water, is removed by the process of sublimation. The often-remembered illustration of sublimation is the evaporation of dry ice, where solid carbon dioxide sublimes to form gaseous carbon dioxide without ever becoming a liquid. In the sublimation portion of the drying step, the amount of water is reduced, often from 90 percent of the original volume, to a level of water that represents only a residual amount of the starting quantity.

The sublimation of the ice must be completed in such a fashion that the integrity of the product characteristics formed during freezing are maintained during the entire process. Freeze-drying a material with retention of initial characteristics requires that the process be completed while maintaining both the frozen and dried product below a critical phase transition temperature. Above this phase transition temperature, a loss of the desired properties established during freezing may occur. Such a loss of the original structure manifests itself as either a melting of the frozen solution or a collapse of the material that remains after the ice has sublimed away (MacKenzie 1975).

In order to prevent degradation of the product due to the presence of any water that may become chemically active, the material needs to be dried to a low level of residual moisture. This entails a third, distinctly different process, referred to as secondary or terminal drying. This process involves the removal of low levels of water to achieve the required dryness. The final level of water to be achieved is based on what is required for the stability of the product (Hsu et al. 1990).

Upon removal of the bulk of the water via sublimation, trace amounts of water are yet remaining. This potentially chemically active "free" water may react with the product, yielding unwanted degradation products and a loss in potency upon storage. The mechanism for finally removing residual amounts of "bound" water is desorption.

Desorption is principally temperature dependent in the presence of low amounts of sorbed material. Therefore, temperatures used may approach or actually exceed ambient temperature.

Although simply stated as three different, interdependent steps of low temperature vacuum drying, each has diverse and complicated aspects. Lyophilization is a technology capitalizing upon basic principles of other sciences, applied in concert as a combination of methods to preserve materials that degrade due to heat or hydration reactions. The preparatory step of freezing, followed by the primary removal of water via sublimation, ultimately achieving a low amount of residual water in the last step, terminal drying, is a unique combination of varied processing techniques. Each warrants careful study, since it is the coalescence of the three that provides a method of preservation for products with unique requirements.

APPLICATION AND PRODUCT DESCRIPTION

Products that are produced by the process of freezing and freeze-drying have a unique combination of physicochemical qualities and processing requirements. Those that require storage in the absence of even trace amounts of water and are heat labile, are an extraordinary class of healthcare products that benefit from lyophilization. A combination of the requirement for delivery of the drug substance where the dried form offers the desired stability and the material is more effective in the liquid form offers an opportunity where both advantages can be achieved.

Types of Lyophilized Products

The requirements of sterile, dry-lyophilized powders are identical to those of ready-to-use liquid preparations. Routes of administration span the range of other sterile pharmaceuticals. For example, methotrexate sodium for injection, an an -inflammatory and antineoplastic agent, is available as a lyophilized powder in strengths of 20 mg, 50 mg, and 1 g per vial. Routes of administration for the non-preserved single-use lyophilized form are intramuscular, intravenous, and intrathecal. How the drug is administered is dependent on the intended therapy regime.

Pharmacologically active compounds that are both heat sensitive and undergo chemical degradation by hydrolysis are those that benefit from low temperature drying where reaction rates are slow and any chemically active water may be removed. Historically, these have often been, but are not exclusively, injectable drug products. The presentations of the drug product may range from a single or unit dose, to a multidose bulk pack for further dispensing in a hospital pharmacy setting for intravenous administration. Single-use preparations are the most prevalent, particularly since the drug substance is not sufficiently stable as an aqueous solution and may be reconstituted using an unpreserved diluent. However, some high-use products, such as the semisynthetic, broad spectrum antibiotic pipracillin sodium, is marketed as a unit dose presentation in 2-, 3-, and 4-gram vials; it is also available as a 40-gram pharmacy bulk vial.

The scope of lyophilized products ranges from antibiotics to fractionated blood products. The active materials may be derived from naturally occurring biological materials, prepared as extracts and purified substances. Some products are novel organic compounds that may be either partially or completely synthesized. A rapid-acting vasodilator, nitroprusside sodium, first provided as a lyophilized powder, is an inorganic salt. The underlying common characteristic for all these types of products is the instability of the material in the presence of water, whether they are a biochemical, organic, or inorganic compound.

Some of the antibiotics include the injectable forms of penicillin, penicillin analogs including ampicillin, and cephalosporins such as cefazolin sodium. Besides antibiotics, other types of pharmaceutical agents include antihypertensives, anti-inflammatory agents, antineoplastics, antivirals, amphetamines, and adrenal corticosteroids. Besides penicillin, one of the earliest pharmaceutical products studied as a freeze-dried preparation was blood products (Flosdorf et al. 1935). The range of purified blood products now encompasses coagulants as serum factors VIII and IX, immune globulin, and serum complexes. New products from biotechnology now on the market include enzymes and proteins produced by recombinant DNA techniques. These products include the recombinant form of the enzyme alteplase as a tissue plasminogen activator and recombinant forms of interferon. Additional biological products are approaching the horizon, addressing a wide variety of health conditions. These encompass new and more effective antibiotics, immune system activators, gene therapies, and novel vaccines. Many of these new products will utilize lyophilization as the preferred method of preservation.

Lyophilized Product Formulations

Both the requirements of the active ingredient and the route of administration influence both the design of product formulation and the development of the lyophilization process. The simplest formulation would contain only the active ingredient itself. An example of this is the antiprotozoal pentamidine isethionate, where each unit dose contains only 300 mg of the drug substance (Williams and Schwinke 1994). Perhaps the most chemically complex and challenging is whole blood. The preservation of blood has been of interest in the early days of freeze-drying and continues to be of interest (Flosdorf et al. 1935; McCoy 1994).

Along with a unique combination of processing steps, lyophilized pharmaceutical preparations also have an extraordinary formulary of adjunct material. A ready-to-use liquid preparation may consist of an excipient for isotonicity, an acid or base for adjusting the pH (perhaps in combination with a buffering system), a solubilizing agent for the drug substance, an antioxidant, or a biological preservative (particularly for multiple-dose presentations of a product).

In comparison, the list of a formulary for a lyophilized preparation may consist of all of those above, although rarely an antimicrobial preservative. Unique to lyophilized preparations, a bulking agent may be used to impart desired physical attributes for the dried material. For biological materials there may also be cryoprotectants used to protect the product during freezing, or lyoprotectants for preserving biological activity during freezing and subsequent drying.

There are a number of agents for controlling the pH, beyond inorganic acids and bases such as hydrochloric acid and sodium hydroxide. Organic acids, such as the amino acids glycine and arginine, may be used. The formulation may utilize buffering systems such as phosphate (monobasic sodium phosphate and dibasic sodium phosphate) and citrate (sodium citrate and citric acid) buffers. Sodium acetate and acetic acid can also be used. Each buffering system has its own specific advantages and influences for a lyophilized preparation.

Excipients used as bulking agents are used in formulations where the therapeutic activity of the agent is high and the dried weight of the active compound is low. In order to provide sufficient dried material where an acceptable cake is achieved, a simple sugar such as sucrose is used as a bulking agent. Mannitol, chemically classified as an alcohol, is perhaps the most-favored bulking agent. Mannitol is often selected because the crystalline form is easily dried and yields an attractive uniform white cake.

Lyoprotectants are used for materials that are sensitive to the dehydration that occurs during freezing and freeze-drying. The mechanism for how lyoprotectants accomplish their unique form of preservation is not clearly understood. The most commonly used lyoprotectants are sugars such as sucrose; maltose has also been used effectively. There is a significant amount of interest in trehalose, a sugar reported to be more effective than sucrose as a protective agent (Rosser 1991).

The particular components of a product formulation are selected with consideration given to both the route of administration and product characteristics. The list of candidates for use in lyophilized parenterals provides greater latitude of acceptable excipients in a product intended for subcutaneous or intramuscular injection than there is for one administered intravenously. The ultimate product design is one that possesses attributes that meet a culmination of product requirements.

Packaging

Lyophilized preparations are most often packaged in type 1 clear flint-molded or tubing vials. For light sensitive products such as deoxycycline hyclate, the glass may be amber rather than clear flint. Although some liquid preparations such as sodium heparin are available in plastic vials, the moisture penetration levels preclude plastics for use in lyophilized preparations.

Rubber septums or stoppers are of unique designs, specific to lyophilized products. The closures not only allow penetration with a syringe or IV spike for reconstitution and subsequent delivery, but also function as vents for water vapor to escape during freeze-drying. Design of the stopper and the geometry of the vents may vary according to a manufacturer's preferred design. Stoppers are available with a single vent, two vents (sometimes called two legged), or three vents. There are a number of critical aspects of the rubber stopper that span compatibility with the drug, vapor permeation, and machinability in processing. Selection of an appropriate stopper design and rubber formulation requires laboratory study and careful consideration of the choices available.

In certain instances the packaging components may have an influence on the effect of the lyophilization process parameters. For example, tubing vials, having a thinner and more uniform wall thickness as compared to molded vials, freeze differently. This may have an effect upon the drying rates during processing, physical appearance of the dried cake, and reconstitution time (Ulfik 1989). Conversely, the geometry of the stopper and the number of vents and their respective effect upon the rates of sublimation have shown there is no statistically significant correlation between stopper design and rates of sublimation (Smith et al. 1989).

MANUFACTURING STEPS

With few exceptions the manufacturing operations for a sterile lyophilized product are identical to those used for ready-to-use liquid preparations. The greatest exception is that the filling operation is interrupted—the product is taken off-line and placed in the lyophilizer to complete the drying step. In this scenario what could be a high speed operation producing large quantities of finished product becomes a limited batch operation. The batch size limitation becomes the size and capacity of the lyophilizer and not the speed of the filling line.

Flow of Materials

The center of a sterile product manufacturing operation is the aseptic processing area. This is reflected in the flow of materials to be processed and the supplies required for this processing toward the aseptic area and, once processed, outward inspection and further packaging activities. Inward flow processes include container and stopper cleaning, sterilization and depyrogenation, bulk solution compounding, sterilization of the product by filtration, and equipment preparation and sterilization. The equipment not only encompasses the filling line equipment but also the trays and rings that are used to transfer the product from the filling line and hold the product in the lyophilizer.

These support activities feed the filling and lyophilization operation where the components come together to form the final product to be lyophilized. With the completion of filling and partially inserting the stopper, the material is loaded into the lyophilizer. At the completion of the freeze-drying process, the containers are sealed within the lyophilizer and then unloaded. The outward flow begins with the capping process by an overseal and continues with the physical inspection of the dried material. The material may then be transferred to a quarantine area until finished product evaluation is completed by Quality Control (QC) and the material is released for further processing, such as labeling and outer packaging.

Preparation of Materials

The packaging containing the product must be clean, sterile, and pyrogen free. The containers may be washed using a sequence of deionized water and steam with a final rinse of water for injection (WFI). The washing equipment may be a batch type, where containers are washed in a rack, or continuous, where containers are washed individually in a continuous washer. For containers washed in a rack, the glassware is then transferred to an oven where it is sterilized and depyrogenated using dry heat at 170–300°C. For large quantity batch operations the washer feeds a dry heat tunnel that is in series with the washer.

Rubber stoppers are processed by first cleaning in a stopper washer to remove particulates and any residues from the stopper manufacturing operation. The washing is often a batch operation where the stoppers are held in a basket that allows water to flow around the stopper. A washer is also used where the stoppers are held in a rotating basket and tumbled. The final step in cleaning is a rinse of WFI. After cleaning the stoppers are siliconized and steam sterilized. For some lyophilized products, particularly those where the formulation yields a dried material with a low weight, the stoppers are further dried to remove additional residual moisture that may still be present after drying in the autoclave or stopper processor. The concern is the desorption of residual water during storage over the shelf life of the product

Filtering and filling hardware is cleaned, prepared, and then sterilized in a steam autoclave. This includes filters and hoses for sterile filtration of the bulk material, pumps, needles and hoses for the filling setup, as well as trays and rings for holding the filled product containers. The autoclave sterilizes the materials using saturated steam conditions at a minimum of 121.1°C and 29.7 psia. The components to be sterilized must accumulate a minimum of 8 F_0 units to assure adequate sterilization. The F_0 value is commonly used as a description of sterilization effectiveness in the pharmaceutical industry (PDA 1978).

Filling

Filling is a crucial step in the preparation of the product for lyophilization. Consider that the fill volume of the starting solution dictates the amount of dried drug substance in the vial. If the fill volume is high, then reconstitution with the directed amount of diluent yields a superpotent product. Conversely, if the fill volume is low, then there is less drug substance in the vial; upon reconstitution the product is subpotent. For this reason fill volume tolerances and frequency of checks in manufacturing are critical. Improvements in accuracy and precision of fill volumes can be achieved by applying the principles of statistical process control (Homes 1991).

The filtered solution is filled into sterile, pyrogen-free containers. The operation may be a manual or high speed operation, dependent on batch size. Automated filling lines may operate anywhere from 30 cpm to 600 cpm. The vials may be placed on a rotary turntable that feeds a filling conveyor. The conveyor then positions the containers at the filling needles for dispensing the solution.

After the solution is filled, the stoppers are partially inserted. This may be part of the filling machine or may be in line with the filler. Stoppers are fed from a bowl onto a track positioned directly over the passing containers. The stopper is then placed on the container, but only partially inserted. This operation is possible at line speeds equal to the filling operation.

The containers are then fed onto a tray loading machine that transfers the containers off the conveyor and onto trays used for loading the lyophilizer. The loading of the vials off the filling line and onto the trays is accomplished by either an arm that pushes the containers onto a tray or by a "pick and place" mechanism. Line speeds equal to the filling operation can be achieved using either type of equipment.

The trays used to hold the filled containers are of a design that allows the bottom to be removed after transfer and loading onto the lyophilizer shelves. The trays are often constructed such that there are rings that enclose all four sides of the tray full of containers. This ring fits within a tray that has only three sides, allowing the ring and vials to slide off the transfer tray and sit directly on the shelf surface. This provides direct thermal contact of the vial and the shelf surface.

Transferring the loaded trays into the lyophilizer may be completed manually or by various mechanical devices. Carts are sometimes used but are less common. Application of robotics devices is being pursued to completely eliminate an operator from the manufacturing operation (Khan 1989).

Lyophilization

Lyophilization equipment for pharmaceutical products ranges from small development and pilot-plant units to large-scale manufacturing systems. The lyophilizers are rated according to both the usable shelf surface area (correlating to the number of containers) and solvent handling capacity. Container holding capacity ranges from less than 1,000 5 cc vials in a development size unit to over 100,000 10 cc vials in a large manufacturing system (Hull Corp. 1994). However, the distinction between the equipment is application and design rather than size and throughput. For example, a pilot-plant unit may be designed for maximum flexibility to process a wide variety of different types of products as would be typical in development and clinical supply manufacturing. In contrast, large manufacturing plants may be dedicated to high volumes of a specific product or type of products. While flexibility is important for development and pilot units, process automation and efficiency are key for manufacturing systems.

All equipment for processing lyophilized products has several minimum design requirements. Since aseptic conditions are necessary, the system must be a sanitary design capable of sterilization. The trend in the industry is to sterilize using saturated steam. Therefore, the equipment is manufactured of stainless steel and capable of pressurization to 34.7 psia and temperatures of 126°C. The system must be easily cleaned. Automated clean-in-place (CIP) is becoming common in the industry for production systems.

It is also becoming more common to design a facility and the equipment to enhance material flow and processing. For example, lyophilizers with a double pass-through system are used for both development and production units for increased flexibility. For units installed in an aseptic area, this allows the material to be removed from the clean room side, eliminating the need for an operator to enter the aseptic area to stopper and remove the finished product. Equipment may be cleaned or repaired from the clean room, thus eliminating the burden to the aseptic area.

The equipment must also include provisions for collecting and maintaining the sublimed water in such a condition that it does not affect the continuation of the process. This is accomplished by use of a refrigerated trap that is optically dense and has sufficient surface area to effectively collect all the water vapor evolved during the drying process at maximum drying rates. It must also have the capacity to hold the total ice load and maintain the ice at temperatures where the corresponding vapor pressure is well below the pressures used for the process—often below $-50°C$ where the vapor pressure of ice is 23 microns (2.3×10^{-2} mm Hg).

The process of sublimation requires an environment of reduced pressures; in an order of magnitude of 10^{-2} mm Hg. In order to reduce the pressure within the system, oil-sealed vacuum pumping equipment that can achieve 10^{-4} mm Hg is used.

Various types of instrumentation are available for controlling and monitoring the operation. Principally, temperature, pressure, and time are the controlled variables for processing. Data such as shelf, condenser, and product temperatures and system pressure are recorded. The possible combination of instruments and the level of sophistication for control are varied and extensive. Instrumentation ranges from simple digital instruments to mainframe computers.

High quality instrumentation is used in the controlling and monitoring systems to provide improved accuracy and precision (Hlinak 1987). Programmable microprocessors are used to control the shelf temperature profile along with completing the sequencing steps for the cycle. This eliminates the need for the operator to constantly attend the process and increases the accuracy and precision of controlling the process variables. Monitoring of the process parameters such as shelf, product, and condenser temperatures and system pressures is also achieved by the process control instrumentation.

Many levels of sophistication are achievable for the process (Nail and Gatlin 1985), ranging from simple control instruments to fully automated process control using a mainframe computer system. The system used is based on the application and the level of automation desired rather than the size of the lyophilizer.

PROCESS AUTOMATION

As a process for the production of sterile pharmaceuticals, lyophilization is a unit operation consisting of a number of processing steps in addition to the lyophilization process itself. The support processes required include cleaning, sterilization, leak testing, and loading/unloading. Each of these support activities can be completed manually or may be fully automated. As in the freeze-drying cycle, there are many advantages to automating these operations. It is important to note that the automation of such a unit operation is more than adding a programmable logic controller (PLC) to replace an operator. One needs to look at the processing techniques, the design of the equipment, and the integration as a unit. This approach allows capitalizing on the advantages of system automation. There are numerous options in the configuration of just the control system alone.

Cleaning

During various phases of operation, the lyophilizer unit may become contaminated. This contamination can occur due to product on the container from the filling line, containers spilled during loading, or product that leaves the container during the drying cycle. There are also circumstances, either due to poor equipment or process design, along with conditions of equipment failure, where processing fluids may enter the chamber. Examples are leaks in the heat transfer system within the chamber or the backstreaming of a vacuum pump oil (Lopez et al. 1982). Cleaning after each batch is completed is necessary for eliminating potential cross-contamination from batch to batch and assuring product purity.

Manual cleaning of the lyophilizer is or ˋ aspect of all the operations that still presents challenges. These challenges include cleaning from within the aseptic area, validation of the manual procedure, and cleaning control and reproducibility. From a clean room management aspect, cleaning from the aseptic area places a significant burden on maintaining Class 100 conditions. Manual cleaning varies with personnel and conditions and raises the question of reproducibility to that of cleaning at the time of validation.

For these reasons a lyophilization system becomes an excellent candidate for automated CIP. However, there are also some special challenges in incorporating this technology in a lyophilization system.

The CIP system must be integral with the design of the lyophilization unit. Adequate wetting and agitation are the two key elements in efficient and effective cleaning. To assure that all the surfaces within the chamber receive effective exposure, spray nozzles must be positioned and sequenced to first wash the entire chamber; then wash each individual shelf by positioning the shelves in front of a set of high-pressure spray nozzles; followed by a second wash of the chamber. With such an arrangement all the interior surfaces are effectively cleaned.

The greatest benefit gained with CIP is when the system is designed from inception with automation of CIP in mind, rather than simply adapting CIP nozzles and piping to a lyophilizer. Paralleling the trend in sterilization, CIP needs to include the chamber and condenser. Assured reproducibility can be achieved when the CIP function is automated. Validation and an effective change control program for cleaning can be better managed by applying process automation to this function.

Sterilization

For the aseptic production of lyophilized products, it has become "state of the art" to sterilize the freeze-drying chamber (PDA 1978). This requires that the vessels be designed to meet pressure code requirements at elevated temperatures. Usually, the vessels are designed and constructed to tolerances similar to autoclaves with pressures to 34.7 psia and temperature to 121.1°C. Because lyophilizer chambers have a considerable amount of weight, the heat-up and cooldown times are substantially longer than the processing times for an actual steam autoclave. Consequently, turnaround times in production become considerable.

The automation of sterilization is an excellent example of the benefits that can be achieved when the automated process is intrinsic to the equipment design rather than an add-on afterthought in equipment construction. If the approach of optimization is incorporated into the concept and design of the system, the achievements are much more significant. The design of the supply piping and the method of steam distribution within the chamber can improve the steam penetration and temperature uniformity. The sterilization process also includes preheating and cooldown through jackets incorporated as part of the structure of the vessel on the outside of the chamber. This planned approach lends itself to automation and allows faster turnaround, yielding increased flexibility in scheduling.

Sterilization of the unit places a special burden on maintaining equipment for dependable vacuum operation. This is particularly true in the processing of parenterals. Any leak of sufficient magnitude to prevent achieving the desired pressure for processing may lead to product sterility concerns. To assure successful processing of the batch, it is becoming commonplace to complete a vacuum integrity test of the equipment prior to loading any product. Vacuum integrity testing may be completed at the end of each batch in addition to or in place of testing prior to loading. The phenomena of outgassing must be taken into consideration while performing a vacuum integrity test. The setting of in-process limits needs to compensate for outgassing. Outgassing of water can occur from ice remaining in the condenser after processing or left over from steam sterilization.

Loading

The greatest source of contamination in an aseptic operation is from human activities within the aseptic area. Substantial gains in sterility assurance can be achieved with automation of the manual tray transfer step in the loading of the lyophilizer. Automated material handling eliminates the need for an operator in the aseptic operation and, therefore, protects the product. In comparison, the processing of some types of products, such as cytotoxic agents, requires protecting the operator from the product.

Often in a freeze-drying operation, the last step at the filling line is traying the product containers for transport and loading into the lyophilizer. The containers are placed on two-part trays where the bottom can be removed when the tray is placed on the lyophilizer shelf, as described earlier. Loading small groups of trays or all the trays for a single shelf is advantageous since the product is maintained at a controlled temperature, within a controlled environment, from aseptic conditions to the primary barrier provided by the lyophilizer.

VALIDATION

The objective of a validation study is to demonstrate the operation's ability to support the process and develop documentation to show that the unit can be maintained within a state of control. Tests that are to be completed, including the procedures and expected results, are established and defined within a validation protocol. The validation includes protocols on installation qualification (IQ), operational qualification (OQ), and process qualification (PQ). Prior to considering validation, the product characteristics and process parameters must be defined.

The first part of implementing a validation study is to complete the IQ and the OQ. Installation qualification is the foundation of the validation study and the first building block upon which the OQ, process validation, and change control program are built.

Installation Qualification

Documentation of a general description, equipment specifications, support utility services, equipment components, and installation of the freeze-drying equipment should be addressed in the IQ protocol. As part of the installation, an initial calibration of all critical instrumentation assures that the instrumentation is in good working condition after shipment and that any data collected for startup qualification of the system is accurate. The support utilities are important to qualify and quantify to assure that adequate supply is available for proper functioning of the system's equipment. A list of common utilities required for a freeze-drying system include the following: electrical, steam, nitrogen, cooling water, compressed air, and adequate drainage.

Operation Qualification

The operational capabilities of the equipment are an important factor influencing the success of the process. The capabilities are established within the IQ and OQ efforts of the validation study. The operational quality of the equipment is paramount because control of the environment is established and maintained by the equipment. This is crucial in lyophilization where the process is completed in a critical environment of controlled pressures and temperatures. The performance of the system needs to be compared to the original equipment manufacturer's specification. Acceptance criteria should be developed that include process parameters and fall within the manufacturer's stated specifications. Following are the types of tests that should be included in an operational qualification:

- Shelf temperature control
- Condenser cooldown
- Vacuum pumping rates
- Vacuum integrity tests
- Sublimation and condensation rates
- Vacuum control
- Steam penetration tests

Completion of qualifying tests in compliance with the written procedures and meeting the acceptance criteria assures that the system will adequately perform to levels required for reliable processing.

Aseptic Process Validation

As with any sterile product, aseptic process validation is an integral part of the validation package. Along with sterilization validation of the lyophilizer itself, media fills are used to demonstrate the level of sterility assurance that can be achieved. The lyophilizer sterilization validation uses biological indicators (BIs) to demonstrate the effectiveness of the procedure. Aseptic filling, transfer, and loading is evaluated using media fills that mimic the operations where human activities are involved.

The media fills are critical because of the transfer and loading processing steps. Unique to lyophilized products, the containers are removed from the filling line, protected under Class 100 conditions, transferred to the lyophilizer, and placed upon the shelves. During this transferring step the containers are still open to the environment, since the stoppers are placed such that water vapor can escape during drying. All of these handling conditions must be challenged using filled containers of media to assure that the operations can successfully provide a substantial level of sterility assurance.

Process Validation

Process validation for lyophilization is done concurrently or retrospectively. Concurrent validation is utilized where the process data is collected and analyzed for product being produced for distribution and use. Retrospective validation is completed through the analysis of historical batch data; the data is gathered from the records of cycles previously run and shown to be under the same conditions.

Concurrent Process Validation

Manufacturing a lyophilized product first requires validation of the procedures and the processes that will be used. This allows for evaluation of the scaleup of new processes in a production environment utilizing production size batches. Extensive data is collected and analyzed. The depth, scope, and quantity of data and its analysis is much more extensive than would be routinely done for a normal production batch. Samples from the validation batches may be placed on a commercial stability program for long-term stability evaluation.

Process parameters and product characteristics are interdependent; therefore, the use of a placebo to emulate actual product runs is often not a reasonable substitution for actual product. Although not totally appropriate as a substitute for product, a placebo formulation may be used to increase the batch size to one that is more representative of actual production if there is a limited amount of actual product available. Actual product at a full or near full batch size is required for the PQ of a large-scale manufacturing operation. With the expense and/or limited quantities of the raw material for new products, PQ is often concurrent with processing of product intended for distribution.

The process and its effect on the product qualities define the significant process parameters that must be quantified. The finished product characteristics become an integral part of the process evaluation and are included in the acceptance criteria. The product data include in-process and finished product testing data. Following are examples of data that may appear in the protocol:

- Shelf loading temperature
- Shelf freezing rate
- Product freezing rate
- Frozen product soak temperature and time
- Shelf temperature ramp and soak functions
- Product temperature
- Condenser temperature
- Vacuum level

These process parameters document that the process was completed within a set of predefined parameters. If the lyophilization cycle was completed within acceptable limits, a predictable product quality would be expected. Some of the measurable characteristics that would demonstrate the product quality may be as follows:

- Product physical form
- Dry cake physical appearance
- Moisture content
- Reconstitution rate
- Product assay
- Purity
- pH

Retrospective Validation

Retrospective validation encompasses the analysis of historical data from product and processes. Data from commercial lots are used to document that the process parameters and product performance are within defined limits and meet the acceptance criteria. This demonstrates that the unit operation can be maintained within a state of control and that acceptable product performance and characteristics are achieved. Analysis includes finished product and product held for stability up to and including the expiration date. This validation approach, therefore, involves batch process data, QC lot release data, and commercial stability program results.

The value of retrospective validation is that the database would be larger than just the three initial batches analyzed in the concurrent validation. Implementation may also be an ongoing basis and is useful in tracking process and product performance.

SUMMARY

Lyophilization is a complicated unit operation, one of the most integrated processes for the preservation of a drug product. The unique element of this technology is that the qualities of the product and its formulation affect the process as well as the process affecting the final product quality.

Opportunities exist for optimizing the integration of the process into an operation for the manufacturing of lyophilized products through a systems approach. Capitalizing on advances in technology leads to achieving greater success in control and quality assurance to meet ever-expanding demands for increased efficiencies and high level of regulatory compliance.

REFERENCES

DeLuca, P., and L. Lachman. 1965. Lyophilization of pharmaceuticals 1: Effects of certain physical-chemical properties. *J. Pharm. Sci.* 54:621.

Flosdorf, E., L. Hull, and S. Mudd. 1935. Procedure and apparatus for preservation in "Lyophile" form of serum and other biological substances. *J. Immunol.* 29:389.

Gatlin, L., and P. DeLuca. 1980. A study of the phase transitions in frozen antibiotic solutions by differential scanning calorimetry. *J. Paren. Drug Assoc.* 34:398.

Hlinak, A. J. 1987. Freeze dryer automation with programmable controllers. *Pharm. Eng.* 7 (6):27–33.

Homes, D. S. 1991. Filling pharmaceutical vials—An SPC analysis. *Pharm. Tech.* 15 (11):32–40.

Hull Corporation. 1994. [Company literature.] 3535 Davisville Road, Hatboro, PA 19047.

Hsu, C., C. Ward, R. Pearlman, H. Nguyen, D. Yeung, and J. Curley. 1990. Determining the optimum residual moisture in lyophilized protein pharmaceuticals. In *Developments in biological standardization, vol. 74. Biological product freeze-drying and formulation* (pp. 255–271), edited by J. C. May, and F. Brown. Basel, Switzerland: Karger.

Khan, S. 1989. Automatic flexible aseptic filling and freeze-drying of parenteral drugs. *Pharm Tech.* 13 (10): 26–33.

Lopez, F. V., I. P. Solis, and F. A. Castro. 1982. Oil foreign particles in freeze-dried injectable powder. *J. Paren. Sci. Tech.* 36:259–266.

MacKenzie, A. 1975. Collapse during freeze drying—qualitative and quantitative aspects. In *Freeze drying and advanced food technology* (pp. 277–307), edited by S. A. Goldblith, L. Rey, and W. Rothmayer. New York: Academic Press.

McCoy, H. D. 1994. Crybio research may lead to increased human blood supply. *Gen. Eng. News* 14:12.

Nail, S., and L. Gatlin. 1985. Advances in control of production freeze dryers. *J. Paren. Sci. Tech.* 39 (1):16–27.

PDA. 1978. Validation of steam sterilization cycles. Technical monograph #1. Parenteral Drug Association.

Pikal, M. J., K. M. Dellerman, M. L. Roy, and R. M. Riggin. 1991. The effects of formulation variables on the stability of freeze-dried human growth hormone. *Pharm. Res.* 8 (4):427–436.

Rosser, B. 1991. Trehalose drying: A novel replacement for freeze-drying. *BioPharm.* 4 (8):47–53.

Smith, E., E. Trappler, and W. Currey. 1989. Effect of stopper design on freeze drying. Presented at the annual meeting of the Parenteral Drug Association, 30 October–1 November, in Hollywood, FL.

Ulfik, L. 1989. Effects of vial on rate of sublimation. Presented in a poster session at the annual Parenteral Drug Association meeting, 5–7 November, in Philadelphia, PA.

Williams, N. A., and D. L. Schwinke. 1994. Low temperature properties of lyophilized solutions and their influence on lyophilization cycle design. Pentamidine isethionate. *J. Pharm. Sci. Tech.* 48:3.

12

Lyophilization Under Barrier Technology

John W. Snowman

Increasing requirements to freeze-dry drugs in a contained environment are driven by three factors.

1. Development in isolation technology improves the possibility of maintaining the integrity of aseptically produced products by containing pollution originating from human operators.

2. Operators must be protected from hazardous materials. Many drugs now being developed are effective in microgram doses rather than the milligrams to which the industry has been accustomed. Some are cytotoxic. Exposure to tiny amounts could endanger operators.

3. Dangerous compounds or patnogens must not be released to the environment.

The flow diagram for a typical filling, lyophilizing, and final packing line is shown in Figure 12.1. Mixing and filtration can be done in contained vessels. Currently, filling of vials, half insertion of rubber bungs, and loading/unloading of the lyophilizer are usually done in an aseptic area. Isolation or barrier technology now allows these operations to be carried out in aseptic conditions, but without the cost and inconvenience of a conventional aseptic area, while at the same time

Figure 12.1. Typical secondary production flowchart showing operations requiring isolation and inputs/outputs that are in product contact.

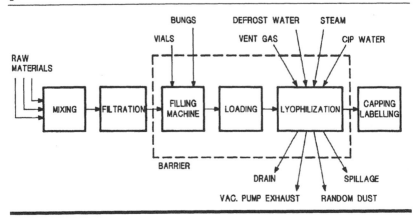

protecting operator and environment from the product. This is the prime application for barrier technology in lyophilization.

A second area of concern must be addressed because lyophilizers generate sporadic movements of air, water vapor moves out of containers of frozen liquid, and there may be product carryover into the ice condenser that is discharged to the environment when the equipment is defrosted. The vacuum pump exhaust is piped to the atmosphere, giving another possibility for environmental contamination.

Finally, it must be appreciated that a lyophilizer is much more complicated than almost any other piece of pharmaceutical hardware. There are many intrusions into the working space that need to be removed for regular maintenance or calibration, exposing maintenance and calibration staff to potentially dangerous quantities of active material.

The fundamentals of lyophilization and some practical features of a modern pharmaceutical lyophilizer are reviewed as a preliminary to the description of how barrier techniques are used.

PHYSICAL PRINCIPLES

The lyophilization process consists of three phases, indicated diagrammatically in the temperature/time and pressure/time diagrams of Figure 12.2:

- Freezing to solidify the liquid material by bringing it below its eutectic zone or collapse temperature.

Figure 12.2. Idealized lyophilization chart.

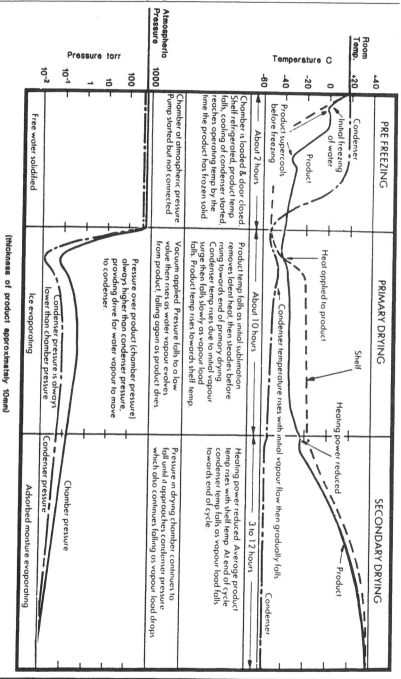

- Sublimation drying to reduce moisture to around 4 percent w/w of the dry product and leave an apparently dry cake of solid residue that is substantially the same size and shape as the frozen mass.

- Desorption or secondary drying to reduce bound moisture to the required final value (normally below 1 percent w/w).

In practice, freezing may not be complete until after primary drying begins and some desorption drying is in progress during primary drying.

Vapor is generated during sublimation from a distinct interface or freeze-drying front that moves from the outer surface of the material. Subliming vapor must pass through the interstices of the substantially dry layers. The shape of the interstices results from that of the ice crystals that form on freezing. If the crystals are small and discontinuous, the escape path for the vapor is limited; if large dendritic crystals are formed, escape is easy and the product can be dried more quickly. The method and rate of freezing are thus critical to the course of sublimation.

Freezing

In some very simple systems both solvent (mainly water) and solute crystallize. Commonly, the solvent crystallizes, but the solute forms an amorphous glass. In the simpler case where both crystallize, cooling from the initial temperature to below 0°C eventually results in the nucleation of ice. The release of heat of crystallization raises the temperature toward 0°C. Crystallization then proceeds at a progressively falling temperature related to the equilibrium melting point of the concentrating solution, as shown in Figure 12.3. As the temperature falls, the solution approaches saturation so that crystals of solute are precipitated.

Eventually, a eutectic point is reached where the material becomes wholly crystalline. When the material contains more than one crystallizable solute, a similar situation exists, but the eutectic point is lower than that of any two components combined.

Where the solutes form a glass, no eutectic is present. In this case the solvent crystallizes in the same way as is shown by the melting and nucleation curves on the left of Figure 12.3, but there is only a single transition line on the right.

Freezing Methods

When liquids are to be lyophilized, the freezing and drying times can be reduced by disposing them in thin layers. A variety of means have been developed for achieving this in freezing baths. One method is by tilting containers to form wedges. Another is to rotate them, either slowly about horizontal axes or rapidly about vertical axes to form hollow cylinders. In the pharmaceutical industry freezing is usually achieved by standing trays of bulk material or vials on the lyophilizer shelves that are then refrigerated.

Sublimation

After freezing the partial pressure of water vapor must be reduced below the triple point pressure of water (about 6 mbar) to allow sublimation to take place. If there is a free escape path for water molecules, the speed of sublimation at the beginning of primary drying is largely limited by the rate at which heat can be transferred to the freeze-drying front.

The eutectic or collapse temperature of the material must not be exceeded during drying, otherwise melting or collapse could occur, giving rise to gross material faults such as melting, shrinking, and puffing. Collapse is the most common failure. It is particularly common if sugars are included in the formulation as cryoprotectants.

Since most of the heat must be transferred through the frozen material, which usually has poor thermal conductivity, the need to avoid melting or collapse means that the thermal gradient, and thus the heat

Figure 12.3. Equilibrium melting temperatures and nucleation curves for solvent and solute.

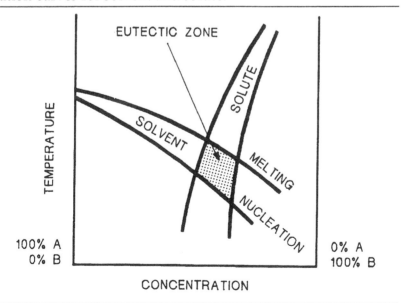

conducted, is small. Later in the process the flow of escaping water vapor, and hence the sublimation rate, is restricted by the increasing depth of the dry part of the product cake.

Heat transfer during sublimation can be increased by spoiling the vacuum-typically by bleeding air into the chamber or by throttling the vacuum pump to maintain pressure around 0.2 mbar. This improves convective heat transfer from the heating shelf to the bottom of the product container.

Typical primary drying time for a 1 cm thickness cake of a "simple" material is between 10 and 20 hours with favorable ice-crystal structure and optimized temperature and pressure conditions in the drying chamber. Drying time varies with thickness approximately by the relationship Time = K (thickness)$^{1.5}$.

Secondary Drying

Secondary drying is the removal of bound moisture which may be water of crystallization, randomly dispersed water in a glassy material, intracellular water, or absorbed water. Primary or sublimation drying accompanied by desorption removes the free water. The remaining bound moisture usually amounts to about 4 percent w/w. The bound moisture is removed by heating the product, typically to between 15°C and 30°C for about one-third of the primary drying time.

The residual moisture content of dry material is a critical factor in storage life. Most freeze-dried materials are hygroscopic and must be stored in sealed containers.

PRINCIPLES OF LYOPHILIZATION EQUIPMENT

The simplest lyophilizer would consist of a vacuum chamber into which wet material could be placed, together with a means of removing water vapor so as to freeze the sample by evaporative cooling, and then maintain the water vapor pressure below the triple point pressure.

The temperature of the sample would continue to fall below the freezing point and sublimation would slow down until the rate of heat gain in the sample by conduction, convection, and radiation was equal to the rate of heat loss as the more energetic molecules sublimed away and were removed.

However, when a material is frozen by evaporative cooling, it froths as it boils. This can be suppressed by low-speed centrifugation during freezing; a more usual approach is to freeze the material before it is placed under vacuum. With small laboratory lyophilizers material is prefrozen inside a flask. The flask is then attached to a manifold connected to the ice condenser.

For production-scale equipment the material in trays, vials, or ampoules is placed on shelves inside the drying chamber. The shelves can be cooled so that the material is frozen by conduction and convection at atmospheric pressure before the vacuum is created.

It is usual to arrange a heat supply to the product support shelves so that, after their initial use for freezing, they can be used to provide heat to maintain the product at a constant low temperature. Figure 12.4 is a simplified representation of a production lyophilizer showing a product drying chamber with shelves on the left and ice condenser on the right.

Modern lyophilizers incorporate refinements to the type of equipment shown in Figure 12.4. The most important are as follows:

- An isolation valve between the product drying chamber and the ice condenser allows for end point determination and simultaneous loading and defrosting.

- The chamber and ice condenser are constructed as pressure vessels to allow for steam sterilization at 121°C or higher.

- Product support shelves are cooled and heated by a circulating intermediate heat exchange fluid to give even and accurate temperatures.

Figure 12.4. Basic industrial lyophilizer.

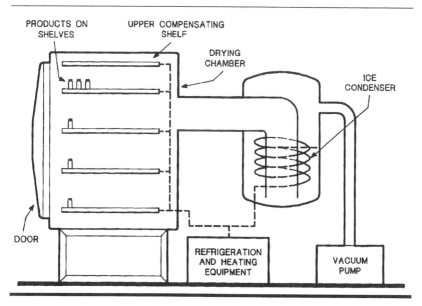

- Additional instruments are used to control, monitor, and record process variables.

- Movable product support shelves close the slotted bungs used in vials and facilitate cleaning and loading.

- The automatic control system contains safety interlocks and alarms.

- Duplicated vacuum pumps, refrigeration systems, and other moving parts enable drying to continue without endangering the product in the event of mechanical breakdown.

- Loading and unloading systems are automatic.

For low value products where safety and possible cross-contamination are not so important, the ice condenser is sometimes placed inside the drying chamber to reduce equipment costs.

REGULATORY BACKGROUND

Over the years bodies concerned with many diverse industries have evolved regulations and codes of practice to promote safe handling and environmental protection. Some regulations of particular interest to pharmaceutical companies drying hazardous materials are recommended below:

- USA Federal Standard Number 209E: *Clean Room and Work Station Requirements Controlled Environment* (IES).

- British Standard 5295

- Code of Federal Regulations CFR Title 21 Parts 600-799.

- "Recombinant DNA Safety Considerations," OECD, 1986.

- *The categorization of pathogens according to hazard and categories of containment,* compiled by the Advisory Committee on Dangerous Pathogens, Her Majesty's Stationery Office

A method of classifying the degree of hazard associated with a material is to adopt the following criteria:

- Is the material hazardous for man?

- Is it a hazard to process workers?

- Is it transmissible in the community?

- Is effective prophylaxis and treatment available?

- Does it damage the environment?

Hazard Groups

Group 1: A material or organism that is most unlikely to cause human disease.

Group 2: A material or organism that may cause human disease and that might be a hazard to workers, but is unlikely to spread in the community or damage the environment. Exposure rarely produces infection and effective prophylaxis or effective treatment are usually available.

Group 3: A material or organism that may cause severe human disease and present a serious hazard to workers or damage to the environment. It may present a risk of spread in the community, but there is usually effective prophylaxis or treatment available.

Group 4: A material or organism that causes severe human disease and is a serious hazard to workers and/or that materially damages the environment. It may present a high risk of spread in the community and there is usually no effective prophylaxis or treatment.

Barrier technology using isolators and special means to prevent environmental damage is essential for handling materials in Groups 3 and 4. It is necessary to decide whether to run the isolator at a positive internal pressure to protect the product or at a negative pressure to protect the operator. It is thought that inspecting authorities will normally insist on a positive internal pressure to protect the product. If manual intervention for maintenance is not required very often, it may be possible to run at positive pressure and then change to negative pressure when intervention is needed.

ISOLATION OF LYOPHILIZATION OPERATIONS

Before reaching the filling machine toxic products will usually be processed in closed vessels so hazards are minimal when standard operating procedures are established. Isolation during lyophilization operations has three main areas of applications:

- Transfer from filling machine to lyophilizer
- Contamination within the lyophilizer
- Protection of the environment from effluents

Early approaches to isolation were attempts to isolate operators from open processes in suits, Figure 12.5. This is cumbersome and

Figure 12.5. Operator in pressure suit.

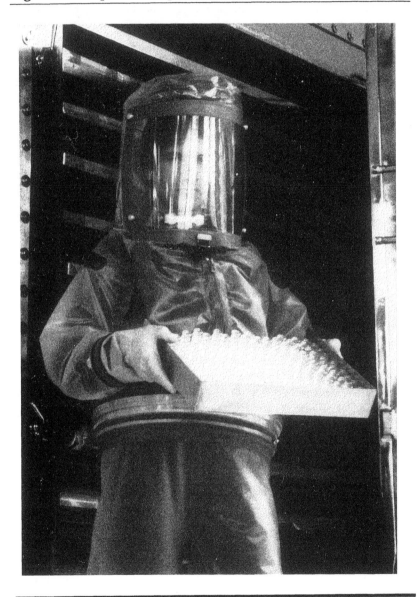

restrictive. Barrier technology provides a means to provide a Class 100 environment for filling and operator protection while transferring from the filling machine and loading the lyophilizer. Isolation of the filling machine alone is treated elsewhere. An isolator for both a small filling machine and a lyophilizer is shown in Figure 12.6. It should be

Figure 12.6. Isolator for small lyophilizer and filling machine.

noted that the use of isolators does not imply that they can be located in uncontrolled areas. Experience indicates that the surroundings should themselves be of a reasonable standard of cleanliness, at least up to Class 10,000 standard. It must be anticipated that regulatory bodies will set standards for areas surrounding isolators.

Transfer from Filling Machine to Lyophilizer

The best approach is to use an automatic or semiautomatic transfer from the filling machine to the lyophilizer. Alternatives are to use a "pick and place" robot for small lyophilizers; or for large production units a cart (Figure 12.7) or a constant level loading system (Figures 12.8 and 12.9) is recommended. The cart illustrated in Figure 12.7 was one of eight in a major installation of freeze dryers. Each cart was power driven and could load/unload half of a freeze-dryer load in one operation. The cart itself was covered with PVC film and carried its own HEPA filter and fan to generate a Class 100 internal atmosphere, thus acting as a mobile isolator. During docking on to the lyophilizer or filling line a door at one end was opened up for vial movement.

Figure 12.7. Loading cart with barrier in place.

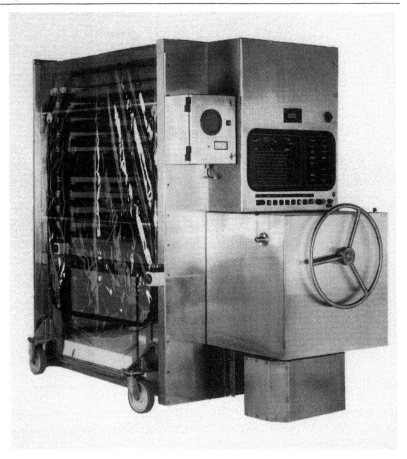

Figure 12.8. Loading frame for constant level loading.

Figure 12.9. Automatic trayless loader for constant level loading.

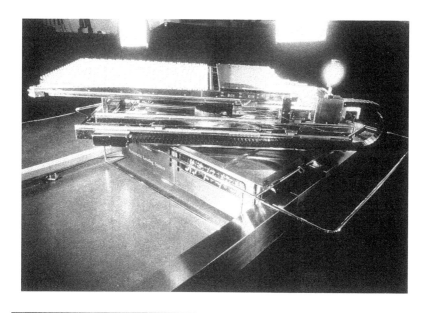

Such doors can be hinged both to left and to right if partial loads are scheduled. The connection to a lyophilizer door is too big to be able to use a double sealed lid as on a rapid transfer box. Isolation during docking depends on a vertical laminar flow over the open area.

Cart systems give better lyophilizer utilization than constant level loaders but have the disadvantage that during docking onto a lyophilizer or filling line, one end must be opened up to vials movement. The connection to a lyophilizer door is too big to use a double sealed lid as a rapid transfer box. Isolation during docking depends on a vertical laminar flow over the open area. Constant level loading systems prevent utilization of the lyophilizer during the whole filling run, but are preferred because they can be permanently isolated. An ideal solution is to use a trayless system for transporting vials, if possible, as the restricted space inside the isolator would require trays to be extracted for cleaning and sterilization.

Figure 12.10 shows a semiautomatic trayless constant level loading system designed for a 20 sq m (220 sq ft) shelf area lyophilizer. Although movements are controlled by a Programmable Logic Controller, half suits and glove access are provided to give access for adjustments. Vials arrive from a filling line and are pushed into one of the two transport frames inside the isolator. When the transport frame

Figure 12.10. Design for semiautomatic trayless constant level loader.

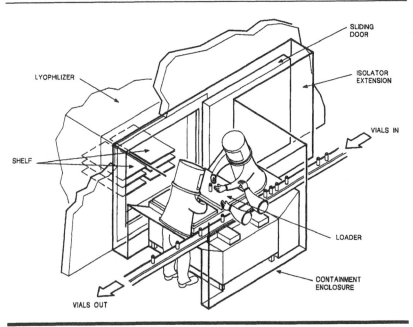

is filled, it is pushed onto the lyophilizer shelf to fill half of it. During that operation arriving vials are fed into the other transport frame. As each lyophilizer shelf is filled, it is lifted up from the loader level and a new shelf is positioned for filling. The process is repeated until loading is complete. The process is reversed for unloading.

If automated loading systems cannot be used for economic or other reasons, product and operator protection can still be achieved with manual systems under isolation technology. Figure 12.11 shows an isolator and lyophilizer with half suit for operator protection mounted in an aseptic area. Ideally, the filling machine isolator should be extended to cover the front of the lyophilizer, with means to move vials from the filling machine into the drying chamber (see Figure 12.6). The alternative is to use separate isolators for the filling machine and lyophilizer with rapid transfer boxes to move vials between them (Figure 12.12). The downside to both of these arrangements is the nuisance value of the vial trays or "fences" that must be used to load/unload the lyophilizer. Yet another alternative is to use sealed trays or "cassettes" that can be opened in the freeze dryer after the door is closed (Figure 12.13). However, all manual methods bring operators into close proximity to liquid product with the consequent hazards. Some degree of automation is strongly preferred, especially for cytotoxic products.

Contamination within the Lyophilizer

Engineering solutions to protect operators and products during transfer from the filling machine to the lyophilizer are relatively easy to

Figure 12.11. Isolator and lyophilizer in clean area.

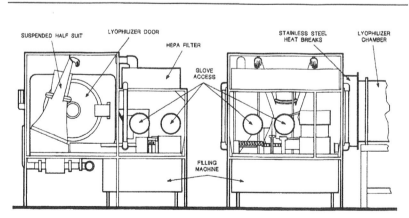

Figure 12.12. Use of transfer boxes to load lyophilizer.

devise. The hazards arising from the lyophilization process itself are less well known. Figure 12.14 is an outline of a typical lyophilizer. It comprises a drying chamber with movable shelves shown with the door open. The door face of the lyophilizer is sealed into the wall of a clean area.

Figure 12.13. Manual loading using cassettes.

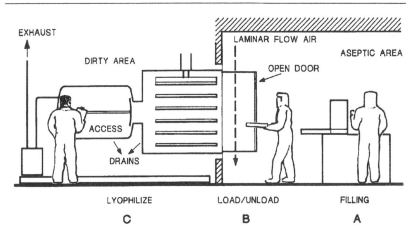

Figure 12.14. Outline of typical production lyophilizer.

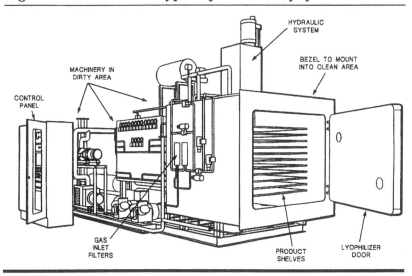

Frequently, a clean-in-place (CIP) system is fitted. The body of the drying chamber, the ice condenser and their associated instruments and machinery are located in a dirty area behind the clean room wall. A separate control panel is fitted in the machinery area. Figure 12.15 shows the same typical lyophilizer diagrammatically and indicates the points presenting hazards.

Minor hazards are created by the removal of fittings and of instruments for calibration. These hazards can be eliminated by detoxification procedures before removal. Detoxification must be es-

Figure 12.15. Diagram of lyophilizer for hazardous products showing main danger areas.

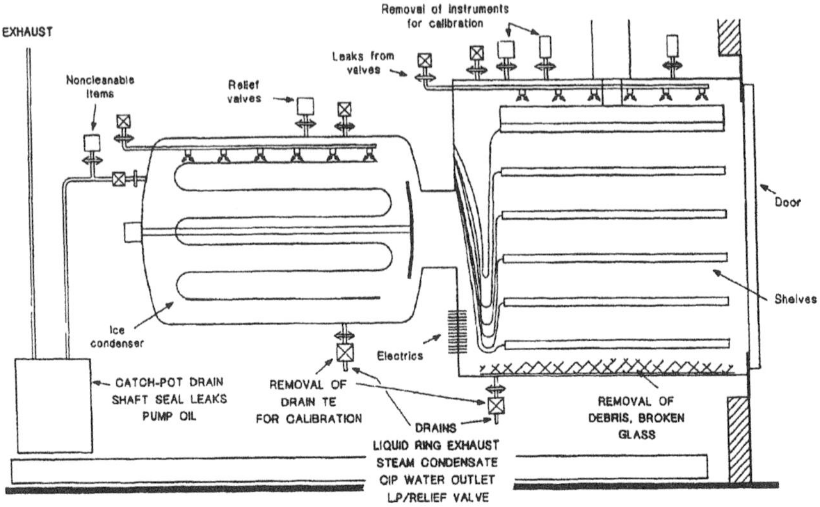

pecially effective for the drain temperature elements (TEs). When noncleanable items must be removed for adjustment or calibration, a safety procedure with a miniature containment cell must be provided.

A major hazard would arise if debris were deposited in the drying chamber. If a lyophilizer is manually loaded, it is almost inevitable that occasional vials will fall down and break. Removal of such debris is a very hazardous operation. Incidents of such a nature are much reduced if automatic loading is used.

Protection of the Environment from Effluents

Minor hazards are created by leaks from relief valves. Leaks are only likely after valves have opened. If the steam sterilization and detoxification procedures are carried out carefully, there is minimal risk. However, for environmental safety the relief valves should be piped to a protected exhaust, such as the one for vacuum pump.

Oil sealed rotary vacuum pumps present several risks. If toxic materials are carried over from the drying chamber and deposited in the oil, then leaks from shaft seals or catch-pot drains may escape. Also, contaminated oil must be disposed of in a safe manner. Leaks from shaft seals can be controlled by using semihermetic rotary pumps or, when rotary boosters are used, special nitrogen gas purge systems for the bearings.

The major danger associated with the vacuum pumping group is of toxic materials being ejected with the exhaust. The presence of oil droplets in the exhaust from oil-sealed rotary pumps makes it more difficult to catch or destroy toxins entrained in the gas. For this reason "dry" vacuum pumps are preferred when freeze-drying toxic materials. If dry pumps are used, no oil is entrained in the exhaust gas streams and several methods are available to prevent environmental contamination. All but one are based on heating. The current possibilities are as follows:

- Incineration upstream of the pump in a labyrinth heated to 110°C. This method was used effectively to contain the Foot & Mouth Disease virus, Type O, BFS 1860, in a vacuum pump exhaust. A trapping efficiency of 99.997 percent was achieved (1 infective unit out for 1.5×10^4 in.).

- Heating in an absorbent block. This method uses a gas reactor column (GRC, Figure 12.16), a device developed for the semiconductor industry. It has the disadvantage that gas flow must be limited to avoid cooling the column, so the GRC should not be used during initial pump-down of the lyophilizer.

- An approach that can be used with a dry pump is to insert a HEPA filter with a "safe change" facility in the exhaust stream.

- Special drain design is needed as conventional "air breaks" cannot be used. The drains must be kept at negative pressure and led to a treatment tank for detoxification.

Design Features of Isolators for Use with Lyophilizers

Clear, flexible PVC (polyvinyl chloride) film is ideal for the upper parts of the isolator. It allows ambient light in and does not engender feelings of claustrophobia when working in a half suit. Normal quality 0.5 mm thick material is used for most parts as it has good stretch. Where good vision is needed, optical quality PVC is used.

If the lyophilizer is to be steam sterilized, it is necessary to ensure that any thermoplastics used in the isolator are not heated beyond their operating range. This can be achieved by fitting a heat break, typically by constructing the isolator with thin stainless steel panels for attachment to the lyophilizer, as shown in Figure 12.17.

The floor of the isolator is vulnerable to damage by dropping things or even by laying heavy, sharp-edged items on the floor. For this reason the floor should be in the form of a rigid tray with the PVC canopy sealed around the edge. One method is by PVC-coated steel strips. The ideal material for the floor is stainless steel. AISI 304 is

Figure 12.16. Gas reactor column.

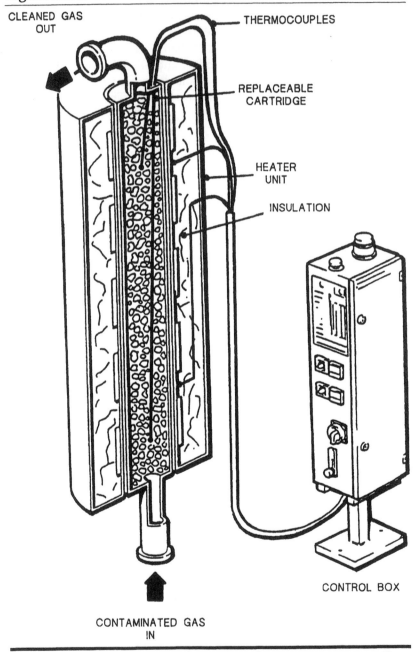

CLEANED GAS OUT

THERMOCOUPLES

REPLACEABLE CARTRIDGE

HEATER UNIT

INSULATION

CONTROL BOX

CONTAMINATED GAS IN

adequate for most purposes, but it may be necessary to specify a higher grade, such as AISI 316L, if aggressive sanitizing agents are to be used inside the isolator.

Figure 12.17. Top view of isolator connected to steam-sterilizable lyophilizer showing position of heat breaks.

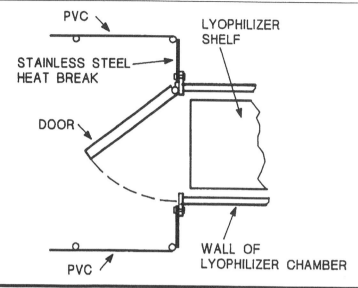

A major problem when using an isolator with a lyophilizer is the space occupied by the lyophilizer door when it is opened. The door should be the double hinged gimbal type shown at b in Figure 12.18, or better still. a sliding door as shown at c.

Figure 12.19 shows the clear area that must be left to accommodate door swing in front of the half suit. Figure 12.20 shows a large lyophilizer with a sliding door suitable for use in an isolator with a layout like Figure 12.18c.

Figure 12.18. Top views of lyophilizer and isolator showing the effect of different door hinges.

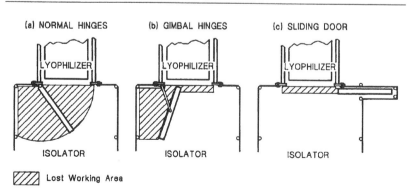

Figure 12.19. Area lost due to lyophilizer door swing.

Ventilation of the Isolator

Ventilation is by a HEPA-filtered fan driving an airflow through the isolator (Figure 12.21). This provides a supply of particle-free air to the unit and also steadily purges particles generated within it toward the

Figure 12.20. Sliding door on an industrial lyophilizer.

Figure 12.21. Sterilization schematic for ventilation.

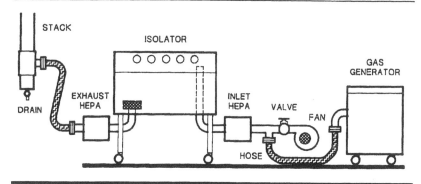

exhaust. The outflow is also protected by a HEPA filter. For sterile applications it is normal practice to run isolators at higher than atmospheric pressure, say 5 mm of water gauge. For applications where the contents of the isolator are hazardous, then it should be run at less than atmospheric pressure by about 5 mm of water gauge. Further protection must be added before exhausting the air to the atmosphere. Air is generally the support medium. Other gases, such as nitrogen or carbon dioxide, can be used.

The flow pattern within the volume of the isolator should be turbulent flow if satisfactory gaseous sterilization will be used. The flow should purge the entire volume, leaving no dead spaces, so that particles are steadily swept away to the exhaust filter. The air distribution system should even out the flow across the isolator.

Half-suit isolators are fitted with fans on both the inlet and the exhaust. This is to eliminate large wall-mounted filters. By using a second fan on the exhaust side, a more compact exhaust filter can be fitted, thus saving space and keeping the walls of the isolator free. A further benefit of the second fan is that the isolator can be run at positive or negative pressure simply by adjusting the fan speed controllers. Also, a positive exhaust drive compensates for back pressure in any exhaust ducting that otherwise might restrict the flow.

Ventilation equipment is best mounted underneath the base tray to give maximum light and vision inside the canopy.

Sanitization of the Isolator

Isolator sanitization can be done by hand spraying. However, something less dependent on the human operator is better; gaseous sterilization is preferred, of which several alternatives are available. The traditional sterilant is formaldehyde solution. Peracetic acid solution is better having lower toxicity and fewer residuals. However, both formaldehyde and peracetic acid can cause corrosion of copper and so should be avoided if electrical circuits would be exposed. Other proprietary solutions have been developed to avoid this problem.

A new possibility is to use hydrogen peroxide (H_2O_2). Proprietary equipment to generate H_2O_2 vapor is available on the market. For best results humidity must be carefully controlled at a specific value and turbulent flow established. The inlet HEPA filter is pre-heated with hot air before mixed steam and H_2O_2 is blown in through it. Dry air can also be imported by the same route.

It should be noted that no gaseous sterilants are effective in penetrating dead legs of pipe ends and ideally these should be avoided. If they must be used then special means must be incorporated to ensure penetration. The seals on transfer boxes have at least one surface that cannot be reached by a sterilizing gas and which must therefore be sterilized by another method, usually 300°C dry heat generated by an electrical trace heater.

Validation

Validation must include calibration of instruments; positioning of spore strips in places where vapor penetration may be poor; and means to confirm that fans are working, HEPA filters are operating correctly, and that all parameters have been tested and recorded. Isolator systems are normally leak tested on installation as part of their validation program. A regular routine for continuous validation will include leak testing and HEPA filter checking. Several methods are available for leak-testing:

Pressure Test

This is a very simple test on the integrity of an isolator. It needs no special equipment and can be run using the isolator as it stands in the case of positive pressure units. The isolator is first inflated to approximately five times working pressure and held there, by closing the stop valves, for 10 or 20 minutes. This allows the canopy to stretch. The pressure is then reset and monitored for 40 minutes. The system can be considered leak-tight if the pressure does not drop more than 3 mm of water gauge in 30 minutes. It should be noted that changes in temperature, atmospheric pressure, and movements of sleeves and half-suits can affect the results, since very small differential pressures are involved. The isolator's own manometer or a separate calibrated instrument can be used for this test.

Smoke Test

The smoke test requires filling the isolator with smoke from a suitable generator. Once filled with smoke, the isolator is then inflated to an overpressure of 10 mm of water gauge and a particle detector is used to check seams, joints, ports, and so on for leaks.

Freon Test

A can of freon is placed inside the isolator, which is then raised to 10 mm of water gauge. A short burst is then sprayed from the can. An electronic halogen detector is then used to look for leaks around the system.

13

Aseptic Packaging and Labeling

Ram Murty

The subject of aseptic processing, packaging, and the labeling of sterile drug formulations is extremely interesting due to inherent challenges associated toward maintaining their physical, chemical, and microbial attributes. A wide variety of commercial preparations are covered under this subject, with varying degrees of complexities, which influence the professional sphere in delineating factors associated for maintaining uniform quality standards consistent with applicable regulations.

A basic understanding of human physiology; the composition, methods of manufacture, storage, and distribution of various formulations; and a knowledge of all relevant regulations and guidelines covering these formulations are essential to drive the science and technological aspects from the current state to further sophistication. It may be recognized that this whole exercise is to distribute sterile drug formulations of acceptable quality, purity, and safety standards for the consuming public. The susceptibility for microbial contamination of these formulations is of utmost concern to industry, regulators, and the consuming public. Microbial contamination can be caused by bacteria, yeast, or fungi—all of which are extremely versatile in their metabolic activities. Moreover, the uncontrolled metabolic reactions of various microbes and other prevailing conditions during manufacture, distribution, and storage can present imminent health hazards due to various toxic by-products, such as fever-producing endotoxins, which will render them unsuitable for their intended use.

Current views and future trends in the aseptic packaging of sterile drug formulations are presented in this compilation with the hope that the reader will gain an insight into various aspects for preserving the integrity of various commercially available preparations.

DEFINITIONS

The term *aseptic* is defined as preventing infection; free or freed from pathogenic microorganisms.

The term *preserve* is defined as to keep safe from injury, harm, or destruction; to keep or save from decomposition; to keep alive, intact, or free from decay; to can, pickle, or similarly prepare for future use.

The term *drug* is defined as articles intended to cure, mitigate, treat, or prevent disease and articles intended to affect the structure or any function of the body.

The term *label* includes all written, printed, or graphic matter on the immediate container. Everything required to be on the label must be on or legible through the outside container.

PACKAGING AND LABELING REGULATIONS

Regulatory issues associated with sterile drug formulations require a review of the statutory provisions affecting these products. These are regulated primarily under the authority of the Federal Food, Drugs, and Cosmetic (FDC) Act passed in 1938 and enforced by the Food and Drug Administration (FDA), an agency of the Department of Health and Human Services.

Food, Drug, and Cosmetic Act

Congress enacted the FDC Act in order to protect consumers from unsafe or deceptively labeled or packaged products by prohibiting the movement in interstate commerce of adulterated or misbranded foods, drugs, devices, and cosmetics.

Drug Listing Act

The purpose of the Drug Listing Act (1973) is to provide the Commissioner of Food and Drugs with a current list of all drugs manufactured, prepared, propagated, compounded, or processed by a "drug" establishment registered under the Federal FDC Act. Therefore, the registration requirement for all drug products is mandatory.

Labeling

The labeling of pharmaceuticals generally falls under pharmacopoeial guidelines. The drug standards division of the United States Pharmacopeia (USP) requires the labeling of inactive ingredients for topical, ophthalmic, and parenteral preparations and is seeking to expand this requirement to all other drug dosage forms. These guidelines are enforceable under the FDC Act. Some of the resources for preparing a parenteral drug label include the following:

- Title 21—Code of Federal Regulations (Table 13.1)
- Parenteral Drug Label Contents Checklist
- Innovator's Label

Regulations on Containers

The authority for the regulation of the pharmaceutical container is derived from the FDC Act. Section 501(a)(3) defines a drug or device as being adulterated if "its container is composed, in whole or in part, of any poisonous or deleterious substance which may render the contents injurious to health." Section 502(i)(1) indicates that the drug is misbranded "if it is a drug and its container is so made, formed, or filled as to be misleading." Section 505(b)(1)(D) requires "a full description of the methods used in, and the facilities and controls used for, the manufacture, processing, and packaging of such drug."

Good Manufacturing Practices Regulations

In general, persons engaged in the processing, packaging, or holding of a drug product for administration to humans or animals are also subject to 21 CFR parts §210 through §226, current Good Manufacturing Practices Regulations (cGMPs) (1978), with the authority derived from section 501(a)(2)(B) of the FDC Act. The section provides that a drug shall be deemed adulterated

> *"if the methods used in, or the facilities or controls used for, its manufacture, processing, packing or holding do not conform to or are not operated or administered in conformity with the cGMP regulations to assure that such drug meets the requirements of this ACT as to*

Table 13.1. Parenteral Drug Label Contents Checklist

Subject	21 CFR
Manufacturer, packager, distributor	§201.1
Directions for use	§201.5 & §201.100(b)
Ingredient information	§201.10(a)&(h)
Quantity of ingredient per unit	§201.10 (d)
Established generic/brand name	§201.10 (g)
Small label exemption	§201.10 (i)
Expiration date and storage conditions	§201.17
Net quantity	§201.51
Dosage statement (see package insert)	§201.55
Rx legend	§201.100(b)(1)
Dosage	§201.100(b)(2)
Route of administration	§201.100(b)(3)
Quantity of active ingredient	§201.100(b)(4)
Inactive ingredients	§201.100(b)(5)
Lot number	§201.100(b)(6)
NDC number	§207.35(b)(3)
Additive statement for LVP solutions	§310.509(f)
Habit-forming drugs	§329.10(c)
Antibiotic labeling	§432.5
Controlled substance schedule symbol	§1302.04

safety and has the identity, strength, and meets the quality and purity characteristics, which it purports or is represented to possess."

For example, 21 CFR §211.94, §211.84, and §211.113 addresses the following, which impact the characteristics of drug formulation and microbiological aspects.

§211.94 Drug product containers and closures

(a) Drug product containers and closures shall not be reactive, additive, or absorptive so as to alter the safety, identity, strength, quality, or purity of the drug beyond the official or established requirements.

(b) Container closure systems shall provide adequate protection against foreseeable external factors in storage and use that can cause deterioration or contamination of the drug product.

(c) Drug product containers and closures shall be clean and, where indicated by the nature of the drug, sterilized and processed to remove pyrogenic properties to assure that they are suitable for their intended use.

(d) Standards or specifications, methods of testing, and, where indicated, methods of cleaning, sterilizing, and processing to remove pyrogenic properties shall be written and followed for drug product containers and closures.

§211.84 Testing and approval or rejection of components, drug product containers, and closures

(d)(6) Each lot of a component, drug product container, or closure that is liable to microbiological contamination that is objectionable in view of its intended use shall be subjected to microbiological tests before use.

§211.113 Control of microbial contamination

(a) Appropriate written procedures, designed to prevent objectionable microorganisms in drug products not required to be sterile, shall be established and followed.

(b) Appropriate written procedures, designed to prevent microbiological contamination of drug products purporting to be sterile, shall be established and followed. Such procedures shall include validation of any sterilization process.

§211.130 Packaging and labeling operations

(a) Prevention of mixups and cross-contamination by physical or spatial separation from operations on other drug products.

(b) Identification of the drug product with a lot or control number that permits determination of the history of manufacture and control of the batch.

(c) Examination of packaging and labeling materials for suitability and correctness before packaging operations, and documentation of such examination in the batch production record.

(d) Inspection of the packaging and labeling facilities immediately before use to assure that all drug products have been removed from previous operations. Inspection shall also be made to assure that packaging and labeling materials not suitable for subsequent operations have been removed. Results of inspection shall be documented in the batch production records.

PACKAGING AND LABELING GUIDELINES

United States Pharmacopeia/National Formulary

The United States Pharmacopeia/National Formulary (USP/NF) has extensive sections on pharmaceutical containers and provides characteristics thereof and specifications and control procedures therefrom (USP XXIII/NF XVIII 1995). Likewise, each monograph has a packaging requirement that outlines the performance characteristics of the pharmaceutical container pertinent to drug product (e.g., tight, light-resistant containers). These performance characteristics are defined under the General Notices—Containers. The important sections of the USP/NF that relate to types of pharmaceutical containers are presented in Table 13.2. Table 13.3 describes various USP tests for sterile products.

Food and Drug Administration

The packaging guidelines issued in the final form in 1987 present very specific recommendations as to the information that must be supplied in new drug applications (IND/NDA submissions) concerning the package and its components (FDA 1987). Pertinent information from the guideline is presented below.

According to Chapter 1, if a drug is listed in the USP monograph, it is considered misbranded if the package does not follow those recommendations unless packagers receive explicit permission for the deviation.

Chapter 2 of the guidelines deals with containers and closures. The section on containers covers four types: (1) Parenteral, (2) Nonparenteral, (3) Pressurized, and (4) Bulk. The following information will cover some of the specific details for parenteral and bulk containers.

Table 13.2. General Notices—Preservation, Packaging, Storage, and Labeling

Containers	
Tamper-resistant packaging	Light-resistant container
Well-closed container	Tight container
Single-unit container	Hermetic container
Unit-dose container	Single-dose container
Multiple-dose container	Multiple-unit container

Source: United States Pharmacopeial Convention, USP 23/NF 18, 1995.

Table 13.3. General Tests and Assays

<1>	Injections
<71>	Sterility tests
<85>	Bacterial endotoxins test
<87>	Biological reactivity tests, in vitro
<88>	Biological reactivity tests, in vivo
<151>	Pyrogen tests
<381>	Elastomeric closures for injections
<661>	Containers
<671>	Containers—permeation
<771>	Ophthalmic ointments
<785>	Osmolarity
<788>	Particulate matter in injections
<1151>	Pharmaceutical dosage forms

The numbers provided in square parentheses correspond to the USP sections.

Source: United States Pharmacopeial Convention, USP23/NF 18, 1995.

Parenterals

The following information is required for glass ampoules and vials:

- Name of manufacturer(s)
- Glass type, USP
- Physical description (size, shape)
- Chemical resistance
- Light transmission, if applicable
- Compatibility with contents
- Leaching/migration tests
- Sampling plan
- Acceptance specifications

The following information is required for plastic vials:

- Name of manufacturer(s)
- Type of plastic
- Composition; method of resin manufacture
- Method of container manufacture, with a full description of analytical controls
- Physical description (size and shape)
- Light transmission, if applicable
- USP tests—biological, physicochemical, and permeation tests
- Vapor transmission test, if appropriate
- Additional toxicity test data
- Compatibility, which includes leaching and/or migration tests
- Sampling plan
- Acceptance specifications

The items needed for cartridges, prefilled syringes, and large-volume parenterals (LVPs) are the same as described above for glass vials and for plastic vials.

The request for the composition and the method of manufacture of plastics is very sensitive information for plastic manufacturers, who originally resisted supplying this information. However, the FDA was adamant on this point because plastic manufacturers use many additives and there is always the possibility that these additives could be absorbed by the drug-a matter of great concern to the FDA.

Bulk Containers (Sterile Products)

The guidelines for bulk containers are very similar to those described under the nonparenteral section, except that bulk containers used for temporary, short-term drug storage are exempt from the reporting requirements. Readers are encouraged to refer to the guidelines for specific details.

The last section of Chapter 2 deals with closures, including caps, liners, inner seals, and elastomeric stoppers for parenterals. Also included is information requirements that are logical and specific for closures: The seal mechanism used for caps, including torque data; a description of inner seal, if used; special information when an elastomeric closure is used; and special information about medicine droppers, if used.

Chapter 3, suitability tests of package components for intended use, deals mainly with plastics. Section 1 discusses physical, chemical, and biological characteristics of plastic containers and explains the importance of detailed elucidation of these characteristics relative to the use of a particular plastic container/closure for a particular drug. Section 2 covers specifications and tests that are needed to assure that each batch of plastic meets the specifications. Examples of tests suggested are infrared spectra, thermal analysis, melt viscosity, molecular weight, degree of crystallinity, and film thickness. Biological testings and tests for elastomeric closures are also discussed.

Integrity challenges after multiple-entry penetration of vial stoppers are required. It is important to note that the FDA will not accept compatibility tests that are not conducted in a real-life situation. The testing must be done on the actual container/closure and the actual drug in question. For example, it is not sufficient simply to test the container material in standard test solutions to assess leaching and migration factors.

The adhesive and ink section of Chapter 3 deals with tests to determine whether or not the organic components of these materials will migrate through the plastic container wall or through elastomeric components.

Chapter 4 deals with Investigational New Drug (IND) applications. The container/closure to be used for a new drug in the investigational stage must be described for approval in the same detail as is needed for the final container. This allows for the fact that the container chosen for use during clinical trials may be quite different from the final container that will be used when the drug is commercial. In general, the FDA prefers that the containers used during the investigational process provide maximum protection and be as inert as possible; in other words, they prefer glass.

Chapter 5 deals with changes in packaging for drugs already on the market. Generally, if a container change is contemplated, the packager must file for an amendment or submit a supplement and provide information on the following:

- Container type (glass to plastic)
- Container style (bottle to blister)
- Closures/liners
- Components of plastics
- Suppliers/fabricators
- Packaging facilities

In cases where the packager uses polyethylene and proposes a change in resin supplier or a change from one type of high density polyethylene (HDPE) or low density polyethylene (LDPE) to another type of HDPE or LDPE, the FDA will allow such changes without prior approval, provided that the container/closure system originally adopted and the testing protocols for it have already been approved, it is demonstrated that the new container is equivalent to the old by use of tests described either in the NDA or in the USP, and stability studies originally performed to establish the expiration date have been expanded to include the new container to be used.

STERILE DRUG FORMULATIONS

A review of various commercially available sterile drug formulations indicates that a degree of complexity exists among the individual formulations. A drug formulation contains an active drug, a substance that stands alone or in combination with a mixture of inactive ingredients or excipients. These excipient ingredients are intended as bulking agents, antioxidants, preservatives, solubilizers, wetting agents or emulsifiers, buffering agents, tonicity modifiers, oleaginous vehicles, lubricants, suspending agents, chelating agents, local anesthetics, specific stabilizers, and so on, depending on the individual formulations.

Physiology and Routes of Use

A basic understanding of the principles of human physiology and the application of drug formulations is important in the selection of packaging components. Various drug formulations and their routes of usage are summarized below:

- *Topical:* Direct contact with skin, wound, and/or burned area

- *Systemic:* Injection—intravenous, intramuscular, or subcutaneous

- *Oral:* Ingestion by mouth (sublingual) and/or gastrointestinal

- *Intranasal:* Solutions and nebulizers of medicaments for nasal use

- *Transdermal:* Ointments, lotions and liniments, mists, patches, and topical liquids for the skin

- *Ophthalmics:* Solutions, suspensions, and ointments for the eye

- *Otics:* Liquids and suspensions for the ear

- *Rectal:* Liquids and suspensions for rectal administration

- *Douches:* Liquids for cleansing or disinfecting the surface of the body cavity

- *Gargles:* Cleansing the eye, throat, vagina, and nasal passages

It may be visualized that the parenteral route is common for sterile drug formulations. There is greater recognition by the regulators and the consuming public of sterile drug formulations. For example, the administration of drugs intended to have systemic effect by inhalation vapors and liquid and solid mists is well established.

All the routes used for systemic access have demanding requirements that can be fulfilled by complex and precisely structured and formulated drug products. For such precision products preservation of the original pristine state of the product is a critical packaging need.

Classification of Sterile Pharmaceuticals

Commercially available sterile pharmaceuticals may be classified into five groups based on their intended uses.

1. Injectables, (solutions, suspensions, lyophilized powders)

2. Ophthalmics (solutions, suspensions, lyophilized powders)

3. Biologicals (solutions, suspensions, lyophilized powders)

4. Vaccines (solutions and suspensions)

5. Miscellaneous (solutions, suspensions, and dry powders)

Routes of Administration

Sterile pharmaceuticals are used by various routes of administration (USP 1994; FDA 1986) depending on the medical necessity. Table 13.4 exemplifies the routes of administration that are recognized in medical practice. In view of their intended application for critical medical necessities, the establishment of appropriate packaging specifications will assure the overall quality of the finished sterile dosage form.

Ideal Requirements for Sterile Drug Formulations

A basic understanding of the ideal requirements of formulations leads to a determination of their packaging needs. Sterility, apyrogenicity, particulate, potency, and stability for these formulations are discussed.

Table 13.4. Routes of Administration—Sterile Pharmaceuticals

Implantations	Intrasynovial
Injections	Intrathecal
Inhalation	Intrathoracic
Intraarterial	Intratracheal
Intraarticular	Intrauterine
Intracardiac	Intravenous
Intracartilaginous	Intravesical
Intradermal	Irrigation
Intralesion	Ophthalmic
Intralymphatic	Perfusion
Intramuscular	Perfusion, Biliary
Intraocular	Perfusion, Cardiac
Intraperitoneal	Subcutaneous
Intraspinal	Transtracheal

Sources: United States Pharmacopeial Convention, 1994; Food and Drug Administration, *Drug Establishment Registration and Drug Listing Information Booklet*, 1986.

Stability

The first consideration in the formulation of a product is to provide an effective and safe treatment of specific conditions. A second concern is that the ingredients maintain their effectiveness from the time of manufacture to the moment of use. Assurance of effectiveness requires the absence of both physical and chemical changes. Physical changes include settling out of a suspension. Whereas, chemical reactions are attributed to the complex molecular nature of the formulation; biological activity depends not only on the specific constituents but often on the exact configuration within the molecule. Tests for biological activity include optical isomerization, discoloration, and precipitation. Degradation is influenced by moisture, permeation of oxygen, and exposure to light through the packaging components. Remedial measures to combat degradation include the complete elimination of light, avoidance of trace contaminants, and an application of antioxidants that do not interact with the packaging components.

Purity and Sterility

Purity and sterility are required attributes of commercial formulations and are achieved by the diligent application of well-established technology during product manufacture. Assuming that a formulation has been developed to achieve the effectiveness and safety claims, the total operation must be designed and controlled to assure product variability; all forms of biological and chemical contamination are kept within acceptable limits.

Thus, the industry is infused with strong quality control emphasis. Raw materials and in-process and finished products are continually checked for approval/rejection. Inspection at all stages during manufacturing and recordkeeping requirements are vital for success.

The specific problem of microbial contamination is much more serious than the stability problem, especially for the manufacture of liquid formulations. Historically, microbial contamination has led to recalls of products as diverse as baby lotions and milk of magnesia antacids. *Pseudomonas aeruginosa*, *Salmonella species*, *E. coli*, and *Staphylococcus aureus*, are commonly isolated microbes in various formulations. Liquids can be ideal growth promotion media for bacteria, as can be many raw materials (e.g., natural products). Bacteria also flourish in the difficult-to-clean parts of processing equipment. Therefore, thorough equipment cleaning and monitoring procedures are essential. Microbial monitoring of process water is essential since this is one of the primary sources for contamination. Routine testing is recommended for bacteria (coliform counts or total bacteria counts) for products intended for rectal, urethral, or vaginal administration.

Antibacterial preservatives are added to the formulation (e.g., alcohols, phenols, hexachlorophene, mercury compounds, and benzoic acid esters). While the problem of the so-called nonsterile products has received renewed interest, the requirements for parenterals have been stringent. The fact that contaminated drug product could cause serious occular infection is well documented. One bacillus, *P. aeruginosa*, is especially virulent and can cause blindness in 24–48 hours. The most likely source of contamination is from the reuse of containers or droppers that have contacted infected areas. Antimicrobial additives will help but are not foolproof since, for example, viruses are not affected.

The preferred method is to deactivate the microbes by heat, steam, high or low energy radiation, and gases as well as through alternative aseptic filling techniques.

Drug-Excipient Considerations

The term *excipient* has usually been defined in association with pharmaceutical formulations. The dictionary definition of an excipient is "an inert substance (such as gum arabic or starch) that forms a vehicle as for a drug." Excipients are a necessary component of pharmaceuticals, enabling the delivery of medicinals in a variety of dosage forms.

Stabilizing agents, antimicrobial agents, and bulking agents are added to preserve the properties of formulations during their intended shelf life. Formulations must be packaged in the selected container/closure system and provided to the consuming public in that system. The product during the course of its manufacture, storage, distribution, and consumption should adequately be protected from extraneous contaminants that influence the physical, chemical, and microbiological attributes of the preparations. Therefore, the selection of proper packaging materials and appropriate controls during their preparation will ensure a high degree of consistency and uniformity in the finished products. Over the years, the FDA has introduced firm policies and provided guidance to drug manufacturers in cGMPs.

Categories of Excipients

A brief review of excipients (*Physicians' Desk Reference* 1994) used in parenteral formulations indicates that twelve categories are employed according to their function.

1. Solvents, water immiscible

2. Solvents or cosolvents, water miscible

3. Surfactants

4. Antioxidants

5. Antimicrobial agents

6. Antifoaming agents

7. Isotonic agents

8. Chelating/complexing agents

9. Buffering agents

10. Suspending and/or dispersing agents

11. Sustained release agents

12. Miscellaneous

THE CHANGING PHARMACEUTICAL INDUSTRY

The pharmaceutical industry has grown comfortably familiar with traditional parenteral dosage form package systems. Glass ampoules, stoppered glass vials or bottles, plastic ophthalmic solution dropper tip bottles, and prefilled syringes have been successfully manufactured for decades. The integrity of these systems has been validated by years of experience dedicated to fine-tuning both the package and its assembly. But the introduction of novel package designs along with a changing regulatory environment may force the industry to reevaluate its approach to package development.

A plethora of package designs and materials of construction is now available to the pharmaceutical manufacturer. Package designs have become more complex to make for easier preparation and administration of parenterals. Examples of new delivery systems include the following:

- ADD-Vantage™ vial/flexible container

- Unit dose insulin syringe cartridges

- Portable intravenous infusion systems

- Wet/dry combination packaging

- HYPAK™ Liqui/dry dual-chamber prefilled system

- Redi-vial dual chamber

- Wet/wet combination systems

In addition, packaging materials have also continued to evolve to include elastomeric closures coated with more inert, lower particulate-shedding polymeric materials; thermoplastic closures; and newer plastics for vials, bottles, and flexible containers. Elastomeric closures coated with Teflon® and/or purcoat and containers fabricated from newer resins such as Daikyo Resin CZ may exemplify recent developments in this area (*Technical Literature*).

Techniques are being developed for sterilizing packaged product and package components consistent with the advances in the package design. For example, gamma irradiation, electron-beam (E-beam) sterilization, and hydrogen peroxide gas may become more viable options. Advances in steam sterilization equipment have made possible air-over-steam cycles and high or ultrahigh temperature cycles, which permit more controlled sterilization of packages prone to blowout and more rapid sterilization or marginally thermal stable chemical compounds.

The FDA has now provided an added incentive for more intensive research via a recently published proposal that all new parenteral products be terminally sterilized or that sufficient data be provided to demonstrate that terminal sterilization is not a workable option (*Use of Aseptic Processing* 1991). Several other countries have already mandated terminal sterilization requirements.

Package integrity is a simple concept, but one that is not easily measured or validated. Leakage is a quantitative term mathematically described as the amount of gas capable of passing through a seal under carefully defined conditions of temperature and pressure. In the same way that nothing is absolutely clean or pure, all packages exhibit some degree of leakage. Verifying package integrity is, therefore, a matter of defining the leakage specification limits and selecting appropriate test methods for detecting leakage at these limits. Leakage specifications should be conservative enough to guarantee sterility of the parenteral package, and to ensure satisfactory product stability and package performance throughout the shelf life, but not so stringent as to make integrity testing and verification prohibitively difficult and expensive. When choosing a leak test method, the sensitivity of the equipment or technique must be weighed against other factors, such as reliability, speed, cost, ease of use, repeatability, safety, and data processing capabilities. This subject has extensively been reviewed by Guazzo (1994).

It may be visualized that the pharmaceutical industry has historically taken a liberal approach to package integrity validation. If nothing appeared to leak through the seal, if the samples tested were sterile, and if product stability was satisfactory, package integrity was considered validated. This approach has been found to be insufficient due to a variety of factors: today's more stringent regulatory environment, the rise in complex parenteral packaging and product delivery devices, and the growing number of products sensitive to even minute leakage of gases or moisture. To meet these challenges a more comprehensive product/package development approach is recommended. The approach begins by precisely defining the functions and expectation of the package in terms of integrity, stability, and functional performance. Then package materials and component designs can be selected; followed by the optimization of package filling, assembly, and processing operations. The integrity and functional performance of the package should be challenged at the limits of component dimensional specifications as well as filling and processing parameters. Leakage tests of the finished product should only be considered as a final check of a thoroughly validated product/package system.

PACKAGING CONSIDERATIONS

Packaging must consistently meet the agreed-upon specifications, such as level of protection. It is also critical that packaging material not be a source of contamination. For aseptic filling the package must be produced and stored under conditions that preserve sterility.

Packaging—Liquids

The packaging needs for liquids are quite demanding. The physical properties of liquids are altered with inadequate packaging components. The evaporation of solvents in a formulation increases viscosity and alters the concentration of a particular ingredient, which can lead to precipitation of some components. This may be cited as an illustration of why the selection of packaging components is very critical. Likewise, sterility assurance is extremely important. In addition, it is well documented that drug solutions are sensitive to oxygen and gases that have deleterious effects. The appearance of particulates in sterile solutions is not desirable. Finally, the proper selection of multiple-unit containers that permit withdrawal of successive portions of the contents without altering the strength, quality, or purity of sterile products is very important.

Packaging—Solids and Semisolids

Sterile solids are represented by a distinct class of sterile products that are intended for reconstitution at the time of administration. Semisolid preparations are intended for topical and ophthalmic uses. They are oleaginious bases and a number of sterile preparations belong to this class. The requirements that arise most often in this description of packaging need are as follows:

- Product containment

- Product protection

- Product dispensing

- Product purity and sterility

The sterility requirement is one of the most frequently cited for various drug formulations. In addition, the stability of these formulations is largely dependent on the conditions of storage in a given container/closure system. Therefore, the selection of the proper container/closure is extremely important in preserving the properties of drug formulations during the intended storage.

Characteristics of Containers

The definitions of a container in the regulations are not clear and consistent. The definition in the USP General Notice:

> *The container is that which holds the article and is or may be in direct contact with the article. The immediate container is that which is in direct contact with the article at all times. The closure is a part of the container.*

This definition has utility since it focuses on what is in direct contact with the article and does not artificially separate container and closure. What is essential is that the container should protect the drug from the normal environment in which it will be stored before and during use (i.e., from the effects of water vapor and other gases [permeation] or light [photodegradation]). Also, any interactions between the container and the preparation (sorption, leaching) must remain below a level that would adversely affect the safety, efficacy, or stability of the drug. Significant to this is the nature of the dosage form and the route of administration. The major potential problems between drug products and containers are summarized in Table 13.5.

Qualification of Containers

In approaching the problem of regulatory control of pharmaceutical containers, it is obvious from the earlier discussions that we are dealing with an unusually complex subject. Variations exist in the types of glass, paper, or metals used as containers; however, this in no way compares with the ever-expanding family of plastic materials with respect to their range of physical and chemical properties. Also, different properties in plastics can be obtained not only by alterations in the molecular weight and geometry of the polymer but by the addition of plasticizers, fillers, lubricants, and other chemical substances. These

Table 13.5. Potential Problems—Containers

Problem	Description	Example(s)
Sorption	Loss of article components to the container	Sorption of benzalkonium chloride by nylon
Desorption (Leaching)	Loss of container components to the article	Leaching of DEHP from PVC tubing
Permeation	The transport of water vapor or gases through the container	Water vapor through LDPE in ophthalmics Nitroglycerine from tablets through plastics
Radiation degradation	The loss of articles and/or container through radiation	Photodegradation of retinoids UV absorption by polymers
Container modification	The alteration of container properties due to contact with the article	The stress cracking of polyethylene on prolonged contact with some liquids at elevated temperatures

Source: American Association of Pharmaceutical Scientists, Workshop on Pharmaceutical Packaging: Issues and Challenges for the 90's, 1993.

have modifying effects on the ultimate container. Laminates of different materials present further complications.

If the selection of an original container is complicated, then the selection of an alternate may be equally or more complex, depending on the situation. The degree of qualification testing for the alternate will depend on the route of administration of the drug product and the past history and industrial acceptance of the package material. Materials that meet well-known industry standards and are used in their normally accepted applications and are not overstressed normally require less testing than innovative materials.

Some considerations in evaluating a pharmaceutical packaging component include: a characterization of the material(s) of construction, the compatibility of the packaging component with the environment, the route of administration, and the suitability of the packaging component for its intended use (i.e., performance characteristics).

The category of parenteral dosage forms represents the highest risk level since the introduction of hazardous contaminants into general circulation is relatively rapid and complete. Glass containers for parenterals are well defined in the literature and are classified in the USP as to the type and recommended use. Pharmacopeial Monographs for parenterals generally specify Type I glass. If the alternate is plastic, then full, physicochemical, material characterization is usually necessary for both the resin and the container; toxicity studies (i.e., subacute extracts) and cell cultures are usually necessary on the finished container. Container characterizations and toxicity studies for the container may require pre- and poststerilization. The information that might be needed for an NDA/ANDA packaging section of a plastic parenteral container might include, but is not limited to, the information outlined in Table 13.6.

Container/Closure Integrity

There are numerous statements in the earlier discussion that define containers for injection. The regulatory definition of container is not precise. A comparative review presenting the container/closure standards in four major pharmacopeias was reported (Dabbah and Paul 1992). Leak testing of ampoules has been described in the literature. The physical and mechanical properties of the vial/closure system that affects the seal integrity was first described in the literature in 1983 (PDA 1983). Determination of container closure integrity by various physical methods such as vacuum retention, vacuum chamber, internal pressure, dye immersion, seal force testing, and so on have been summarized. Also, microbial methods have been described to assess the integrity of the container/closure system.

Package integrity is a measure of a package's ability to keep the product in and to keep potential contaminants out. While this is a relatively simple concept, it is not precisely measured or validated. The product necessitating containment consists of the liquid or solid parenteral drug product, as well as the gas headspace in the case of a package sealed under vacuum or with an inert gas. Potential contaminants include microorganisms, pyrogens, other chemicals or materials and particulates. Various tests are used by the pharmaceutical industry to measure parenteral product package integrity (Guazzo 1994). However, many of these tests are insensitive and qualitative. Validation of package integrity is rarely determined by evaluating packages representing the full range of package component or production assembly variables. Only sterility of the packaged product is required by the FDA as verification of package integrity. Yet package integrity is a critical requirement of all parenteral products.

Table 13.6. Criteria for the Selection of Plastic Containers

Part I. Product Information

1. The base polymer, the formula, and configuration of the container
2. The manufacturer's product code, name, address, and DMF number with an authorization to access the DMF. In the DMF the holder should supply the exact formula of the plastic container and include any appropriate specifications and their attendant test procedures.
3. Additional treatments such as washing, coating, and sterilization

Part II. Polymer Characterization

1. Multiple Internal Reflectance (MIR), thermal analysis by differential scanning calorimetry, density, melt index, modules of elasticity
2. USP physicochemical tests—plastics (USP 23/NF 18 <661>).

Part III. Container Characterization: Pre- and Poststerilization

1. Infrared spectroscopy, turbidity, extractables, heavy metals, thermal analysis, other polymer parameters
2. Water vapor and gaseous permeability tests
3. Permeability to microorganisms
4. Toxicity tests include acute systemic toxicity and intracutaneous reactivity

Part IV. Packaging Compatibility (Stability Studies)

1. Interaction of the drug product with the container including absorption, adsorption, permeation, and leaching
2. Testing of the container and drug product under worst-case scenario of actual production (Testings should include IR, UV, TLC, GLC, LC, and TGA)

Part V. Quality Control Testing

1. Certificate of analysis from the vendor
2. Incoming identification tests, such as thermal analysis or infrared spectroscopy
3. Conformance to drawing or master sample
4. Performance tests

Source: American Association of Pharmaceutical Scientists, Workshop on Pharmaceutical Packaging: Issues and Challenges for the 90's, 1993.

The apparent lack of advancement in parenteral package integrity validation and test method development may be due to industry's success at providing adequate sterile packaging. However, novel packaging designs and the need to minimize moisture pickup or gas headspace loss for very sensitive products challenge the traditional standards of integrity.

Fortunately, there is a large body of knowledge (belonging to other scientific disciplines) on the subject of leakage scattered throughout the literature. By gleaning a general understanding of leakage concepts, it may be possible to more logically design, assemble, and validate integral parenteral product packaging.

In considering the choice of a container and the tests associated with its approval for use in the packaging of pharmaceuticals, the overriding objective is the maintenance of the quality of the preparation during storage and administration. With drugs the biological results are paramount; other aspects of the container, such as appearance, should not replace safety and efficacy from their primary position. The risks emanating from interactions between containers and drug formulations vary from the trivial to the dangerous, from a reduction in the intensity of a flavor to the administration of an unknown toxic substance. In this potential for interaction, the nature of the dosage form and, thereby, the route of administration plays a significant role. It is quite obvious that regulatory agencies would pay far more attention to plastic containers for injections than those used for oral dosage forms.

Parenteral drug products in plastic containers require an approved NDA as a condition for marketing a parenteral in a plastic container (21 CFR §310.509). The regulation most specifically addresses LVPs—parenterals with a capacity of 100 ml or more. Safety concerns regarding LVPs are the incompatibilities of the drug product with the plastic container and the leachability of certain plastic ingredients. New plastic containers for LVPs should be qualified by NDA submission. Applications submitted after NDA approval may then be submitted as an abbreviated new drug application (ANDA).

The Center for Drug Evaluation and Research (CDER) policy does not require NDAs for new plastic containers for all small-volume parenterals (SVPs), parenterals with a capacity of less than 100 ml. Safety concerns regarding SVPs also include the incompatibilities of the drug product with the plastic container and the leachability of certain plastic ingredients. Small-volume parenterals packaged in plastic containers can be accepted as ANDAs. Thus, integrity testing will become a subject of greater scrutiny; the establishment of test procedures as well as the validation of such procedures will be challenging for pharmaceutical scientists.

Sealing Integrity Tests

Hermetically sealed ampoules are routinely tested for proper sealing through a leaker test. The leaker test is intended to detect an incomplete seal, capillary pores, or tiny cracks in ampoules. The procedure is described in the Federal Specifications from the Defense Personnel Support Center (DPSC) (1976). However, literature reports indicate that although ampoules have undergone dye testing, the ampoules exhibited leakage because the morphology of normal vitreous glass differed due to the flame sealing of ampoules by a process known as devitrification (Levine 1986).

In the case of glass/rubber closure systems, the integrity can also be verified by the leak test method described by the DPSC. The test requires that the container be placed in an inverted position for 2 hours at room temperature and then for 4 hours at 49°C ±3°C. If the containers exhibit no evidence of leakage during or at the completion of test, they are considered to be acceptable.

Ophthalmic containers (e.g., metal tubes, plastic tubes, jars, foil, etc.) holding semisolid products shall be subjected to the following test:

> *The filled immediate container, with seals applied when specified, shall be selected at random from each lot; exterior surfaces thoroughly cleaned, and stored for 8 hours (in a horizontal position for tubes and foil containers, and inverted position for jars) at a temperature of 60°C ±3°C. No leaking, except that quantity that could come only from within the crimp of the tubes or thread of screw caps, shall be evidenced during or at the completion of the test* (DPSC 1976 and USP 23/NF 18 1995).

A second test employed in the validation of container/closure systems requires the filling of containers with a soybean-casein digest medium and precisely processing the filling containers according to normal production conditions. The containers are then incubated and macroscopically examined for microbial growth. This also includes growth promotion tests at each time in the inspection interval.

A closure displacement test as described in the literature (Levine 1986), the results from which can correlate the biological tests, ensures the consistency of the sealing process. The test utilizes a digital micrometer and a spring force tester, whereby it is possible to measure the force required to compress the capped closure at a specified distance. The greater the downward force required, the more rubber closure has been crimped by the metal cap. Thus, one can quantify the capping process during the product development phase and normal production operation. An optimum lower control limit must be established for the closure displacement test.

Terminally sterilized products (via steam) are normally cooled with water prior to unloading from the autoclave. It must be shown that the container/closure system maintains integrity during the cooling process, which can then be verified by adding a stable chemical entity to the cooling water and analyzing the product for the chemical entity.

Engineering and microbiological personnel must work together to examine the closure design to determine whether any potential pathways exist that would allow the ingress of microorganisms. Once the design concept is accepted as feasible, challenges to the design concept may be made by conducting an assortment of physical tests, such as dye/pressure tests or other appropriate tests. The type of closure capping or insertion application and associated forces must also be considered.

The manufacturing process must permit appropriate sidewall/sealing surface contact to occur in the case of rigid container systems such as glass. If plastic and rubber components are used in the closure system, the process must assure appropriate fit to preclude outward or inward leakage. The adequacy of closure system's integrity should include an evaluation of overall design, sealing qualities and characteristics, functionality, product contact, materials of composition, size of the container/closure, and process conditions.

Very little information exists in the literature concerning the evaluation of closure systems by either physical or microbiological methods. Due to the heterogeneity of closure and container systems that exist in the pharmaceutical industry, standardized procedures are not available. The situation is more pronounced with the advent of wide usage of plastic containers. While physical and microbiological tests may not be standardized, physical tests for the most part will yield quantitative measurement that permits one to evaluate the process and design changes on the integrity of closure system.

Various leak testing methods—their levels of sensitivity, advantages and disadvantages, and reported usage—have been summarized (Guazzo 1994; Korczynski 1987).

The selection of the physical method is independent of the end use of the product. However, the microbiological test method selected should have a relationship with the intended use of the product.

Packaging—Selection

The next phase involves the selection of appropriate packaging materials and packaging component designs. Materials are first screened based on physical and chemical compatibility with the product as well as their ability to withstand sterilization. Performance criteria should also include the ability of the material to ensure adequate package integrity. The dimensions of the components must be specified according to appropriate fit, clearance, and interference. Component samples representing multiple component lots should be evaluated at the dimensional specification limits for their impact on packaging functionality and package integrity.

The close relationship between a pharmaceutical preparation and its package is of major concern to the industrial pharmacist. Faulty packaging of dosage forms can invalidate the most valuable formulation. Consequently, it is essential that the choice of immediate container materials for each formulation be made only after a thorough evaluation has been made of the effects of these materials on product stability.

Functions of Drug Packages

Various functions of drug package have been summarized in a recent publication (Jenkins and Osborn 1993):

(i) Containment is the most fundamental function of a drug package to contain the packaged product and not allow it to become part of the environment. Principally, this requires a package that will not leak, that remains impervious to attack by the ingredients of drug formulations and that is strong enough to hold the contents during physical distribution.

(ii) Protection is unquestionably the second most important package function. The product must be protected against physical damage— such as breakage of tablets—against loss of contents or ingredients and against intrusion of unwanted components of the environment such as water vapor, oxygen, liquids, dirt and light.

(iii) Other essential functions include:

(a) The package must provide some way to dispense the contents, either into another container or directly onto to the hand or into the body.

(b) Some provision has to be made for reclosure of the package so the unused contents will not lose their potency or efficiency, become contaminated or represent a hazard to small children.

(c) When the package contents are sterile, this sterility must be maintained, including the sterility of the unused remainder.

(d) The package must present all the information about the drug that is required by law and good therapeutic practices.

(e) The package should help sell over-the-counter products without the need for intervention by a pharmacist.

(f) Although most packages do not perform this function, packages which aid in compliance pay large dividends in reduced health care costs.

(g) The package design must provide evidence of tampering for those products which are so regulated by the FDA.

(h) Packages for prescription drugs and certain OTC products must thwart access by young children.

In addition, drug packaging has some special requirements because of the following:

- Many drugs are harmful or even toxic unless taken in carefully regulated amounts. Unless drugs are taken in the proper amounts at the proper frequencies, they will be ineffective, a patient's health will be at risk. Therefore, the accuracy of labeling information is important.

- The sterility requirement for drugs injected intravenously into the bloodstream or other subcutaneous routes is extremely important.

- A knowledge of the drug-package interactions is very critical.

- Regulations for drugs are more rigid. Shelf-life predictions indicating the stability, provisions for point-of-use repackaging, a requirement for tamper-evident packages in certain classes and packaging devices for a select group of patients (such as geriatrics and pediatrics) all play an important role in the overall success of the drug.

STERILE PACKAGING

The five essential quality criteria for sterile pharmaceuticals are sterility, apyrogenicity, particulates, potency, and stability. The specific processes used for sterilization and the integration of these processes into the filling operation will be covered. The preferred technique is terminal sterilization of the final packaged product as soon as possible after sealing. Where terminal sterilization is not possible because of drug sensitivity to the sterilization conditions, the product is aseptically packaged in equipment where both the product and the package are separately sterilized and then brought together in a sterile environment. For critical products, such as parenterals, aseptic packaging is combined with terminal sterilization. For noncritical products, aseptic packaging is often used even though terminal sterilization is possible.

Terminal Sterilization

Autoclaving with saturated steam under pressure is the most common terminal sterilization method. For large-scale operations, assembled of packages are placed in the autoclave, which is a thick-walled pressure vessel. The periods for heating up, holding at sterilization conditions, and cooling down are regulated to suit the heat sensitivity of the package and the drug and the time required to achieve the required level of microorganism destruction. For especially sensitive products, like dextrose injection, the heating rate is very rapid and cooling is accelerated by water sprays. Autoclaving is used primarily for aqueous liquids. It is ineffective for dry powders in sealed containers and anhydrous oils.

For materials such as petroleum jelly, mineral oils, greases, waxes, and talcum powder, where steam sterilization is ineffective or is too severe, filled packages are subjected to dry heat in a hot air oven. A wide range of inactivation times and temperatures have been established, which depend on the bacteria involved and the humidity. Lower temperature cycles have been developed for some preparations such as sulfonamides, powders with low melting points, and certain oil-based solutions that cannot withstand normal conditions.

Nonthermal methods of sterilization that are commonly employed for medical devices are not suitable for pharmaceuticals, especially liquids, since ionizing radiation adversely effects the drug and sterilizing gases will not penetrate the package.

An exception is the European practice of using ionizing radiation to sterilize some powders, such as penicillin, streptomycin, polyvitamins, and certain hormones. While no drugs are sterilized this way in the U.S., several drug container systems are presently undergoing the FDA approval process. It is predicted that in the future radiation will be used for final sterilization once a drug has been aseptically packaged. These radiation doses will be considerably smaller than those required for primary sterilization, thus avoiding radiation damage to drugs and packaging materials.

Aseptic Packaging

Aseptic packaging requires separate sterilization of the equipment, the product, the package, and the environment. The containers may be supplied to the drug packager, already cleaned and sterilized; or they may be cleaned and sterilized in the aseptic filling line; or the containers may be manufactured under conditions that insure sterility as part of the filling operation, as in the case of blow-fill-seal system.

The sequence of steps for preparing containers that are free of particulates and microorganisms usually includes washing, drying, sterilization, and cooling. For the in-line sterilization of glassware containers are commonly exposed to a laminar flow of very hot air (~350°C) while being passed through a Class 100 tunnel. Such a tunnel contains, per cubic foot of air, less than 100 particles of 0.5 μm or larger in size. Generally, a container temperature of 180–200°C is sufficient for achieving sterility and contact time if the oven is adjusted to account for container size and shape. The air used is recycled, filtered and dried if containers are wet from washing. Upon emerging from the tunnel, containers are cooled with filtered, laminar flow air washed over an area in a stable layer that isolates the area from the surrounding environment and controls the level of airborne particles.

Autoclaving is also used for glassware where lower sterilization temperatures are accepted (115–138°C), since wet heat is more effective in killing microorganisms than dry heat. These lower temperatures permit the use of autoclaves to sterilize more heat-sensitive containers, closures, and packages, such as those made from plastics including nylon, HDPE, polyesters, polypropylenes, and other elastomers.

Plastic containers are sterilized with ethylene oxide by placing the articles in a chamber containing the gas for four hours at 130°F. Upon completion of the sterilization step, gas is removed by vacuum and the containers are stored until residual gas and by-products dissipate. This step is critical since residuals can irritate the skin and mucous membranes.

Chemical sterilization by immersion in liquids is widely used for plastics and paper packages in aseptic filling operations. For the Tetra Brik process, which is a make-and-fill operation where a box is formed from a paper, plastic, and aluminum foil laminate, the laminate is immersed in a bath of 25 percent hydrogen peroxide. In this case the residual sterilization products have been shown to be at an acceptably low level or harmless. Collapsible metal tubes are sterilized with dry heat or by autoclaving. Irradiation sterilization of containers and closures has been widely studied, but is not used except for medical devices.

Sterilization of liquid products prior to aseptic packaging is accomplished by autoclaving, heating and/or filtration. Filters used must have openings small enough to retain microorganisms and yet allow reasonable flow rates. Pressure or vacuum assist the rate of flow. Common filter materials include fused porcelain, sintered glass or metal, and cellulosic or plastic membranes. Typically, coarser first stage, and finer second stage filters are employed with pore sizes of about 0.3–0.5 μm and 0.2–0.3 μm for each stage, respectively.

A typical sterile operation was recently described for filling blown HDPE bottles with liquids (Sharp 1987). Isolated Class 100 areas are maintained for the remote bottle making operation and filling steps, while intermediate Class 1000 areas enclose transition operations such as bottle handling, storage, and cartoning. Bottles coming off the blow molding line are packaged in three layers of presterilized polyethylene bags. As the bottles leave the bottle making area, the outer bag is removed.

The middle bag is discarded during the transition into the filling area. The final bag is removed just prior to loading bottles into the filling operations. By this technique contamination from each previous area is left behind at each transition point. The blow molding equipment is designed so that all parts can be cleaned and sterilized including bottles, molds, and conveyor lines. In the filling operation liquids, sterilized by heat and/or filtration, are brought together with bottles and presterilized caps to complete the aseptic process.

Some of these elaborate bagging procedures and Class 1000 handling areas can be eliminated when the bottle blowing operation is carried out immediately adjacent to the filling and sealing steps. In this case the whole process can be enclosed and fitted with internal sterile air showers so that critical filling point is in an aseptic area and the whole process runs without the need for operator intervention. This isolation of the process from operating personnel eliminates the major source of contamination, according to a recent survey of pharmaceutical companies (Sharp 1990).

SUMMARY

From the foregoing discussion it is evident that packaging and labeling requirements for sterile drug formulations are critical. In view of their intended therapeutic uses in critical medical necessities, this chapter presented relevant regulations and guidelines pertinent to the manufacture and control of sterile dosage forms. Moreover, an understanding of the physiology and routes of administration, as presented, will assist the pharmaceutical scientist in the selection of appropriate container/closure assembly consistent with the current industry standards. It may be mentioned that methods to ensure the integrity of the packages is an important component of quality control assessment for this class of medications.

REFERENCES

Code of Federal Regulations. Title 21: Food and Drugs. Washington, DC: U.S. Government Printing Office.

Dabbah, R., and W. L. Paul. 1992. Container/closure standard requirements in four major pharmacopeias-A comparative review. *Pharm. Forum* 18 (4):3772.

The Drug Listing Act—An act to amend the Federal Food, Drug, and Cosmetic Act. February 1, 1973.

FDA. 1987. *Guideline for submitting documentation for packaging for human drugs and biologicals.* Rockville, MD: Food and Drug Administration, U.S. Department of Health and Human Services.

FDA. 1986. *Drug establishment registration and drug listing information booklet.* Rockville, MD: Food and Drug Administration, Drug Listing Branch.

FDA. 1978. Current good manufacturing practices for finished pharmaceuticals. *Federal Register* 43:45077.

Federal Food, Drug, and Cosmetic Act of 1938, as amended, Secs. 201–702, 21 USC, 321–372.

Federal Specifications. 1976. *Containers, packaging and packing for drugs, chemicals, and pharmaceuticals,* PPP-C-186 C. Philadelphia, PA: Defense Personnel Support Center, Directorate of Medical Material.

Guazzo, D. M. 1994. In *Parenteral quality control,* 2nd ed, edited by M. J. Akers. New York: Marcel Dekker, Inc.

Jenkins, W. A., and K. R. Osborn. 1993. *Packaging drugs and pharmaceuticals.* Lancaster, PA: Technomic Publication Company, Inc.

Korczynski, M. S. 1987. Evaluation of closure integrity. In *Aseptic pharmaceutical manufacturing: Technology for the 1990's,* edited by W. P. Olson, and M. J. Groves. Buffalo Grove, IL: Interpharm Press.

Levine, C. S. 1986. Validation of packaging operations. In *validation of aseptic pharmaceutical processes,* edited by F. J. Carleton, and J. P. Agalloco. New York: Marcel Dekker, Inc.

PDA. 1983. *Aspects of container/closure integrity,* Technical Information Bulletin No. 4. Bethesda, MD: Parenteral Drug Association.

Physicians' Desk Reference, 48th ed. 1994. Montvale, NJ: Medical Economics Data Production Company.

Sharp, J. 1990. Aseptic validation of a form/fill/seal installation: Principles and practice. *J. Paren. Sci. Tech.* 44 (Sept–Oct): 289.

Sharp, J. 1987. Manufacture of sterile pharmaceutical products using blow-fill-seal technology. *Pharm. J.* 239 (6441):106-108.

Technical Literature. Lionville, PA: The West Company.

Use of aseptic processing and terminal sterilization in the preparation of sterile pharmaceuticals for human and veterinary use. 1991. *Federal Register* 56:51354.

USP. 1995. *USP 23/NF 18 and supplements.* Rockville, MD: United States Pharmacopeial Convention.

USP/DI. 1994. Rockville, MD: US Pharmacopeial Convention, Inc.

Workshop on pharmaceutical packaging: Issues and challenges for the 90's. 1993. Arlington, VA: American Association of Pharmaceutical Scientists.

14

Barrier Isolation Technology: A Systems Approach

Jack P. Lysfjord
Paul J. Haas
Hans L. Melgaard
Irving J. Pflug

Barrier isolator technology is the culmination of a long evolution in pharmaceutical manufacturing. Pharmaceutical manufacturing began with open-manual production (Figure 14.1) and progressed to glove boxes. World War II saw developments of both drugs and blood products as well as clean rooms. Sandia Laboratories (Albuquerque, NM) originated the clean room concept, which became common practice for atomic energy applications; microelectronic applications; and pharmaceutical production in the 60s, 70s, and 80s. One of the first steps taken by the pharmaceutical industry in moving toward barrier isolator use was by SmithKline Beecham (Philadelphia, PA) in the mid 1980s, with the application of filling lines with walls built through them to separate the clean room from the gray side maintenance area (Figure 14.2). Dr. Willie Lhoest of Belgium was one of the original promoters of gray side maintenance. Taking this concept a step further encloses the filling line totally and removes the operator from the open vial area. Up to this point in time, the limited number of applications of barrier isolators have not been very user-friendly.

Barrier isolators for sterility testing have been in use for the past several years. The applications of barrier isolation to production systems for pharmaceuticals have been few, but use will be increasing rapidly. The difficult task is to create a sealed enclosure, sterilize it, bring sterile components (such as vials and stoppers) inside, bring

Figure 14.1. Early, manual pharmaceutical filling (*Note: degree of product protection*). Photo courtesy of The Upjohn Company.

Figure 14.2. Clean room pharmaceutical filling with a wall for separating the clean room from "gray side maintenance" area. Photo courtesy of SmithKline Beecham.

sterile product in, and have the complete package leave the barrier isolator without contaminating the interior of the barrier isolator. It is desirable that this be done at an increased sterility confidence level (SCL) while being physically located in a nonclassified, but controlled environment. A continuous process system is actually not sealed, but controlled by overpressure zones. This is a challenge in itself, but validation of the system is an even greater challenge.

It has been known for a long time that people represent the largest source of bioburden or viable particles in the classical clean room operation. The barrier isolator approach separates people from the critical processes that are done with the open vial. It can also separate the operator from hazardous products. Numerous papers have been written demonstrating an improved SCL (reduced contaminants) by keeping people away from open vials prior to stoppering. The Upjohn Company presented data at an FDA open forum in October 1993 that showed media fills in excess of 100,000 without a positive growth were done with an "open" (not sealed) barrier located within a Class 100 clean room. A barrier isolator allows for easier monitoring of parameters as well as providing a consistent environment—an analogy could be made to a product tank or an autoclave.

Reduction in cost, using barrier isolators, is a significant factor given the atmosphere of medical cost containment in the industry. Merck has presented numbers indicating that capital cost savings of 50 percent to 70 percent can be achieved with barrier isolators over conventional clean room facilities of the same capacity. Operating cost savings come from energy savings due to reduced volume of sterile area enclosed, reduced gowning estimated at $50,000 per operator per year, and better utilization of operators. Upjohn expects three times the people utilization when barrier isolators are used since people can cover multiple lines. (Figures 14.3 and 14.4 show comparable facilities, conventional clean room versus barrier isolator approach.)

Operator protection when potent compounds are being filled is of great importance. Barrier isolation is the most cost-effective way of dealing with potent compounds.

In the United States, as in Europe, the objective throughout the pharmaceutical manufacturing industry is to produce product where the SCL is less than 10^{-6} (fewer than one unit in one million units are nonsterile). Only a process that has positive control of the critical points of contamination can be validated to produce a SCL of 10^{-6}. It is impractical to prove a SCL of 10^{-6} by testing the product.

There must be a processing system where the integrity that prevents microbial contamination is controlled in a positive manner. It must be similar to an autoclave sterilization process in that it can be truly validated. It is only when a validated process is carried out

Figure 14.3. Layout for four 300-per-minute vial processing lines in a conventional clean room format. Length = 140 ft; width = 80 ft; total area = 11,200 sq ft; total Class 100 area = 2,464 sq ft.

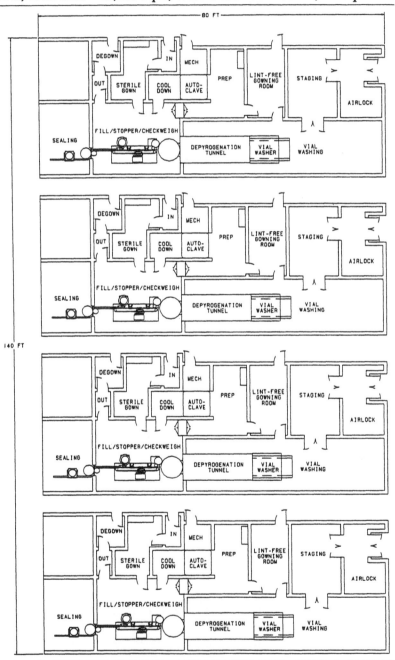

Figure 14.4. Layout for four 300-per-minute vial processing lines utilizing barrier isolator technology. Length = 89 ft; width = 68 ft; total area = 6,052 sq ft; total Class 100 area (prep and barrier only) = 688 sq ft; barrier area only (shaded) = 292 sq ft. (*Note: reduced facility size and reduced Class 100 area compared to Figure 14.3*).

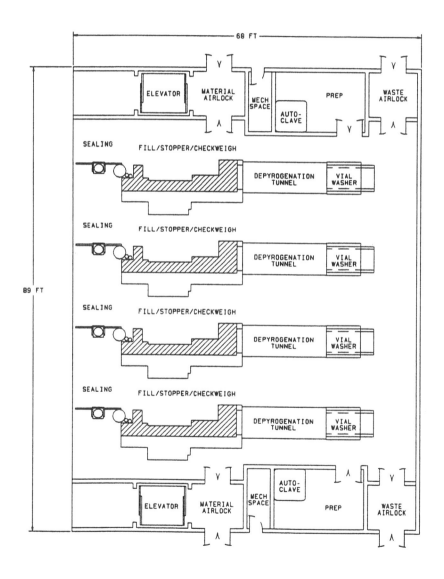

according to the established protocol that the final product will have a SCL of 10^{-6}.

Two very important words are *control* and *validation*. *We cannot validate a process where we do not have control of the process variables.* It is only with a barrier isolator system that one has the control necessary to even consider validating the system to a SCL of 10^{-6}.

In the aseptic assembly of pharmaceutical products in the mid 1990s, containment systems and new, small, barrier isolator systems are used. When a containment-type system (such as gloves, masks, and bunny suits) is used, there is a structure between the sterile area and the nonsterile person or element that will resist but, in general, will not prevent microorganisms from moving from the nonsterile person to the sterile product. There is general agreement in the industry that one cannot truly validate a clean-room production facility that contains a person in protective clothing (in a bunny suit) the way that one can validate an autoclave. To validate to a SCL of 10^{-6} there must be a true, positive barrier isolator to limit the movement of contamination.

A *barrier isolator system* is defined as a system having a positive barrier between the sterile area and the nonsterile surrounding area. It is necessary to distinguish between the different kinds of barrier isolator systems: *flexible barrier isolator systems* are those systems that have flexible components (e.g., gloves, flexible isolators, half suits, and similar systems); *rigid barrier isolator systems* are systems that have rigid walls where pressures can be produced and maintained that can be used as an input for an integrity control system.

The general goals for a barrier isolator system are to

- Protect the product from line operators. This means that we must remove people from the open-vial filling area and maintain an aseptic assembly environment through the use of HEPA[1]– and ULPA[2]–filtered air that will produce a probability of contamination of a product unit with a SCL of less than 10^{-6}.

- Protect the operators from new, very toxic, pharmaceutical products. This means that there must be a positive barrier isolator between the toxic product and the people.

---

[1]A HEPA (high efficiency particulate air) filter is an extended-media dry-type filter in a rigid frame having minimum particle-collection efficiency of 99.97 percent for 0.3 μm thermally generated dioctylphthalate (DOP) particles or specified alternative aerosol, and a maximum clean-filter pressure drop of 2.54 cm (1.0 in.) water gauge, when tested at rated airflow capacity.

[2]A ULPA (ultra low penetration air) filter is an extended-media dry-type filter in a rigid frame having minimum particle-collection efficiency of 99.999 percent for particulate diameters >0.12 μm in size.

- Reduce the overall costs of producing aseptically assembled products.

- Design equipment systems that are flexible and user-friendly.

A parenteral product filling line in a barrier isolator is a system that is so designed and constructed that the manufacturing and packaging operation will produce pharmaceutical products with a microbiological SCL of 10^{-6}. To do this requires that the following three activities be carried out successfully.

1. The system has the ability to transfer and deposit product in the package and close the package so the SCL of the unit of product is 10^{-6}.

2. An active, continuous control system is in place so that if there is a problem, it will first alert the operators and, secondly, will cause production to stop if critical control points necessary to produce or maintain a SCL less than 10^{-6} fail or move out of specification.

3. There is an acceptable validation and certification program. Validation is the carrying out of a series of tests that substantiate that the specific process or activity will produce the required result; certification is assembling the results and accumulated reports of the validation program in a package where the responsible persons in the company review and then approve or sign off on the program.

To produce an aseptically assembled product that has a SCL of less than 10^{-6} in a system using barrier isolation technology requires the following parts and inputs to the barrier isolation enclosure.

1. The product, as it arrives for packaging, must have a very low microbial load so that after packaging, the product unit SCL will be less than 10^{-6}.

2. Packaging in glass vials

   a. Glass vials will arrive at the barrier isolator from the discharge end of a continuous-flow glass sterilization-depyrogenation unit.

   b. Closures must have a SCL of less that 10^{-6} when they enter the isolator.

   c. There is a rigid enclosure unit of stainless steel and glass, with opening lockable doors for assembly and disassembly of equipment. These doors can be sealed so that the isolation enclosure becomes, in effect, a low pressure vessel. The inside of the enclosure is sterilized prior to start-up. It will remain sterile because it will be pressurized with HEPA– and/or ULPA–filtered input air.

3. Microbial control and validation of the barrier isolator

   a. Validate that through wipe down, clean-in-place (CIP), decontamination, and/or sterilization before start-up, the microorganisms inside the barrier enclosure are killed. Such validation provides assurance that at start-up there is less than one viable microorganism per 10 square meters of surface and that there is a validated HEPA– or ULPA–filter system in place that will provide air to the barrier isolator with a very low viable particle count per cubic meter of air.

   b. A validated control system will alert operators and stop production if the barrier isolator integrity is breached during production.

   c. A validated particulate measurement-control system will continuously sample the air at the container opening level and will alert operators and halt production if the particulate count in the air at the filling nozzle/top of open-vial level exceeds the established control point.

When these requirements are examined, we see that the first item is part of the pharmaceutical product manufacturing operation; we assume that is being taken care of since, at the present time the general assumption is that the product has a SCL of 10^{-6} when it enters the manufacturing area. The second item is also taken care of in today's manufacturing system, where glass containers, sterilized in a sterilization-depyrogenation tunnel, arrive at the isolator in a sterile condition; closures are presterilized and, therefore, arrive at the isolator in a sterile condition.

The following major topics areas follow:

1. The System
2. Filler Design
3. The Barrier Isolator
   A. Materials Selection and Compatibility
   B. Interface Issues
   C. Handling Freeze-Dried Products
   D. Particulate Control Considerations
   E. Barrier Isolator Internal Condition Control and Monitoring

4. Clean-in-Place of Barrier Isolators
5. Sterilization of Barrier Isolators
    A. Atmospheric Steam/Hydrogen Peroxide Sterilization System
6. Validation Considerations

THE SYSTEM

A systems approach to barrier isolator technology is comprised of the filler, the barrier isolator, and the sterilization process along with control of critical parameters that allow the system to be operated under control and be validated.

FILLER DESIGN

The filler design forms the foundation for the system and is the basis for the potential success of the system from an ergonomic standpoint. The concept for the design of the filler must be resolved first. Whether the line is rotary or linear (Figure 14.5) must be resolved along with indexing motion vs continuous motion. A rotary machine is typically much wider and has more potential for restricted access when compared to a linear system. Speed usually dictates whether the line has indexing or continuous motion. Reliability, redundancy, and particulate concerns increase when talking about barrier isolators at higher SCLs; continuous motion can be done with a lighter duty machine, is gentler to sensitive protein products, and results in lower particulate generation due to reduced glass vial "clatter." The best approach is a linear continuous motion filling system.

The total system elements need to be considered. A production filling line in a barrier isolator, at improved SCLs, would have difficulty accepting glass in a batch process. An upstream vial washer and depyrogenation tunnel would be utilized upstream from the filler.

Often, an accumulator disc is utilized as a capacitor between the tunnel and the filler. A disc works well for this function, but does not give first in first out (FIFO) motion of vials. This gives a variable open time for a given vial prior to stoppering or a process that varies from vial to vial. Accumulation is necessary, but it must be done with FIFO flow in mind.

Filling needs to be done accurately, repetitively, and cleanly without drips. The fill mechanism must operate with low shear to prevent damage to protein-based products. Vials cannot tip and should be

Figure 14.5. Top view of rotary versus linear vial processing equipment for ergonomic considerations.

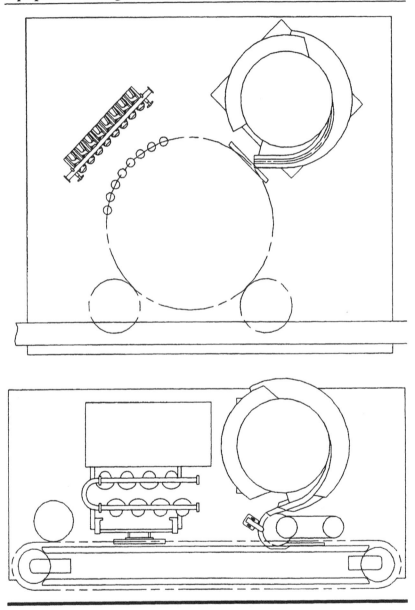

separated to prevent glass-to-glass particle generation. The filler should have rinse-in-place and steam-in-place (SIP) capability. Rinse-in-place is perhaps a more appropriate term than CIP when talking about rubber or plastic components at the molecular level.

With people removed from the interior of the barrier isolator, automated checkweighing is necessary to determine what fill volume each nozzle is actually producing.

A stoppering system that is capable of running both plug and slotted stoppers in appropriate sizes is necessary. The system should be capable of dealing with stoppers with reduced siliconization levels.

- Silicone is a particulate and transfers to the product.

- Protein products degrade with silicone.

The overall system must be easy to reach across and must have easy access through doors (with interlocks) for changeover or maintenance in the enclosed area. It is not realistic to enclose a system and never open the barrier isolator.

Other considerations for the system are as follows:

- Removing people from open vial area.

- Desired operating speed.

- Container sizes to be handled.

- Fill volume range in each container.

- Product cost.

- Product characteristics—foaming, viscosity, and so on.

- Unique process for product.

- Minimal enclosed volume inside the barrier isolator.

- Minimal mechanisms inside the barrier isolator.

- Barrier isolator should have CIP capability on the surface interior for potent compounds.

- System tolerance of various sterilization methods.

- Stopper reservoir.

- Easy maintenance.

- Ergonomics in every decision.

An example of the evolution of filler development for liquid products is shown in three phases. Phase I is a rigid barrier isolator placed over a conventional fill, checkweighing, stopper machine (Figures 14.6 and 14.7). Phase II removes many mechanisms from the enclosed volume and the width of the enclosure drops by 20 inches to 33 inches

Figure 14.6. Top view of conventional vial filling, check-weighing, and stoppering equipment with a barrier isolator installed. (TL Barrier Isolator, Phase I, 1991–early 1992)

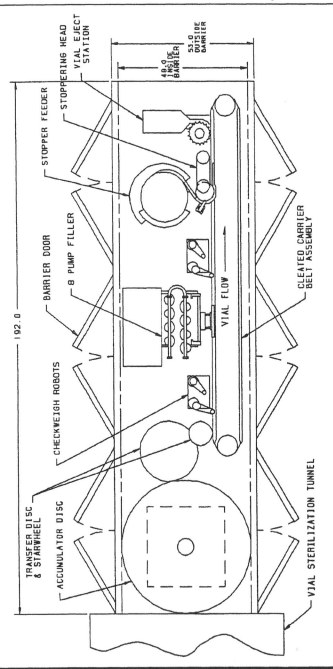

Figure 14.7. Section view of Figure 14.6 through filler. (TL Filler, Phase I) (*Note: outside width is 53 inches*)

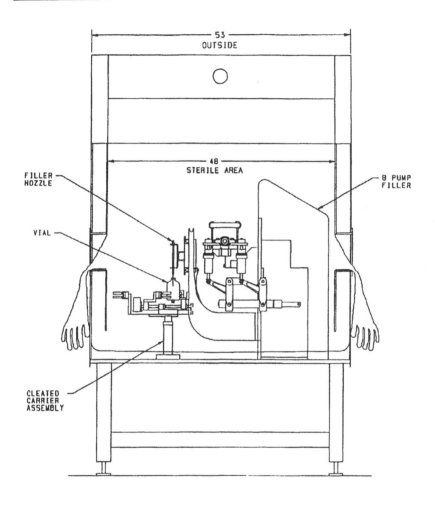

(Figures 14.8 and 14.9). Phase III (Figures 14.10, 14.11, and 14.12) is radically different, leaving only necessary components inside the enclosure, reducing volume, providing drains for CIP, and again reducing width by another 10.5 inches to 22.5 inches at the glove ports. TL Systems greatly appreciates the input of personnel from Eli Lilly & Company (Indianapolis, IN), The Upjohn Company (Kalamazoo, MI), and the Merck Manufacturing Division (West Point, PA) in this evolution.

Figure 14.8. Phase II top view of improved vial filling, check-weighing, and stoppering equipment that has been redesigned to move most of the fill mechanism below the tabletop and to narrow the equipment for ergonomic reasons. (TL Barrier Isolator, Phase II, mid 1992)

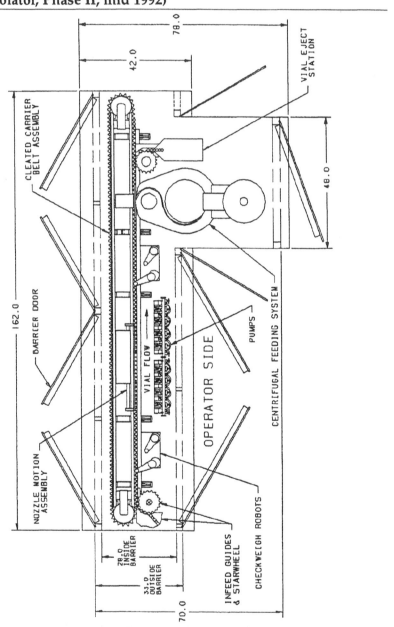

Figure 14.9. Section view of Figure 14.8 through filler. (TL Filler, Phase II) (*Note: outside width is 33 inches*)

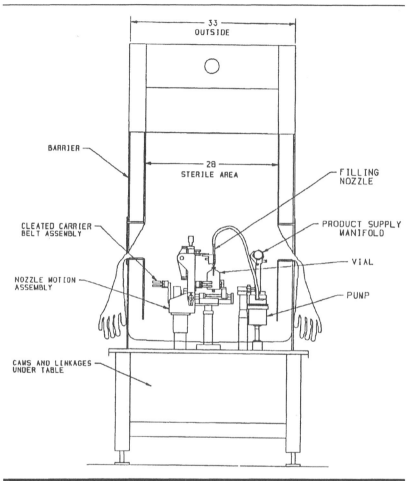

THE BARRIER ISOLATOR

The design of the filling machine forms the basis for the barrier isolator that is mounted to it. A rigid-walled barrier isolator is desired with construction of stainless steel and glass to minimize material that could absorb the sterilant used. Ergonomic considerations need to be taken into account to allow for easy use by the maximum number of potential operators. The design of the enclosure must be designed to properly interface and provide access to critical filler components. A study between the filler supplier, the barrier isolator suppliers, and the ultimate user is necessary to determine where

Figure 14.10. Top view of Phase III. Dramatic redesign with same functions as in Figure 14.6 and Figure 14.8. MAFS (Mini Aseptic Filling System)™ (patent pending)

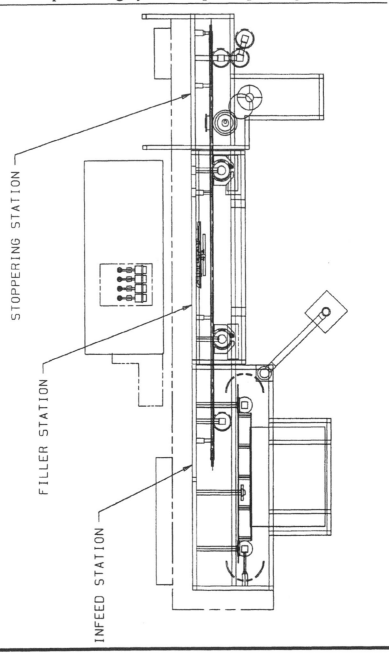

Figure 14.11. Section view of Figure 14.10 through filler. (TL Filler Phase III; MAFS™, patent pending) (*Note: width is 22.5 inches*)

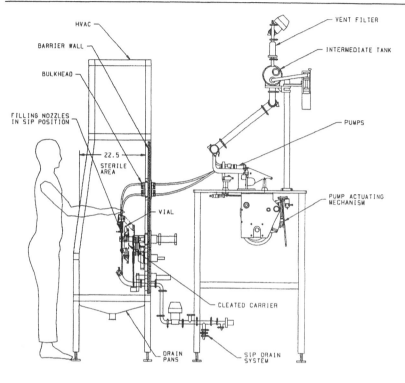

glove ports or manipulators are needed to allow for reliable operation of the system. Ergonomic modeling with computer simulation is a good place to start this evolution. (Figure 14.13 shows a computer model of accessibility.) A second model is necessary to look at airflow distribution. (A computer model of this is shown in Figure 14.14.) A third model that is necessary is construction of an actual mockup system to ensure no oversight occurs. This can be cardboard or plywood and Plexiglas™. The payback on this effort is manyfold. (An example is shown in Figure 14.15.)

Materials Selection and Compatibility

Filler and barrier isolator materials must be compatible with the CIP/SIP processes. With respect to the use of hydrogen peroxide (H_2O_2) as the sterilant, either in vapor form or in combination with

Figure 14.12. View of MAFS™ from operator side without barrier isolator (patent pending).

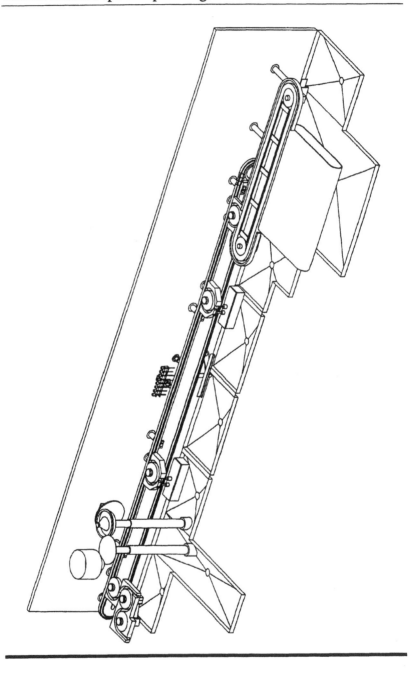

Figure 14.13. Computer simulation of ergonomic aspects of fill area.

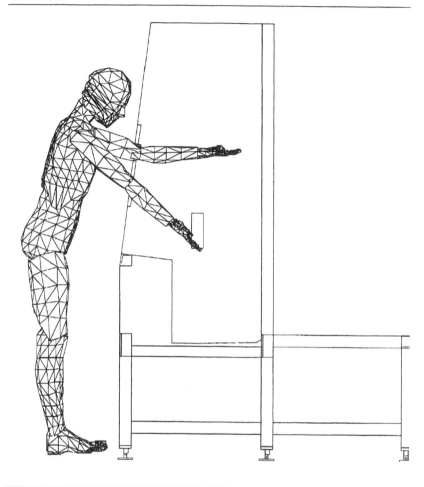

steam, there are a number of materials that are fully compatible with the use of the sterilant. These materials are also, in most cases, impervious to the penetration of H_2O_2. This is particularly true of metals. Plastics and elastomers would normally be held to minimal content to avoid any absorption and desorption questions with respect to the sterilant. In the case of the steam H_2O_2 system, the materials must also be designed to repeatedly withstand 100°C temperatures and 100 percent humidity conditions. See Table 14.1 for a chart of material compatibility with H_2O_2.

Figure 14.14. Computer simulation of airflow in fill area.

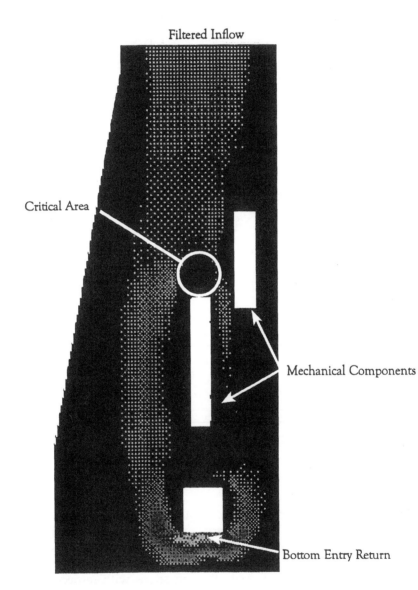

Figure 14.15. Cardboard and Plexiglas™ mockup for ergonomic evaluation.

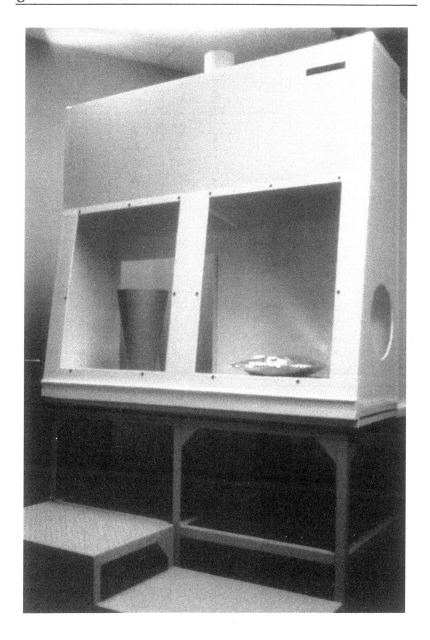

Table 14.1. Material Comparability with Hydrogen Peroxide

	Material	1%	2%	3%
Plastics	ABS Plastic	—	A	—
	Acetal (Delrin™)	C	D	D
	CPVC	—	A	A
	Epoxy	—	C	B
	Hytrel™	—	—	—
	LDPE	—	A	C
	NORYL™	—	A	A
	Nylon	—	C	D
	Polycarbonate	A	A	A
	Polypropylene	A	A	B
	PPS (Ryton™)	—	A	A
	PTFE (Teflon™)	A	A	A
	PVC	A	A	A
	PVDF (Kynar™)	—	A	A
	Buna N (Nitrile)	—	D	D
	EPDM	—	A	B
	Hypalon™	—	D	D
	Kel-F™	—	A	B
	Natural Rubber	—	B	C
	Neoprene	C	D	D
	Phar Med	A	A	A
	Silicone	A	A	B
	Tygon™	—	B	B
	Viton™	A	A	A
Metals	304 stainless steel	B	B	B
	316 stainless steel	A	B	B
	Aluminum	A	A	A
	Brass	D	—	—
	Bronze	—	B	B
	Carpenter 20	—	C	B
	Cast Iron	—	C	B
	Copper	D	—	D

Continued on next page

Continued from previous page

	Material	1%	2%	3%
	Hastelloy-C™	—	A	A
	Titanium	—	A	B
Non-metals	Carbon Graphite	—	C	C
	Ceramic Al$_2$O$_3$	A	—	—
	Ceramic Magnet	—	A	A

Ratings: A = No effect—Excellent
B = Minor effect—Good
C = Moderate effect—Fair
D = Severe effect—Not recommended
— = No data

Data from published material and exposure observation.

Trademarks: Delrin, Freon, Hypalon, Hytrel, Teflon, Viton are registered trademarks of E.I. du Pont de Nemours & Co.; Hastelloy-C is a registered trademark of Cabot Corp.; Kel-F is a registered trademark of 3M Co.; Kynar is a registered trademark of Pennwalt Corp.; NORYL is a registered trademark of General Electric Co.; Ryton is a registered trademark of Phillips Petroleum Co.; Tygon and Phar Med are registered trademarks of Norton Co.

Interface Issues

Several issues need to be addressed with any barrier isolator used with a high speed parenteral filling line. First and foremost, commodity entrance and product exit have to be properly engineered. The largest volume of material coming into a parenteral filling line is glass. The glass typically exits a depyrogenation tunnel with a pressure differential between the barrier isolator and the tunnel itself. The opening at the end of the tunnel may be as large as the entire width of the infeed tunnel, or there may be a transfer within the cooling zone of the tunnel to feed the glass single file into the barrier isolator. The main concern here is establishing the flow required to maintain a pressure differential between the filling area and the tunnel exit.

The next major consideration is the entry of stoppers and other commodity products. Some lines have been designed to interface the barrier with a batch stopper processing system. These involve the bulk transfer of large quantities of sterilized stoppers into the barrier isolator through a sterilized or decontaminated interface port. Another method of transporting either commodity materials or filler components into the barrier enclosure is through a rapid transfer port (RTP). Such ports allow for an aseptic or sterile transfer of material from the outside environment to the inside of the barrier isolator without exposure to contaminates. (An RTP interface is illustrated in Figure 14.16.)

Recent concern has been raised relative to the line (ring of confidence) interface formed by the inner and outer door seals of this type of transfer. Work is currently under way to provide a fully sterilizable line interface between these components. (An example of a fully automated transfer port that can be dry heat sterilized is shown in Figures 14.17 and 14.18).

Further examples of transfer mechanisms include the double-ended autoclave and the high-intensity ultraviolet pass-through tunnel. The ultraviolet pass-through tunnel allows for surface sterilization of prepackaged components required to be transferred into the barrier isolator. (An example of the UV pass-through is shown in Figure 14.19).

Figure 14.16. Rapid transfer port with transfer container.

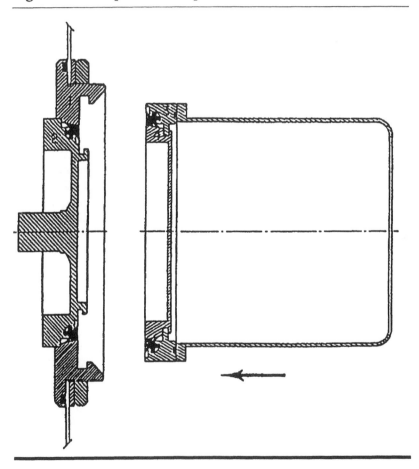

Figure 14.17. Rapid transfer port that can be dry-heat sterilized. Illustration courtesy of Central Research Laboratories (patent pending).

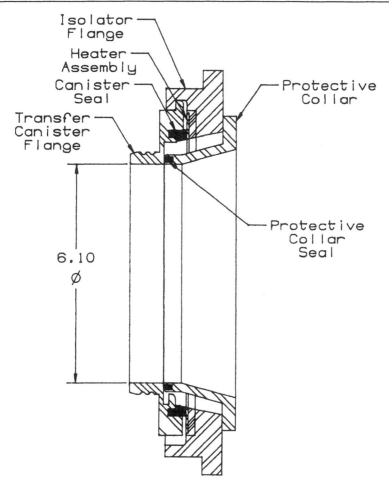

The product exit is the final consideration for interface with the barrier isolator. The product exit may be through a transfer star wheel system, or a small tunnel that feeds the finished product single file into the packaging section of the line.

Handling of Freeze-Dried Products

Products to be freeze-dried (lyophilized) can be handled in two ways. Vials can be tray loaded onto specially protected carts that would

Figure 14.18. Rapid transfer port cycle for connection, sterilization, opening insertion of protective collar, and transfer of components. Illustration courtesy of Central Research Laboratories (patent pending).

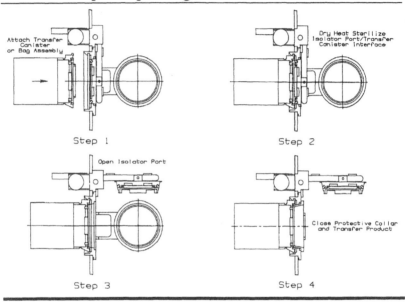

Figure 14.19. UV pass-through for component transfer.

attach to the output end of the barrier isolator utilizing the standard mechanical interface (SMIF) pod concept, as is used in the microelectronics industry. The other approach is to use a freeze dryer with an automated loading system in an uninhabited Class 100 clean room that mates up to the barrier isolator.

In the case of potent compounds and other materials that must be isolated from operators, it is possible to build bidirectional flow tunnels where the pressure inside the initial enclosure is positive with respect to the outside world and yet flows counter to a flow from the outside environment through a higher level negative located in the middle of the entry and exit tunnel (Figure 14.20).

Particulate Control Considerations

Particulate control is also critical in barrier isolator design. The most important consideration is the filtration of the outside air that is used as makeup air for the barrier isolator. All filter banks should have challenge and checking ports available for testing. Since viable particles may be present, the general technique used in barrier isolators is to recirculate the inside air through the HEPA filter, and to add filtered makeup air as required, to maintain the positive pressure inside the

Figure 14.20. Pressure diagram when potent compounds are considered.

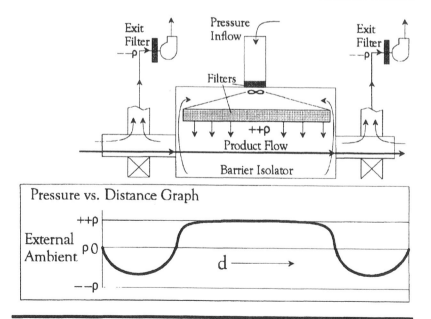

barrier isolator. (This is generally done in a fashion similar to Figure 14.21.)

In addition, the makeup air system is usually provided with one or two roughing filters to extend the life of the HEPA or ULPA filter used as the final filter on the makeup air. The probability of a particle passing through from the makeup air system into the isolator enclosure and finally through the recirculation filter on the isolator enclosure to the product environment below is shown in Figure 14.22. This is not to say that this is the probability of having a particle of any type at the level of the product. This is the probability of a particle making its way through the multiple levels of filters to reach the sterilized, inside critical area. The probability of any particle reaching the critical area must be factored to include the probability of a viable particle reaching that area. Using the known references, the probability of a viable particle as opposed to an inert particle, is on the order of 10^{-5}. So, the probabilities shown in Figure 14.22 must be reduced by that probability, in order to correctly define the probability of a particle of a viable nature reaching the inside of the barrier from its external environment.

Barrier Isolator Internal Condition Control and Monitoring

The internal barrier isolator conditions are both controllable and measurable. This lends itself to a much higher level of repeatability than

Figure 14.21. The airflow into and inside the barrier isolator.

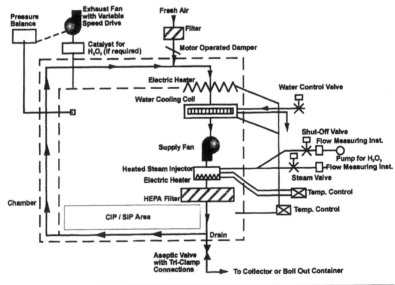

Figure 14.22. The level of particulates at different points in the barrier isolation airflow system and associated probabilities.

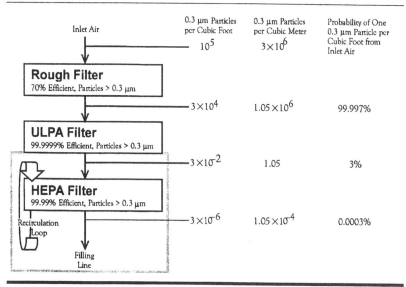

those systems that have random human presence and intervention. Key among the internal conditions is the internal pressure maintenance or the pressure within the barrier isolator, as opposed to its surrounding conditions. Incoming flow requirements of makeup gas into the system are based upon both a cross-sectional open area and the pressure differential between the barrier inside and its surrounding area. Typically, pressure differentials, as used in the past, between filling sites and surrounding areas have been on the order of magnitude of 0.05 of an inch water column. The following calculations show the flow through an open orifice of large size due to this magnitude of pressure differential.

Calculation of the velocity required to maintain a specified pressure differential across an opening starts with the Bernoulli equation for steady flow:

$$p = \left(\frac{V^2}{2g}\right)\left(\frac{\delta}{5.192}\right) = \left(\frac{V}{1096.7}\right)^2 \delta$$

where p is the pressure in inches of water column, V is the velocity in ft/min, and δ is the density in lb/ft^3.

$$V = 1096.7\left(\frac{p}{\delta}\right)^{0.5}$$

Air at standard temperature and pressure (STP) is .075 lb/ft³, therefore

$$V = 4 \times 10^3 \left(\frac{p}{\delta}\right)^{0.5}$$

$$Q = V \times A$$

Example: If the pressure difference across the wall of the sterile enclosure is 0.05 inches and A is 0.25 ft², then

$$V = 0.5 \times 4 \times 10^3 = 8.94 \times 10^2 \text{ ft/min}$$
$$Q = 8.94 \times 10^2 \text{ ft/min} \times 0.25 \text{ ft}^2 = 223 \text{ ft}^3/\text{min}$$

This flow (Q) is with containers present and approximates the gross airflow requirements to maintain the pressure differential between the sterile enclosure and the ambient environment.

One of the questions raised with respect to barrier isolators is the ability to stay at a positive pressure condition when gloves are rapidly removed from the enclosure. Figure 14.23 shows an experiment that was conducted with the removal of two full-size gloves from an

Figure 14.23. The barrier isolator pressure near the exit when two hands were withdrawn from the glove ports. Glove volume = 404 cu. inches; chamber volume = 102,214 cu. inches.

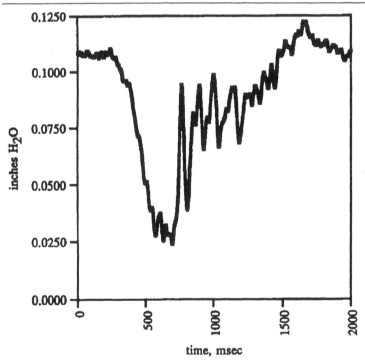

enclosure with normal makeup air fed into the enclosure. The pressure measurement was made close to the exit hole of the enclosure and shows that at no time, even with rapid removal, was there a pressure reversal for the inside of the enclosure transition from positive to negative.

CLEAN–IN–PLACE OF BARRIER ISOLATORS

Any barrier isolator must be able to be opened. Following the opening of a barrier isolator, a CIP and SIP process must be completed to render the inside of the barrier sterile for filling operations. The CIP cycle may also be done to decontaminate potent drugs prior to opening the barrier isolator.

The CIP process can be operated with several different approaches. The best approach is spray balls with full pattern coverage. Filter protection with some type of screen is required to prevent wetting of the HEPA filters. Rounded corners in all cleanable areas and proper drain location assist in completing the removal of material from within the barrier isolator.

STERILIZATION OF BARRIER ISOLATORS

Sterilization in place of the barrier isolator can be accomplished through a number of different methods. The method currently used on sterility test isolators has largely been a room temperature vaporized H_2O_2 sterilant. Despatch Industries and TL Systems have recently developed a sterilization process consisting of saturated steam at atmospheric pressure with H_2O_2 for use in barrier isolators. Other methods, such as ozone, ethylene oxide, or raw steam, have also been used.

Atmospheric Steam/Hydrogen Peroxide Sterilization System

The atmospheric steam with H_2O_2 sterilization system was developed as part of the systems approach to barrier isolation that would provide a robust sterilization process that could be completed in two hours or less. The sterilization cycle is described in Figure 14.24.

The test apparatus used to gather basic data on the destruction of microorganisms is shown in Figure 14.25. D values as a function of H_2O_2 concentration are shown in Figure 14.26. D values for four bacterial spores are shown in Figure 14.27. Figure 14.28 shows the process time with H_2O_2 at various concentrations. A prototype barrier isolator of 50 cubic feet in volume was constructed to do thermal imaging and other microbiological tests (Figure 14.29). The coldest location was

Figure 14.24. Steam hydrogen peroxide sterilization process cycle.

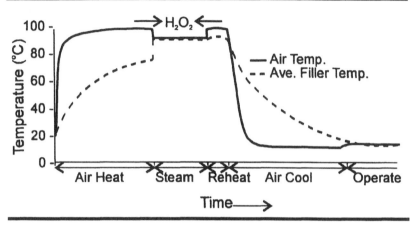

Figure 14.25. Diagram of atmospheric steam hydrogen peroxide test apparatus.

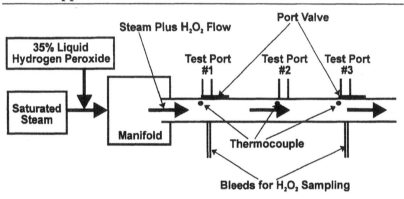

used to place planchets with *Bacillus stearothermophilus* spores to evaluate the process. This is shown in Figures 14.30 and 14.31. With the system cycle of dry preheat, steam, H_2O_2 with steam, steam, evaporation, and cooling, total SIP time would be two hours or less and would be very effective on *Bacillus stearothermophilus* spores with a concentration of H_2O_2 from 0.25 percent to 1 percent.

It is important as part of the cycle to verify the sterilant concentration. A method provided to do that is shown in Figure 14.32 (patent pending). This method works with the steam H_2O_2 sterilant to give a repeatable indication of concentration. It is also necessary to ascertain residual levels and several pieces of equipment are available to accomplish this. Following the completed sterilization cycle, removal of

Figure 14.26. *D* value vs. hydrogen peroxide concentration for spores on planchets in the steam plus H_2O_2 system: Spore and planchet received heat treatment immediately before testing, which is similar to the condition that will exist when steam plus H_2O_2 is used to sterilize a barrier enclosure.

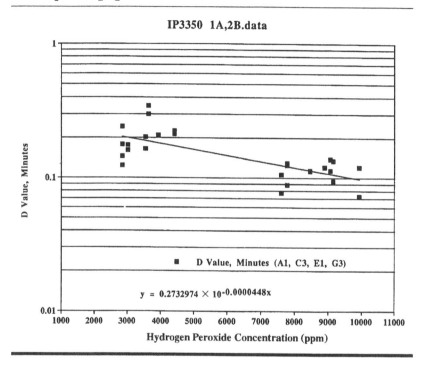

IP3350 1A,2B.data

D Value, Minutes (A1, C3, E1, G3)

$y = 0.2732974 \times 10^{-0.0000448x}$

Hydrogen Peroxide Concentration (ppm)

all sterilant from the enclosure needs to be well documented. The normal level to be reached before initiation of filling, using H_2O_2 would be less than 1 ppm residual.

Validation Considerations

In order to ensure that the conditions seen by the process lines are repetitive, conditions must both be controlled and monitored. The controlled conditions include the pressure differential between the inside of the barrier isolator and the outside. This can be done by utilizing a pressure transducer feeding to a variable frequency drive package, as illustrated in Figure 14.33.

The barrier isolator CIP/SIP cycle also needs to be controlled. The CIP cycle control would include the length of time and fluid volume passed into the enclosure during the spray operation. The SIP cycle

Figure 14.27. *D* **values for four species of spores subjected to steam plus 2500 ppm hydrogen peroxide in the tube system.**

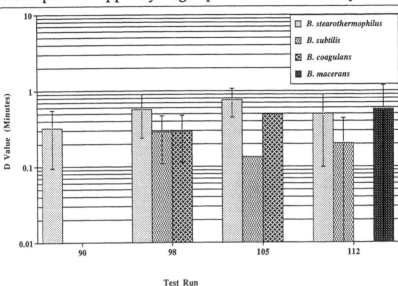

Figure 14.28. Process time for a 12-log cycle microbial control process using atmospheric steam plus hydrogen peroxide at various concentrations.

	Process Time (Minutes)
750 ppm	2.4
2500 ppm	5.7

will depend on which sterilant is used. Most sterilants require preconditioning, which involves either temperature or humidity control or both. Following the preconditioning, sterilant flow and time need to be controlled. The sterilant removal cycle also needs to be a controlled parameter.

The physical locking of the access doors into the enclosure must be assured during and following the sterilant cycle. Verification of enclosure integrity can be accomplished through air pressurization and decay testing, with all openings closed. The measured variables during the operation of the barrier isolator enclosure would include pressure differential to the outside environment, sterilant flow, quantity and duration for the sterilization cycle, and sterilant concentration.

For the sterilant concentration, as noted earlier, Despatch Industries has developed an instrument for monitoring the concentration of

Figure 14.29. Prototype barrier isolator for thermal imaging and running microbiological tests.

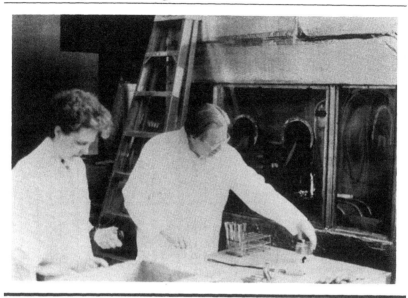

Figure 14.30. Planchet location diagram. Vertical longitudinal cross-section through the prototype barrier isolator enclosure showing the inlet for the steam or steam plus hydrogen peroxide, the HEPA filter bank, the location of the inoculated planchets used in evaluating the system, and the exhaust area.

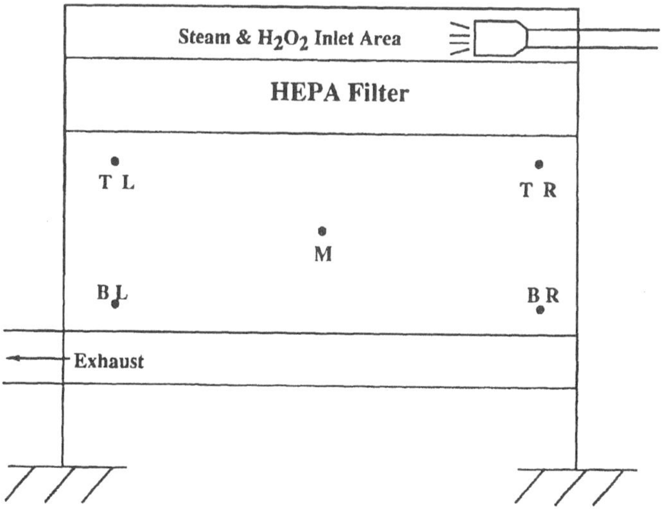

Figure 14.31. Test results on planchets from a prototype barrier isolator.

Results of a series of tests where *Bacillus stearothermophilus* spores PB27CT deposited on stainless-steel planchets were subjected to a steam hydrogen-peroxide (H_2O_2) atmosphere at an H_2O_2 inflow rate of 150 ml/m² of isolator cross-section area for 6 and 12 minutes.

In the initial experiments, there were two replicate planchets at each location. Halfway through the project, we increased to three replicate planchets per location. (Empty boxes are conditions not tested.)

| Test ID | Initial Number | 900 ml H_2O_2/m² 150 ml H_2O_2/min m² for 6 min | | | | | | 1800 ml H_2O_2/m₂ 150 ml H_2O_2/min m² for 12 min | | | | | |
| | | BL[1] | | | BR | | | BL | | | BR | | |
		P1[2]	P2	P3	P1	P2	P3	P1	P2	P3	P1	P2	P3
CM2196	5.33E+6	1820	2850		5	453		0	0		0	0	
CM2204	6.00E+6	597	14340		0	0		0	0		0	0	
CM2211	4.92E+6							0	0		2	2	
CM2217	1.49E+7	323	359		0	0		0	0		0	0	
CM2225	1.45E+7	1507	2977		0	27		0	25		0	0	
IP2232[3]	1.15E+6	0	0		0	0		0	0		0	0	

Continued on next page

Continued from previous page

Test ID	Initial Number	900 ml H_2O_2/m^2 150 ml H_2O_2/min m^2 for 6 min						1800 ml H_2O_2/m_2 150 ml H_2O_2/min m^2 for 12 min					
		BL[1]			BR			BL			BR		
		P1[2]	P2	P3	P1	P2	P3	P1	P2	P3	P1	P2	P3
IP2240[4]	1.54E+7	0	2251		0	0							
IP2245[4]	1.39E+7	7152	11474	17860	280	3895	6885						
IP2253A[5]	1.27E+7	3	5	535	0	1	54	0	12	0	0	0	0
IP2253C	1.27E+7	1	1	19	1	29	304						
IP2260	1.34E+7	63	1911	5615	359	1089	1214						

[1] Data are for the bottom ⬚ the barrier isolator column. BL is the bottom left and BR is the bottom right. See Figure 14.5.
[2] P1, P2, and P3 are replicate planchets.
[3] In Experiment IP2232, the N0 was 1.15E+6, whereas in adjacent experiments N0 was one log higher.
[4] In Experiments IP2240 and IP2245, the 900 ml of H_2O_2 was delivered in 12 instead of 6 minutes.
[5] In Experiment IP2253, the 6 minute experiment was repeated two times: A and C.

Figure 14.32. The hydrogen peroxide concentration measuring system (patent pending).

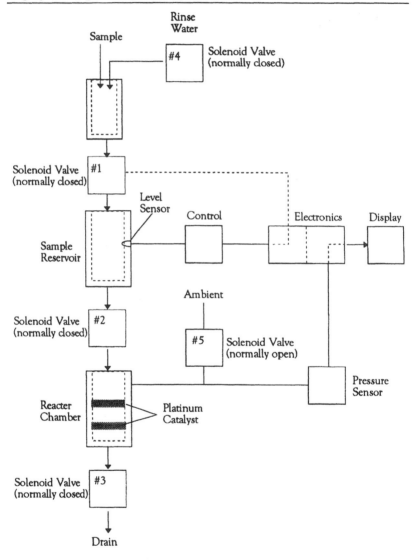

H₂O₂ during the steam/H₂O₂ cycle that has been shown to be consistent with titration measurements. It is also quick, taking under one minute to complete. A machine for doing this is described in Figure 14.32.

Measured variables may also include particulate levels at the filler elevation. Experiments have shown that following the opening of a

Figure 14.33. The pressure balance control system.

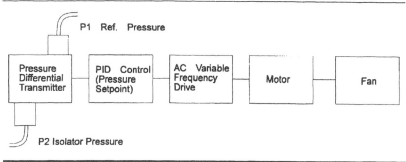

barrier isolator system, the makeup and recirculation air brings the particulate level at the filler elevation to within Class 100 conditions in less than 10 minutes of sealed operation. Class 10 levels are achieved within 30 minutes of closing the system. These measurements were made while operating the barrier isolator in an uncontrolled environment with particulate levels of 10^5 0.5 μm particles per cubic foot. Figure 14.34 shows the barrier isolator on which these tests were conducted.

Figure 14.34. The barrier isolator operating in the unclassified environment.

Figure 14.35. The complete system: washer, tunnel, and barrier isolator with filler, checkweighers, and stoppering.

SUMMARY

It is a giant step to go from the present Class 100 clean room technology to the new system shown in Figures 14.35, 14.36, 14.37, and 14.38. The major issues are ergonomics, particulates, and increasing the product SCL from 10^{-3} to 10^{-6}. Simply placing a barrier isolator over an existing conventional filler will not achieve the goals.

A holistic approach is required when designing a pharmaceutical product filling line in a barrier isolator to achieve the important goals of

- A product SCL of 10^{-6}.

- Reduced overall costs

- Protection of operators from potent compounds

- A flexible and user-friendly system

It is truly a paradigm shift to design, build, control, validate, and operate a system such as this with the absence of people from the open vial area.

Figure 14.36. Tunnel exit accumulation into infeed tunnel.

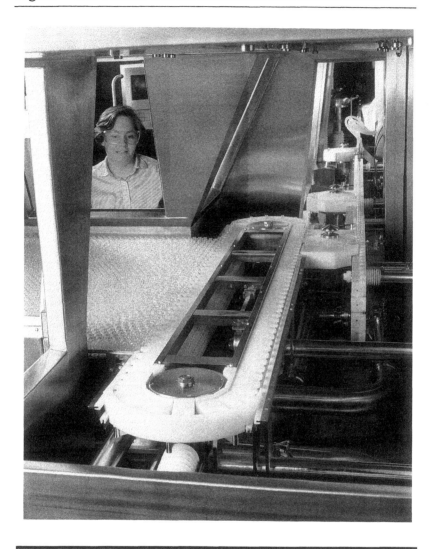

Figure 14.37. Stoppering area.

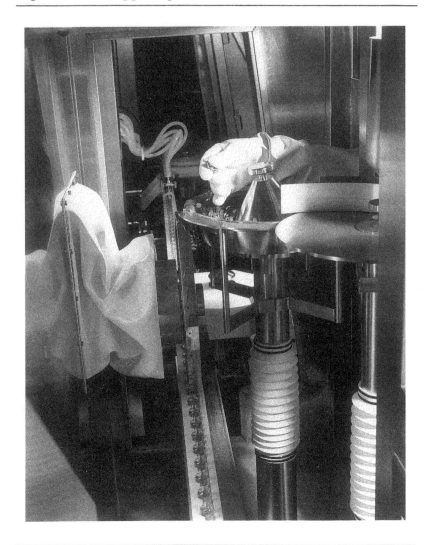

Figure 14.38. Mechanical access for maintenance from back side of barrier isolator.

RECOMMENDED READINGS

Akers, J. E., and J. P. Agallaco. 1993. Aseptic processing—a current perspective. In *Sterilization technology: A practical guide for manufacturers and users of health care products.* New York: Van Norstrand Reinhold.

Akers, J. E., and C. Wagner. 1994. *Barrier technology.* Third International PDA Congress and Workshops, 14–18 February, in Basel, Switzerland.

Block, S. S. 1991. *Disinfection, sterilization and preservation,* 4th ed., pp. 85–128. Philadelphia: Lea and Febiger.

Bradley, A., S. P. Probert, C. S. Sinclair, and A. Tallentire. 1991. Airborne microbial challenges of blow/fill/seal equipment: A case study. *J. Paren. Sci. Technol.* 45 (4).

Casamassina, F. J., J. W. Hulse, R. P. Tomaselli. 1993. Controlling medical product contamination. In *Sterilization technology: A practical guide for manufacturers and users of health care products.* New York: Van Norstrand Reinhold.

Curran, H. R., R. R. Evans, and A. Leviton. 1940. The sporicidal action of hydrogen peroxide and the use of crystalline catalase to dissipate residual peroxide. *J. Bacter.* 40:423–434.

Federal Standard 209D. 1988. *Clean room and work station requirements.* Washington, D.C.

Farquharson, G. 1994. Aseptic filling in network of rigid isolators. Third International PDA Congress and Workshops, 14–16 February, in Basel, Switzerland.

Frieben, W. R. 1993. Presentation at FDA Open Conference on Sterile Drug Manufacturing, 12 October, in Bethesda, MD.

Haas, P. J., H. L. Melgaard, J. P. Lysfjord, and I. J. Pflug. 1993. Validation concerns for parenteral filling lines incorporating barrier isolation techniques and CIP/SIP systems. PDA Second International Congress, 22–27 February, in Basel, Switzerland.

Hoffman, G. 1992. Presentation at Barrier Isolation Technology Conference, 26–27 August, in Minneapolis, MN.

Killick, P. F. 1992. Facility design—Isolation technology. Fourth International Congress of Pharmaceutical Engineering, 8–10 September, in Vienna, Austria.

Klapes, N. A., and D. Vesley. 1989. Vapor-phase hydrogen peroxide as a surface decontaminant and sterilant. *Appl. Environ. Microb.* 56:503–506.

Leaper, S. 1984. Comparison of the resistance to hydrogen peroxide of wet and dry spores of *Bacillus subtilis* SA22. *J. Food Technol.* 19:695–702.

Lewis, J. S., and J. R. Rickloff. 1991. Inactivation of *Bacillus stearother-mophilus* spores using vaporized hydrogen peroxide. Unpublished paper, presented at the 1991 ASM Meeting, in Dallas, TX.

Loy, L. H., and J. F. Melanhn. 1993. Current stopper processing methods and handling techniques. Barrier Isolation Technology Conference, 19–20 August, in Minneapolis, MN.

Lysfjord, J. P., P. J. Haas, H. L. Melgaard, and I. J. Pflug. 1993. The potential for use of steam at atmospheric pressure to decontaminate or sterilize parenteral filling lines incorporating barrier isolation technology. Presented at PDA Spring Meeting, 10 March, in Philadelphia, PA. (To be published in the *Journal of Parenteral Science and Technology.*)

Melgaard, H. L. 1994. Barrier isolation design issues. ISPE Barrier Isolation Technology Seminar, 12–13 May, in Philadelphia, PA.

Melgaard, H. L. 1989. The historic and current atatus of dry heat sterilization and depyrogenation. Presented at ISPE EXPO '89. (Unpublished report by Despatch Industries, Minneapolis, MN.)

Melgaard, H. L., and I. J. Pflug. 1993. Nature and quality of the air leaving the filters at the top of a barrier isolator. Unpublished.

NASA. 1968. *Standard procedures for the microbiological examination of space hardware.* National Aeronautics and Space Administration Document No. NHB 5340.1A. Washington, D.C.: Government Printing Office.

Peck, R. D. 1988. What will the new federal standard 209C mean? *Pharm. Eng.* 8 (2):17–21.

Peterson, A. 1994. Barrier isolator filler design issues. ISPE Barrier Isolation Technology Seminar, 12–13 May, in Philadelphia, PA.

Pflug, I. J. 1992. Microbiological testing program of the TL/Despatch barrier system. Barrier Isolation Technology Conference, 26–27 August, in Minneapolis, MN.

Pflug, I. J. 1990. *Microbiology and engineering of sterilization processes, rev. 7th ed.* Minneapolis, MN: University of Minnesota.

Pflug, I. J., A. B. Larson, and H. L. Melgaard. 1992. Barrier isolation system. (US Patent Pending.)

Pflug, I. J., H. L. Melgaard, C. A. Meadows, J. P. Lysfjord, and P. Haas. 1993. Rigid isolation barriers: Decontamination with steam and steam hydrogen peroxide. In *Sterilization of medical products*, Vol. VI, pp. 115–132. Somerville, NJ: Johnson & Johnson.

Pflug I. J., H. L. Melgaard, S. M. Schaffer, J. P. Lysfjord. 1994. The microbial kill characteristics of saturated steam at atmospheric pressure with 7,500 and 2,500 ppm hydrogen peroxide. Presented at PDA Spring Meeting, 10 March, in Chicago, IL. (To be published in the *Journal of Parenteral Science and Technology.*)

Rickloff, J. R. 1988. The development of vapor phase hydrogen peroxide as a sterilization technology. A report presented at HIMA Conference on Sterilization in the 1990's, 30 October–1 November, in Washington, D.C.

Rickloff, J. R., and P. A. Orelski. 1989. Resistance of various microorganisms to vapor phase hydrogen peroxide in a prototype dental handpiece/general instrument sterilizer. Presentation at the 89th Annual Meeting of the ASM, in New Orleans, LA.

Sinclair, C. S. 1993. Predictive sterility assurance for aseptic processing. Presented at the Kilmer Memorial Conference on the Sterilization of Medical Products, 13–15 June, in Brussels, Belgium.

Toledo, R. T., F. E. Escher, and J. C. Ayres. 1973. Sporicidal properties of hydrogen peroxide against food spoilage organisms. *Appl. Microb.* 26:592–597.

Whyte, W. 1994. The influence of clean room design on product contamination. *J. Paren. Sci. Technol.* 38 (3): 103–108.

15

Hydrogen Peroxide Vapor Sterilization: Applications in the Production Environment

Leslie M. Edwards
Robert W. Childers

INTRODUCTION TO HYDROGEN PEROXIDE VAPOR STERILIZATION

History

Although sterilization using gaseous agents has been practiced for over four decades, hydrogen peroxide vapor (H_2O_2) did not begin to gain acceptance as a sterilant until the 1990s. The sporicidal effects of H_2O_2 vapor, which were first discovered in 1979/1980, were overshadowed for years by the microbicidal effectiveness of liquid H_2O_2 solutions.

The first data on the sporicidal effects of H_2O_2 vapor that was available to the scientific community was contained in the patents that were published in the 1980s. This data suggested that sterilization could be affected after a few hours of exposure to sterilant vapors.

AMSCO acquired the rights to the patents and began to experiment with the technology. AMSCO discovered that "flash vaporization" could be utilized to produce higher concentrations of vapor than would exist naturally in the vapor space above liquid H_2O_2 solutions. Data obtained combining this vaporization method and multiple-exposure pulses in a single, subatmospheric pressure sterilization cycle indicated that sterilization could be accomplished with exposure times of less than one hour.

AMSCO submitted a sterilant registration application to the EPA in 1984 under the guidelines of FIFRA and then submitted a 510(k) for equipment utilizing the sterilant to the FDA in 1989. Equipment utilizing the process became commercially available in 1990.

Hydrogen peroxide vapor is quickly displacing ethylene oxide, peracetic acid (both mist and vapor), chlorine dioxide, glutaraldehyde, and formaldehyde vapor (generated from paraformaldehyde crystals by heating); it is becoming the gaseous sterilant of choice in many applications. Hydrogen peroxide vapor can affect sterilization at near ambient temperature conditions while using low concentrations (500 to 6000 ppm) of sterilant at pressures ranging from a vacuum to near atmospheric pressure. The residual by-products of H_2O_2 vapor are water vapor and oxygen.

Hydrogen peroxide vapor has been shown to kill a wide range of bacteria, bacterial spores, viruses, molds, and fungi. *Bacillus stearothermophilus* spores have demonstrated the highest resistance to the vapor of all the organisms listed in Table 15.1.

Theory

Sterilization Mechanism

The mechanisms for H_2O_2 vapor sterilization are as yet undefined, but literature suggests that the formation of hydroxyl free radicals and their subsequent exposure to the microorganism leads to its destruction (Sintim-Damoa 1993). It has been suggested that these highly reactive free radicals attach themselves to membrane lipids, DNA, and cell components (Ingraham 1992), although the specific mechanisms by which each occurs and which component reaction is the key to sterilization is unknown.

Studies on H_2O_2's effect on DNA suggest that the hydroxyl radical attacks pyrimidine bases and causes the DNA chain to break. Countering this effect in both eukaryotic and prokaryotic cells are several mechanisms that repair the DNA. Studies on *E. coli* by Pollard and Weber (1967) as well as those by Carlsson and Carpenter (1980) suggest that these cells would require a tremendous presence of hydroxyl radicals to demonstrate a significant loss of cell viability.

Kawasaki et al. (1970) demonstrated that enzymatic destruction was instrumental in contributing to a loss of cell viability. This study, and another by Heinmets et al. (1954) on *E. coli*, imply that the destruction of the enzyme required for cell metabolism brings about the loss of cell viability. It has been suggested that methionine (the initiating amino acid of protein synthesis) concentrations are decreased by an oxidative exposure to H_2O_2.

Table 15.1. Microorganisms Evaluated for Their Resistance to Hydrogen Peroxide Gas

Bacterial Spores	Viruses	Bacteria	Fungal Spores, Molds, and Yeasts
Bacillus cereus	Adenovirus 2	*Brevibacterium acetylicum*	*Aspergillus niger*
Bacillus macerans	Herpes simplex type I	*E. coli*	*Aspergillus terreus*
Bacillus pumilus	Influenza A2	*Lactobacillus casei*	*Candida parapsilosis*
Bacillus stearo-thermophilus	Polio type I	*Mycobacterium smegmatis*	*Fusarium oxysporum*
Bacillus subtilis	Rhinovirus 14	*Norcardia species*	*Penicillium chrysogenum*
Clostridium sporogenes	Vaccinia	*Proteus vulgaris*	*Rhodotorula glutinis*
		Pseudomonas aeruginosa	*Saccharomyces cerevisiae*
		Pseudomonas cepacia	
		Serratia marcescens	
		Staphylococcus aureus	
		Staphylococcus faecalis	
		Staphylococcus faecium	

Note: *Bacillus Stearothermophilus* is the most resistant microorganism tested to date.

Hydroxyl free radical exposure has been shown to cause a cell to lose its selective permeability by rupturing the cell wall (Miller 1969). Electron micrographs by Polyakov et al. (1973) supported this assertion by documenting cell wall damage to *E. coli* as well as granule formation in the cytoplasm. Other studies on *Bacillus subtilis* have demonstrated spore coat sustenance, but spore cortex degradation with exposure to H_2O_2 (Bayliss and Waites 1979). These studies suggest that physical damage to the cell wall and spore cortex are probable major modes of action for H_2O_2 attacks on microorganisms.

Factors Influencing Microbicidal Effectiveness

There are four major processing parameters that affect the inactivation of microorganisms by H_2O_2 vapor: vapor concentration, percent saturation (100 times the saturation ratio), temperature, and exposure time. The presence (or absence) of air or other inert gases does not affect the ability of the vapor to inactivate organisms; however, it can assist in, or obstruct, the delivery of the sterilant to the sites to be sterilized. In flow-through systems the air acts as a carrier to deliver the sterilant vapor to the sites to be sterilized. In systems with "dead legs," a vacuum is utilized to deliver the sterilant vapor.

Selectively increasing the concentration of the sterilant has been shown to directly increase the microbial inactivation rate for all vapor concentrations $(500 < C < 6000 \text{ ppm})$ tested thus far (Figure 15.1). The rate of microbial inactivation has also been shown to increase for a given H_2O_2 vapor concentration as the ratio of the actual concentration to the maximum allowable (dew point) concentration increases (Figure 15.2). The literature (Schumb et al. 1955) indicates that the presence of water vapor tends to stabilize the H_2O_2 vapor, which may explain why the rate of microbial inactivation increases for a given H_2O_2 vapor concentration as the water vapor concentration is increased.

Increasing the temperature without increasing the concentration of the H_2O_2 vapor decreases the overall microbial inactivation potential because it reduces the percent saturation and increases the rate at which H_2O_2 vapor breaks down into water vapor and oxygen. However, increased temperatures can be accompanied by higher vapor concentrations.

An increase in exposure time results in a corresponding increase in microbial inactivation if the other process parameters are kept constant. This is very easy to accomplish when utilizing a steady state flow-through, or recirculating, process that maintains the sterilant concentration at a specific level by continuously replenishing decomposing sterilant vapors with a fresh sterilant supply. This type of process was used to generate the data contained in Figures 15.1 and 15.2.

Figure 15.1. *D* values for H_2O_2 vapor at 80+ percent saturation versus H_2O_2 concentration.

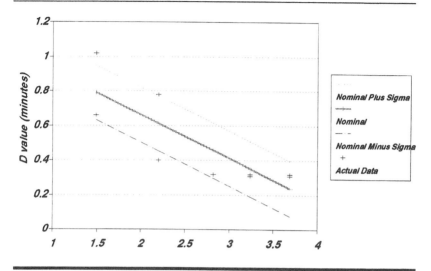

Figure 15.2. *D* values for a 1.6 mg/ℓ H_2O_2 vapor concentration versus percent H_2O_2 saturation.

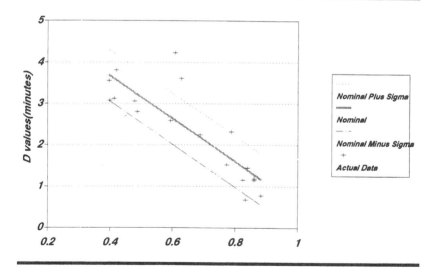

Subatmospheric pressure processes require a multiple-exposure pulse sterilization cycle similar to that shown in Figures 15.3(a) and 15.3(b), whereby doubling the exposure time requires doubling the number of exposure pulses. Merely doubling the exposure time during

Figure 15.3(a). Typical VHP®DV series H₂O₂ vapor steriliza-tion cycle—dry phase and leak test.

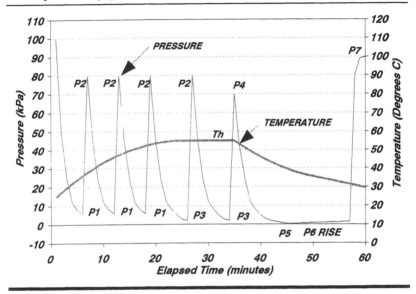

Figure 15.3(b). Typical VHP®DV series H₂O₂ vapor steriliza-tion cycle—sterilize and aerate phases.

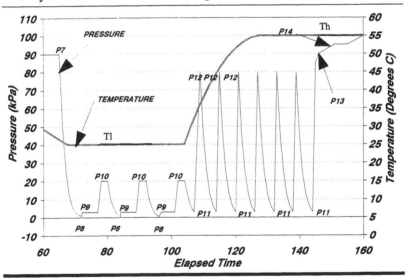

a single pulse will not double the microbial effectiveness of the pulse because the hydrogen H₂O₂ concentration can decrease significantly in a few minutes.

Sterilization Equipment and Processes

Commercially Available Systems

Even though H_2O_2 vapor technology is well established, only a hand-full of products have been fully commercialized. AMSCO has the greatest number of commercialized products as of this time, their first being the Flow Through VHP®1000 Generator (Figure 15.4), which is

---

Figure 15.4. AMSCO VHP®1000 biodecontamination system for surface sterilization using H_2O_2 gas.

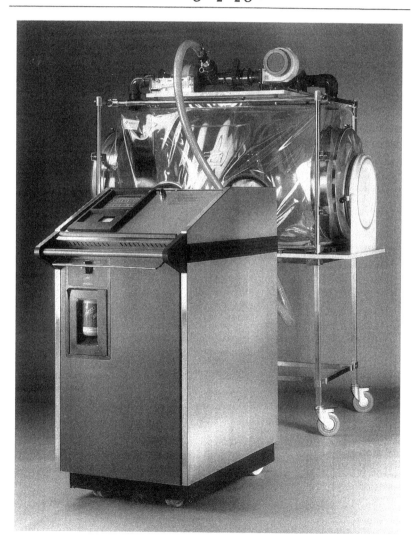

an atmospheric pressure surface sterilization system. It is used most often in "batch mode" processes, such as safety cabinet sterilization, pharmaceutical product quality control (QC) sterility testing, and air lock/pass-through sterilization in the pharmaceutical production environment. Its uses are expanding to include "continuous flow" processes as well.

The LiquiBox Pacesetter™ 2000 system (Figure 15.5) was built in conjunction with AMSCO as part of a technology licensing agreement. This FDA–approved food industry low acid aseptic packaging system contains a pressurized sterile tunnel through which the container spout and closure device must pass prior to and during the filling operation. As the container passes down the tunnel, the spout (and closure device) are sterilized using H_2O_2 vapor and are aerated before the container is filled and capped.

In 1992 the first VHP®DV 1000 (Figure 15.6) was delivered to The Upjohn Company as part of a corporate partnership for technology development involving the sterilization of freeze dryers. Since that time other companies have also retrofitted their large production-sized freeze dryers with this system. The primary advantages of this system

Figure 15.5. The LiquiBox Pacesetter™ 2000 aseptic packaging system for low acid foods utilizes a sterilization tunnel with H_2O_2 vapor.

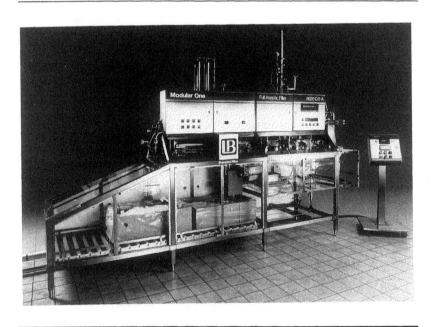

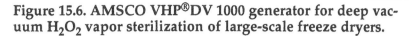

Figure 15.6. AMSCO VHP®DV 1000 generator for deep vacuum H_2O_2 vapor sterilization of large-scale freeze dryers.

are the rapid cycle times and, more importantly, the ability to sterilize the evacuated freeze dryer while at, or near, room temperature. The reduced temperature excursions will have a strong positive effect on the lifespan of the average freeze dryer (less metal fatigue when compared to steam sterilization) and will result in fewer leaks during lyophilization. The pharmaceutical product will have a reduced probability of being contaminated as a result.

The smaller version of the VHP®DV 1000, the VHP®DV 30 (Figure 15.7), is used primarily for pilot-sized freeze dryers. Because of

Figure 15.7. AMSCO VHP®DV 30 generator for deep vacuum H₂O₂ vapor sterilization of smaller-scale freeze dryers.

regulatory restrictions and labeling laws on products processed using ethylene oxide, as well as the thermal stresses applied by steam sterilization, the VHP®DV 30 promises to be a successful system. Generally, it is retrofitted into existing pilot-size freeze dryers, but it can be built into the AMSCO/Finn-Aqua GT Series freeze dryers (Figure 15.8).

The STERRAD™ Sterilization System (Figure 15.9) from Advanced Sterilization Products, a division of Johnson & Johnson, Inc. uses a patented process for its general purpose sterilizer involving the generation of a low temperature hydrogen peroxide–based plasma.

Figure 15.8. AMSCO/FINN AQUA GT10–VHP H$_2$O$_2$ sterilizable freeze dryer.

The STERRAD™ Sterilization System received its 510(k) clearance in October, 1993.

Plasma is a fourth state of matter as distinguished from solid, liquid and gas. There is a widespread occurrence of plasma in nature (principally in outer space), and increased use of the plasma state in various applications of modern technology. An example of a naturally occurring plasma is the aurora borealis or "northern lights."

The STERRAD™ System sterilizes objects using a cool, dry and environmentally safe process. Because of a complete absence of toxic

Figure 15.9. Surgikos STERRAD™ hospital general purpose H₂O₂ plasma sterilizer.

residues, aeration is not required. A single cycle takes just about an hour and the primary end products of the process are oxygen and vaporized water.

Peroxide Plasma as a Sterilizing Agent.   Items to be sterilized are placed in the chamber, the chamber is closed and a vacuum is created. An aqueous solution of hydrogen peroxide is vaporized within the chamber, surrounding the items to be sterilized.

A radiofrequency-induced electrical field accelerates electrons and other particles, which collide with each other. These collisions initiate reactions which generate hydroxyl-free radicals, hydroperoxyl-free radicals, activated peroxide, UV light, and other active species. Free radicals and other species in the plasma cloud interact with cell membranes, enzymes, or nucleic acids to disrupt the life functions of microorganisms.

At the end of the cycle, the active components recombine to form oxygen and vaporized water. There are no toxic residues, so aeration is not required. Because the primary end products of the STERRAD™ System sterilization process are water vapor and oxygen, there are no harmful emissions into the atmosphere.

Applications

The STERRAD™ Sterilization System sterilizes metal and nonmetal objects as well as many sophisticated heat- and moisture-sensitive items. By using low-temperature peroxide plasma, the STERRAD™ System sterilizes without the degrading effects of steam, which can dull sharp objects such as microsurgical instruments. The low-humidity environment produced by the STERRAD™ System makes it possible to process moisture-sensitive devices without damaging sophisticated electronic circuitry.

Because the 100 liter capacity (about 4 cubic feet) STERRAD™ System offers a rapid turnaround time (total cycle time is around one hour), it can have a throughput equal to that of a much larger ethylene oxide sterilizer. This high throughput can help reduce inventory and minimize storage space. As with other sterilization systems, the sterilized items can be stored in protective packaging until needed. The STERRAD™ Sterilization System and all of the other hydrogen peroxide sterilization system discussed are not designed to sterilize cellulosics or standing liquids since the sterilizing agents would tend to be absorbed and or condensed into these materials.

The STERRAD™ System technology helps preserve environmental integrity. There are no emissions harmful to workers or the environment because the primary end products of the sterilization process are oxygen and water vapor.

The unique patented hydrogen peroxide cassette delivery system enhances user convenience and safety. The cassettes automatically advance through each sterilization cycle, eliminating the need for tanks, valves, or generators.

The STERRAD™ System has a small footprint and can be placed wherever it is needed within a facility. It requires only a standard 208 volt electrical connection. No external vents, drains or other utility hookups are needed. The system is designed with self-diagnostic functions for ease of use.

The first healthcare market product from AMSCO, the VHP®100, is intended for use in the sterilization of endoscopes. The VHP®100

(Figure 15.10) utilizes a combined vacuum and flow-through system in order to penetrate the lumens within the endoscope. A general purpose, low temperature hydrogen peroxide vapor sterilizer is expected to be offered by AMSCO to directly compete with the ASP STER-RAD™ System.

Surface Sterilization Process of the VHP®1000

The AMSCO VHP®1000 usually utilizes a "closed-loop" flow-through approach to the sterilization of sealed enclosures, eliminating the need

Figure 15.10. AMSCO VHP®100 endoscope sterilizer.

to dispose of sterilant-laden carrier gas as in an "open-loop" process. It is superior to a wipe-down process because it is repeatable and validatable and there is no risk of a breach of containment by the introduction of a human. No chemical entities are introduced into the sterile environment to assist aeration. The H_2O_2 vapor sterilization cycle consists of four major steps (Table 15.2):

Dehumidification Phase. The enclosure is first dehumidified so that the process always begins with the same amount of water vapor present. Also, with the water vapor level decreased, a higher concentration of H_2O_2 vapor may be circulated throughout the enclosure, decreasing the time required for sterilization. Generally, the enclosure absolute humidity is decreased to 4.6 mg/ℓ (20 percent RH at 25°C), although the VHP®1000 allows for cycles at 2.3 mg/ℓ (10 percent RH) and 6.9 mg/ℓ (30 percent RH). Expediency and practicality are the reasons that 4.6 mg/ℓ (20 percent RH at 25°C) is generally chosen.

Table 15.2. Cycle Phases and Required Data for Theoretical Cycle Development

Cycle Phase	Data Required
Dehumidify	Enclosure volume
	Initial enclosure relative humidity (RH)
	Goal relative/absolute humidity (RH/AH)
Condition	Enclosure volume
	Sterilant inlet hose temperature
	Desired sterilant concentration (determined in sterilize phase calculations)
Sterilize	Enclosure volume
	Minimum enclosure temperature (during condition and sterilize only)
	Maximum enclosure temperature
	Goal RH/AH (from dehumidify phase calculations)
Aerate	Enclosure volume
	Sterilant concentration (from sterilize phase calculations)
	Desired residual H_2O_2 vapor concentration

Dehumidification to lower humidity levels can be very time-consuming and may be impractical, especially for larger enclosures. The VHP®1000 utilizes a fixed-capacity desiccant dehumidification system that requires periodic regeneration so larger auxiliary dehumidifiers (continuously regenerating desiccant wheel systems) are often used because they allow for more rapid enclosure dehumidification and continuous operation. Since cycle turn around time is of primary importance in the production environment, there is a growing need for larger capacity vapor generators and/or modular systems that can be interfaced to large desiccant wheel systems.

Condition Phase.   Hydrogen peroxide vapor is introduced during this phase at a high rate in order to rapidly increase the vapor concentration to the desired level. The purpose of this phase is to decrease the overall cycle time by reaching the desired steady state concentration in the shortest time period possible. The non-VHP®1000–related determinants of condition phase time are the size of the enclosure and the temperature of the inlet hose (or piping) to the enclosure. If the concentration of the inlet vapor during the condition phase is too high, condensation could occur within the inlet hose.

Sterilization Phase.   The H_2O_2 vapor concentration within the enclosure is maintained during the sterilization phase, providing a consistent environment with reproducible sterilization results in a preprogrammed time period. The major determinants for sterilization phase times are the concentration of H_2O_2 vapor that can be maintained within the enclosure, the percent saturation, and the desired sterility assurance level (SAL). In general, a 10^{-3} to 10^{-6} SAL is utilized.

Aeration Phase.   Upon completion of sterilization, H_2O_2 vapor is decomposed into its component parts of water and oxygen ($2H_2O_2 \rightarrow 2H_2O + O_2$) by recirculating the enclosure air/vapor mixture through a catalytic converter. This water vapor and oxygen gas may remain in the enclosure or may be flushed with dry nitrogen or other inert gas. The change in oxygen concentration within the enclosure is minimal since only 16.479 percent by weight of the total vapor (35 percent H_2O_2, 65 percent H_2O) introduced is turned into oxygen and the remaining 83.53 percent is water vapor.

Figure 15.11 contains a computer simulation of a VHP® 1000 H_2O_2 vapor sterilization cycle.

The desiccant system in the VHP®1000 is capable of adsorbing approximately 6–8 pounds of water vapor before a regeneration cycle (removal of moisture from the reusable desiccant within the VHP®1000) is required. The process for regeneration requires 18 hours and may be

Figure 15.11. Computer simulation of a typical VHP®1000 H₂O₂ vapor sterilization cycle.

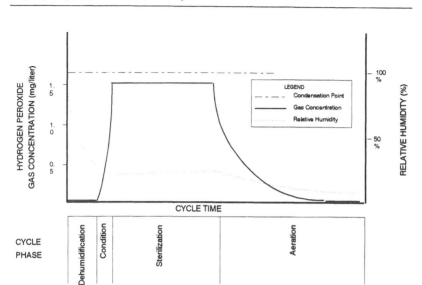

set to run automatically following the last sterilization cycle of the week or it can be run overnight.

It is anticipated that higher capacity, second generation VHP® generator systems will become available, employing drying systems capable of continuous operation. AMSCO is reported to be working on a VHP®2000 and Microflow, a European manufacturer of contamination control equipment, intorduced the Hy Per-Phase generator in late 1994. The preliminary literature from Microflow indicates that the Hy Per-Phase generator can produce an air flow that is up to 50 percent higher than the VHP®1000.

Subatmospheric Pressure Process of the VHP® DV Series Generators

The cycle chart contained in Figures 15.3(a) and 15.3(b) detail the four phases of a typical sterilization cycle for the AMSCO VHP® DV Series Generators.

Dry Phase.   The dry phase typically contains two consecutive series of about 5–10 dry pulses. Each pulse consists of an evacuation to a preset vacuum set point followed by an air admit (vacuum break) to a pressure that is typically in excess of 500 torr. The first series of evacuations is to about 5–10 torr. The second series of evacuations is typically down to approximately 1 torr. Moisture is withdrawn from the drain lines and evaporated in the chamber and condenser during this phase.

Leak Test.   The optional leak test consists of an evacuation to a set point (typically 1 torr) followed by a preset waiting period in which the vacuum level in the chamber and condenser is monitored. The pressure rise from a leak and/or the evaporation of any remaining moisture is compared to a preset maximum value.

Sterilization Phase.   The sterilization phase consists of an even number (typically 4–8) of sterilize pulses. Each pulse begins with an evacuation to a preinjection vacuum set point (typically 1 torr). A measured amount (by weight or volume) of H_2O_2 is vaporized and then introduced into the chamber, condenser, and accompanying piping in a controlled manner. The exact amount of vapor that is introduced is thus preset and is determined based on the temperature of the freeze dryer as well as the internal volume of the freeze dryer. After a short (2–8 minute) exposure time, air is controllably admitted to raise the total pressure inside the freeze dryer to a vacuum set point (typically between 100 and 300 torr). A second short (1–4 minute) exposure time completes the sterilize pulse.

Aeration Phase.   The aeration phase is made up of 10–25 pulses. Each pulse typically consists of an evacuation to a vacuum set point that is below 10 torr, followed by an air admit (vacuum break) that raises the chamber pressure above 500 torr. The vacuum system may continue to withdraw air from the system during the "air admit," creating a flow-through aeration. Residuals are removed from the chamber and the condenser during this phase.

GENERAL APPLICATIONS IN THE ASEPTIC PHARMACEUTICAL MANUFACTURING ENVIRONMENT

Emerging Technologies and a Regulatory Perspective

Hydrogen peroxide vapor is now emerging as a leading technology in aseptic processing not only due to its proven microbicidal effectiveness and positive environmental safety aspects, but also due to the emergence and acceptance of absolute barrier technology and other advanced aseptic processing technologies by regulatory agencies.

Recent U.S. FDA statements during a two-day open meeting on October 12–13, 1993, recognized the potential for the use of aseptic technologies that are designed to minimize human intervention, such as blow/fill/seal, robotics, and barrier systems. The FDA has indicated that they will support the use of advanced aseptic processing techniques for products that cannot withstand terminal sterilization, provided adequate data regarding SALs achieved in the process is presented. This may be a revolutionary breakthrough for aseptic processing in terms of U.S. regulatory acceptance, which promises to increase the use and development of these and other advanced aseptic processing techniques worldwide, well into the 21st century. European regulatory authorities have accepted barrier technology for many years and the resulting industry data supports high SALs (or rather "low probability of contamination") as evidenced by long-run media fills (>30,000 vials) and environmental monitoring.

Barrier Systems

The basic concept behind barrier technology is to maintain the user's physical presence in the operation as required (via glove ports or "half suits") while the operator remains biologically removed from the process. Common sense dictates, and numerous studies have confirmed, that the elimination of human contact with the sterile product or its surrounding environment is the single most important factor for decreasing the probability of contamination in the system.

The basic construction of a barrier system is a rigid- or flexible-walled enclosure made from stainless steel, glass, or plastic. HEPA–filtered air supplies (and often HEPA exhausts) provide airflow characteristics within the barrier that are compatible with the product and/or processes. These barriers may be absolute (fully sealed) or partial (nonsealed but operated at a positive pressure in order to prevent the inward migration of microbial contaminants). Both barrier types have contributed to the decreased probability of contamination in filling lines and other production and processing equipment. Additionally, absolute barriers can also provide operator and environmental protection when dealing with human or animal pathogens as well as cytotoxins.

Barrier systems cover a wide range of applications in the aseptic manufacturing environment. Most noted uses for sterile environments are vial, syringe, and ampule filling lines; accumulator tables; stopper feeders; interface isolators for dry heat sterilizers; autoclaves and lyophilizers; and equipment and product transfer isolators. With the complexity of these items and the material flow characteristics of the system, operator accessibility to the process and the maintenance of sterility during product transfers are critical. Therefore, custom built isolators designed for the specific process are the norm in the isolator industry.

Barrier System Sterility Assurance Level Determination

Figure 15.12 contains experimentally derived curves for a blow/fill/ seal machine manufactured by Automatic Liquid Packaging at Fisons Pharmaceutical. The test data was generated by varying the spore

Figure 15.12. Extrapolations of experimentally derived curves for probability of system contamination for form/fill/seal systems using a HEPA–filtered air shower and barrier technology. Adapted from the results of Bradley et al. (1990), with the addition of a proposed barrier system and predicted results.

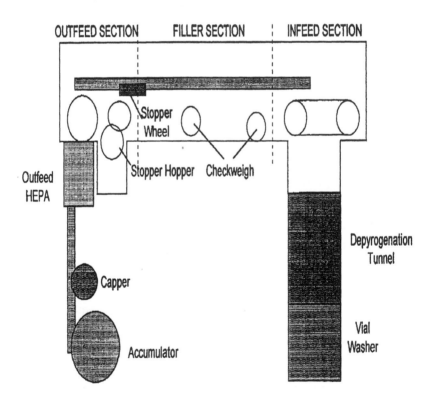

concentration in the environment. A one-log reduction in probability of contamination (SAL) was demonstrated using this technique when a HEPA–filtered air shower was introduced over the filling mandrels (Bradley et al. 1990). This technique of increasing the spore aerosol challenge concentration in the environment surrounding a sterile barrier would be a very practical method to verify the SAL that is achieved (Leo 1993) since verification by media fill would be impractical if the SAL is as low as predicted.

Other Applications

A large number of other H_2O_2 sterilize-in-place (SIP) applications are also being developed, including powder mills and stainless steel tanks that were previously steam sterilized. The major advantage in using H_2O_2 vapor in these applications is the low temperature, nonpressurized sterilization that decreases metal fatigue and wear on gaskets and seals.

Small rooms are often sterilized using H_2O_2 vapor, both in a laboratory environment and in production facilities. These rooms may be transfer areas for large equipment, small clean rooms, or other rooms adjacent to the aseptic core in conventional aseptic facilities.

Although 8–16-hour cycle times are common today, significantly shorter times are possible with the appropriate construction material selection and auxiliary air handling systems. The advantage of H_2O_2's environmentally friendly decomposition characteristics makes it a preferred method over other technologies, such as formaldehyde vapor and other gaseous sterilants. Also, of the commercially available gas delivery systems, few have the extensive documentation that accompanies the microprocessor control of the H_2O_2 vapor generators that monitors and controls key cycle parameters. This level of control and documentation allows these systems to be used in a GMP–controlled environment.

UTILIZATION OF HYDROGEN PEROXIDE VAPOR STERILIZATION IN FILLING LINES

At this writing there are only a few enclosed filling lines that are validated and in operation using H_2O_2 vapor sterilization. A number are under construction or are in various phases of design or validation. Other technologies, such as peracetic acid and H_2O_2–based liquid systems, have successfully been utilized for long-run media fills. Hydrogen peroxide gas, though, appears to be the technology of choice for the development of new aseptic filling systems and is replacing the other technologies due to its consistency, safety, and environmentally friendly qualities.

When a new aseptic filling project is being designed, the flow of materials and division of processes need not be the same as in the past. Sterile material transfer into, and out of, sterile Class 100 (or better) isolators is now possible. These isolators, which can be portable or stationary, are equipped with glove ports and half suits and need only be housed in high classification areas (100,000 or greater) or even unclassified, but monitored environments. The large areas that were previously dedicated to gowning and related activities are no longer required. Operators can enter or exit glove ports or half suits in a matter of seconds. The combination of barrier (isolation) technology and H_2O_2 vapor sterilization not only reduces the total floor space requirements and the initial capital outlay but also results in reduced operating costs (utilities, gowning, labor). The barriers, and the equipment they contain, can be relocated fairly easily if changes occur in the manufacturing process.

System Design Considerations

The entire filling operation, including all enclosures and process equipment, should be collaboratively designed by the customer, the architect, the engineers, the process equipment vendors, the enclosure manufacturers, and the H_2O_2 vapor sterilization equipment vendor. This cooperation is essential to provide a high quality system that considers all necessary design parameters. A properly staffed and managed program is essential to the project's success. This will insure that all parties are educated regarding the various technologies and are prepared to make the appropriate design decisions.

A block diagram of the major filling system components and/or functions is contained in Figure 15.13. There are three main functions represented: the introduction of filling components (glassware, stoppers, product, etc.), the main filling function (including stoppering, capping, etc.), and the exit of the finished product (for lyophilization, quality control testing, labeling, packaging, etc.). A sterile environment must be maintained throughout all phases of these steps.

Material Considerations

The materials of construction for all system components can affect the overall system operation. There are three different aspects of "material compatibility" that must be considered. The materials must be evaluated with respect to their chemical resistance to H_2O_2 vapor, their

Figure 15.13. Generic vial filling system utilizing barrier technology and hydrogen peroxide vapor sterilization.

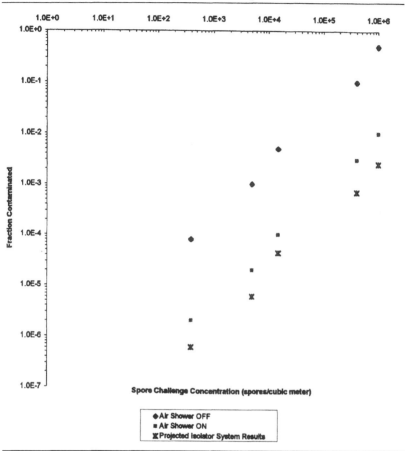

ability to act as a catalyst to the breakdown of H_2O_2 vapor, and their H_2O_2 vapor absorption/adsorption characteristics. Stainless steel, glass, silicone rubber, and Viton® are commonly used materials that are acceptable from all three perspectives (Table 15.2). Materials that should be avoided include cellulosics (paper), nylon, latex, and butyl rubber. Many plastics are not "attacked" by the low vapor concentrations present during sterilization, but they will not outgas afterwards as rapidly as stainless steel or glass.

The primary material used in the construction of the filling machine and other major system components is stainless steel. Rigid isolators frequently have a stainless steel main structure, glass walls, and aluminum-framed glass fiber HEPA filters. PETG, polycarbonate, or acrylic are often substituted for the glass, but will likely require increased aeration times. Flexible wall isolators typically have a thick, clear, PVC canopy.

The filling machine vial guides are typically made from abrasion-resistant materials such as Delrin®, ultrahigh molecular weight (UHMW) polyethylene or Teflon®; the accumulator table conveyor belt is typically made of high density polyethylene (HDPE). Components made from these materials are still in use after 300 H_2O_2 vapor sterilization cycles. The higher density materials tend to adsorb/absorb less than the lower density materials and are less permeable.

Sealing materials that can be continuously, or repeatedly, exposed to H_2O_2 include Teflon®, Viton®, and silicone rubber. Most elastomers are candidates for product use (stoppers, septums, etc.) since they will only be exposed once for a fairly short time. Refer to Table 15.3 for a list of some materials that have been tested for both chemical resistance and absorptivity. It is important to note that different formulations of the same material (e.g., Delrin®) may demonstrate widely different material compatibilities. This disparity is more likely due to the filler materials used as opposed to differences in the characteristics of the elastomer. The authors recommend that individual material testing be performed if there is any doubt about the suitability of construction materials.

Closure Transfer and Sterilization

Stoppers, caps, and other closures may be bulk packaged and presterilized by autoclaving or irradiation prior to use on the production line. Thus, only the outside surface of a plastic bag, outer wrap, or other closure must be sterilized in a transfer station. The bag is aseptically opened in a sterile isolator and the stoppers conveyed via a vibratory feeder to the filling isolator. Other methods of transfer may include passing supplies through a double door and an autoclave interface isolator and into the attached filling line isolator. A canister with a rapid transfer port (RTP) may also be loaded with product, sterilized, and attached to the filling line isolator; the RTP port is then opened and the sterile contents aseptically passed into the filling line isolator.

Sterilization of the Autoclave Interface Isolator

An autoclave interface isolator is an isolator that is sealably attached indirectly to the chamber of a double door autoclave. When the autoclave door is opened, carts of sterile product can be brought directly into the isolator, which serves as a staging area for the transfer of sterile materials into the filling machine isolator. Over the past five years vertical and horizontal sliding door autoclaves have become

Table 15.3. Hydrogen Peroxide Vapor Sterilization Material Compatibility

Item	Preferred Materials	Materials Dependent on Application[a]	Materials to Be Avoided
Isolator walls	316 stainless steel 304 stainless steel	PETG Lexan Acrylic Fiberglass Polyvinyl choride	Cellulosics Carbon steel
HEPA filters	Glass fiber media Stainless steel and aluminum frames	Plastic and nickel plated steel frames	Cellulosic media Wood frames
Seals	Viton[b] Silicone rubber	EPDM Buna N	
Component parts miscellaneous	Ceramics Norprene[b] Teflon[b] Kynar[c] Zirconium	CPVC Hypalon[b] Natural Rubber Neoprene[b] Delrin[b] Polypropylene Polyethylene Epoxy Anodized Aluminum Polyamide Polyurethane	Nylon

[a]Results may vary widely dependent on application and composition of any given material. It is recommended that the sterilization equipment vendor be provided with material lists and samples of selected materials for compatibility testing.

[b]Delrin, Hypalon, Neoprene, Norprene, Teflon, and Viton are registered trademarks of E.I. duPont de Nemours & Co.

[c]Kynar is a registered trademark of Pennwalt Corporation.

increasingly popular for this application since they do not intrude into the space within the isolator. All of the mechanisms for opening and closing the sliding door are typically located in a plenum that should also be sterilized. Communication can be established between the autoclave controller and that of the H_2O_2 vapor generator to synchronize operation, assure operator safety, and provide for movement of the door, thereby guaranteeing exposure of the door seals and all other moving mechanical components located within the plenum.

Vial Transfer and Accumulator Sterilization

Glassware must be both sterilized and depyrogenated prior to entry into the filling operation. The vials are fed directly or are accumulated and fed from a sterile, enclosed accumulator table. A 100 percent recirculated airflow will produce a nearly particulate free environment and will provide an even distribution of the H_2O_2 vapor sterilant. The accumulator table is operated (empty) during sterilization in order to expose all of its surfaces to the H_2O_2 gas. The interfaces between adjoining isolators are of major importance when considering the sterilization of the system. The design of an acceptable seal that will not disrupt the smooth flow of the vials across it is challenging. The project manager is urged to pay close attention to detail as door closures and interfaces are designed and/or selected.

Syringe Tray Entry and the Continuous Sterilization Tunnel

Syringe tray entry to a filling machine presents a unique challenge due to the size of the containers and the rate at which they must be supplied to a high speed line. In a "batch mode" the outer wraps of many syringe trays must first be sterilized and then removed in a sterile environment in order to provide the necessary supply. This approach requires a great deal of storage space in a staging area, which is generally undesirable.

Evans Medical, Ltd. (Speke, England) has utilized a continuous syringe tray sterilization system that provides a steady supply of trays to match the filling requirements. Figure 15.14 shows the basic line, which starts at a staging area where a first outer wrap is removed in a Class >100,000 area and then placed in the conveyor system. As the tray passes down the conveyor, it is warmed (to permit a more rapid sterilization without condensation), sterilized in a tunnel (3) filled with H_2O_2 vapor, and then passed through a two-stage aeration process and into the filling isolator (6). In the filling isolator an operator removes and discards the final outer packaging before placing the tray into the filling machine. The warming isolator (2) as well as the primary (4) and secondary (5) aeration tunnel air handling systems have in-line catalytic converter systems that capture and destroy any migrating H_2O_2 gas. The Evans system, which can be refined further, is viewed as a clever and practical solution to the continuous processing of syringe trays.

Figure 15.14. Block diagram of a continuous sterilization tunnel system for syringe filling in use at Evans Medical, Ltd., Speke, England (designed by Total Process Containment, Surrey, England). The numbered zones have the following functions: (1) Removal of outer wrapping of syringe trays and entry onto the conveyor; (2) Primary heating of the tray; (3) Sterilization tunnel for the outside of the syringe trays; (4) Primary aeration zone; (5) Secondary aeration zone; (6) Material transfer and buffer area to the filler; (7) Filling machine isolator; and (8) Product output isolator.

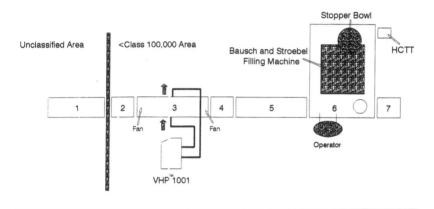

The Filling Machine Isolator

A single H_2O_2 generator can be attached in tandem to two or three isolators, if the isolators are equipped with the proper inlet and outlet connections, if proper vapor distribution is maintained, and if the total volume does not exceed 500–800 cubic feet. For larger enclosures and larger multi-isolator systems, two generators are used to reduce the overall sterilization cycle time.

The filling machine isolator, which is similar to the accumulator isolator described above, is usually a Class 100 (or better), laminar flow system. Materials are carefully selected and the design of the isolator and the filling machine are optimized for rapid and effective surface sterilization. Only the active filling components and the syringe supports and guides are located within the isolator. All other components are external for ease of servicing. The filling apparatus is operated empty during sterilization to assure that all parts are exposed to the sterilant vapors. Care is taken to insure that the least exposed surface obtains the required sterilant dosage to provide the desired SAL.

The design of a filling machine should minimize the number of moving parts in the sterile area and locate most of the major mechanical items below the deck of the filler, thereby leaving only the critical components in the sterile area. The sealing of the deck at the mechanical interfaces may present an engineering challenge, but it will simplify the validation of the system.

The suction system used for stopper placement and other functions can also be beneficial for distribution of the vapor to some of the critical areas. A small catalytic converter should be placed in series with the suction system vacuum pump to prevent the discharge of any H_2O_2 vapor.

Direct Transfer to Quality Control (QC) Sterility Testing

An added benefit of barrier technology usage to both the manufacturing and the QC function is the ability to perform a rapid, sterile transfer of the product directly to a QC transfer isolator or a QC suite. This is an improvement over loading transfer isolators with product, sterilizing the surfaces of the product containers with H_2O_2 vapor and using an RTP to aseptically transfer them into the QC suite. There is also no longer any concern over the possibility of a false negative resulting from sterilant penetration into a nonsterile (contaminated) product while the surfaces of the product containers are being sterilized.

Sterile Product Transfer to the Lyophilizer

If a sterile-filled product requires lyophilization, it should be aseptically transferred from the filling line and into the freeze dryer to avoid contamination. This can be accomplished using automated (or manual) loading equipment that is contained in a sterile isolator. The lyophilizer interface isolator can be permanently attached to both the filling line and the lyophilizer or it can be individually attached using an RTP and relocated as required. In the event that the product is not capped when it exits the lyophilizer, the interface isolator can be used to aseptically transfer the product to the next manufacturing operation.

Validation and Sterility Maintenance

There are five general steps required for the sterility validation on an atmospheric pressure enclosure. These steps, which are described below and summarized in Table 15.4, provide a basic framework for proving efficacy and optimizing performance. The actual means of performing these tests, and even the general approach, will vary between companies. The steps are not necessarily performed serially,

Table 15.4. Five Basic Validation Steps for Surface Sterilization

Validation Step	Description
Cycle development	Develop a theoretical cycle first, based on estimates of enclosure temperatures, then adjust for actual conditions based on studies in other steps.
Temperature distribution	Place thermocouples throughout the enclosure, both on air and surfaces, to determine key temperatures.
Vapor distribution	Utilize chemical indicator strips to redirect airflow patterns and adjust sterilization phase time based on results. Focus efforts here in order to minimize repeated biological testing and decrease overall validation time.
Biological challenge	Inoculate and/or use packaged biological indicators (BIs) to challenge critical areas in the enclosure. Also useful to perform a D value study to verify the vapor concentration present and SAL to be achieved.
Aeration verification	Verify removal of the H_2O_2 vapor and sample both air and materials for residuals. Product effect testing may or may not be appropriate.

but are part of an iterative process of optimizing the system performance by minimizing total cycle time while maximizing the SAL.

Cycle Development

The VHP®1000 *Cycle Development Guide* can be used to determine theoretical, first-pass cycle parameters. The guide is based on a computer simulation of a "flowing mixer." There are examples in the guide that illustrate the cycle parameter development process. The first-pass parameters can be plugged into an actual computer simulation that is incorporated in a spreadsheet program that runs on Quattro Pro™. The parameters that the user must supply include enclosure volume, enclosure temperature, airflow rate, initial humidity, liquid sterilant concentration (typically 30 percent or 35 percent), liquid sterilant delivery rate, sterilant vapor delivery line temperature, and sterilant vapor half-life.

The output from the computerized, theoretical model is only as accurate as the input parameters. Since some of these are not known in advance, they are estimated. Even though the model does not account for the sterilant vapors that are adsorbed onto, or absorbed into, the enclosure during the sterilize phase, the effects from this phenomenon should be minimal after the first few minutes of exposure. This effect can be taken into account during aeration by including a residual level in the aeration air stream.

Once a conservative, first-pass cycle has been determined, it should be run and a temperature profile mapped. If no condensation (fogging) was evident, the injection rate should be increased (typically by 10 percent) and another run made. This step should be repeated until traces of fog are evident. This will determine the saturation limit, or sterilant dew point concentration. Operation should be targeted for at least 80 percent of this upper limit. The expected seasonal temperature variations of the enclosure will determine how close operation can be to this dew point.

Temperature Distribution

In the early stages of the development of the theoretical cycle, it is necessary to run a dehumidification phase cycle with temperature sensors located on the sterilant delivery line and throughout the enclosure. Some of these sensors are placed on the surfaces and others should hang freely to measure enclosure air temperatures. These temperature readings are used to estimate the temperature of the enclosure during sterilization. The sterilant delivery line temperature establishes a maximum sterilant injection rate.

Later on, these same sensors are used to obtain a full temperature mapping of the enclosure during the sterilization phase. Surface temperature data should be taken not only on the walls of the enclosure, but also on areas that could act as heat sinks or sources such as the large masses of stainless steel contained in the frame of a filling machine.

If a cold tool, or product, is present during a sterilization cycle, a thin film of condensation will occur on its exterior surface if it is below the dew point temperature of the vapor. This film will disappear during the ensuing aeration so it may not present a problem, but caution is advised because sterilization efficacy may be compromised.

Vapor Distribution

The chemical indicator is a tool that can be used to qualitatively determine if the distribution of the H_2O_2 vapor is uniform throughout the enclosure. If there are large areas that are not receiving an adequate level of H_2O_2, the low level will be indicated by a slower color change of the chemical indicator. Air circulation can be altered to improve sterilant distribution. The chemical indicators can be used to efficiently locate the areas that may present the greatest biological challenge.

Biological Challenge

The biological challenge of the system is the key measure of the sterilant's efficacy. In general, the time spent on refining the uniformity of the vapor distribution will be reflected by a more rapid progression through biological efficacy testing.

The challenge during biological efficacy testing is to quickly identify the most critical (product-contact) components and the most difficult to sterilize locations in the system. Once these are identified, triplicate runs can be made in which the biologicals are put in place, the sterilization cycle run to completion, and the biologicals removed and incubated. The filling system should be able to perform an "empty fill" or mock run during sterilization, without any product, containers, or stoppers in the system. This action of the conveyors and other moving parts allows for the exposure of all equipment surfaces to the H_2O_2 vapor. If any air is discharged from the system, it must be discharged to an outside exhaust or passed through a catalytic converter.

The detailed methods for testing the system are beyond the scope of this chapter, but a few comments on packaged BIs, inoculated carriers, and direct product inoculation are warranted. Packaged BIs are by far the easiest to use since they are preinoculated, have a long shelf life, and are packaged to readily accommodate aseptic culturing. Inoculated carriers (also known as "coupons" or "planchets") are traditionally stainless steel, but may also be representative samples of other materials from the intended application made into more easily tested carriers. The major advantage of using inoculated carriers is that they represent the actual performance of the sterilization process on the materials to be sterilized. There are numerous disadvantages, including unknown shelf life and poor spore adhesion. Direct surface inoculation is traditionally a difficult method to apply due not only to accessibility issues but also to variations in cleaning, inoculation, and recovery techniques. When properly utilized, this method provides an excellent means for testing system performance.

Upon completion of the sterilization process, sterile media fills are generally conducted with subsequent incubation and inspection of the media filled containers for microbial growth. Media fills provide a means for demonstrating the initial achievement of sterilization by producing a sterile product; they also provide a means to verify sterility maintenance. To test SALs beyond 10^{-3} though, the number of vials required to be filled and inspected for growth becomes unmanageable and costly and may be better tested using methods similar to those described by Bradley et al. (1990).

Aeration Verification

The aeration phase typically reduces the sterilant vapor concentration within the system to below 5 ppm before outgassing becomes evident. Outgassing will tend to maintain the H_2O_2 vapor concentration in the system at the 1–3 ppm level for an extended time period; it is dependent on many factors, including materials of construction, exposed surface area, exposure time, exposure concentration, and the air exchange rate during aeration. Hydrogen peroxide vapor detectors and sensors are used to determine when the residuals are below a specific level.

Aeration of the enclosure until an undetectable level of hydrogen peroxide remains can be time consuming and may be unnecessary. The Occupational Safety and Health Administration (OSHA) states that human exposure must be limited to 1 ppm over an 8-hour time-weighted average (TWA). Stability and potency tests can be used to determine if 1 ppm, or any other target level of residual, has an effect on the product. Thus, an aeration target of 1 ppm may be acceptable if there is no adverse affect on the product.

Aeration time studies should be conducted using full vapor exposure times with an injection rate (vapor concentration) that is 5–10 percent above the nominal value. Aeration should be to a nominal value that is around half of the target value. Thus, for a target value of 1.0 ppm at the end or aeration, the aeration time should be capable of reducing the residual level to 0.5 ppm. Vapor sensors can also be utilized to terminate the aeration when the residual level reaches one half, or some other appropriate percentage, of the target value.

If components, or vials containing product, are sterilized with H_2O_2 vapor, then they should also be tested for residuals. Aqueous extraction techniques can be used to determine both the amount of sterilant that is absorbed by a component and the amount of sterilant that penetrates through a container into the product (water).

Sterility Maintenance and Frequency of Sterilization

The procedures for sterility maintenance in these isolation systems will differ by application and company philosophy. It is likely that sterilization will be required between product lots, after power failures or maintenance procedures, and at specific time intervals. A complete sterilizations record will be retained along with documentation of continuous pressurization with HEPA–filtered air.

The relevance of the technology becomes apparent as we examine this new paradigm, which centers upon the removal of the main source of system contamination, the human. Most pharmaceutical companies will routinely clean and sterilize product-contact isolators on a lot-by-lot basis and maintain the necessary sterilization documentation. It is logical when viewed from the new paradigm, to not resterilize until a different product is introduced into the line or until a specified time interval has passed. Longer filling runs on the order of weeks or months may be used before a new "lot" must be declared if there are no product formulation or storage limitations. Sterility will be assured if sterile transfers are observed, if the initial environments are sterile, and if there are no breaks in the system.

Sterile suction wands can be integrated into clean-in-place (CIP) systems that employ a sterile water for injection (WFI) rinse so that sterility need not be breached during cleanup operations. Standard change parts can be transferred in via a steam autoclave interface or via a H_2O_2 vapor sterilized transfer isolator.

The validation issue can be divided into barrier integrity and transfer methods. Barrier integrity can be assured by utilizing positive pressure systems that prevent the ingress of contaminated air (viable particulates) into the sealed enclosure. This concept was demonstrated by Gold (1993) to be very effective, even for partial enclosures. It is anticipated that future studies of this type for sealed enclosures will utilize a specific external environmental challenge as in the Bradley form/fill/seal system and will show that contaminated air ingress is not a concern (Bradley et al. 1990). Leak rates can automatically be monitored for a sealed system while it is maintained under positive pressure using state of the art analogue pressure sensors and microprocessors.

Transfer methods must provide for a sterile transfer of components and product into the barrier system. Operator error and component malfunction must not be allowed to go undetected if they can compromise the integrity of the sterile transfer process.

UTILIZATION OF HYDROGEN PEROXIDE VAPOR IN LYOPHILIZERS

Introduction to Freeze Dryer Sterilization

A substantial percentage of the drugs and vaccines that are produced annually in the United States require lyophilization during the manufacturing process. It is estimated that two-thirds of the production freeze dryers are capable of being sterilized using either steam or ethylene oxide (EtO). The other one-third is typically sanitized between lyophilization runs using a variety of liquid chemical disinfectants.

An even higher percentage of the new drugs and vaccines under development in research laboratories is expected to require lyophilization. Yet, it is estimated that less than 20 percent of the research and pilot production freeze dryers are capable of being sterilized with steam or EtO.

Industry experts agree that it is a good idea to sterilize between lyophilization runs. There is a need for an affordable, validatable, sterilization technology that could be retrofit onto research, pilot, and production freeze dryers that are incapable of being sterilized with steam or EtO. This would provide the pharmaceutical industry with a choice of either retrofiiting their existing freeze dryers with sterilization capability or purchasing new sterilizable freeze dryers.

Hydrogen peroxide vapor is that technology. Hydrogen peroxide vapor sterilization cycles (Figures 15.3(a) and 15.3(b)) have been developed and validated to operate at near ambient temperature under subatmospheric pressure conditions. Vapor generators are commercially available for small freeze dryers having less than 30 cubic feet of total internal volume and for large freeze dryers having up to 1000 cubic feet of total internal volume.

These generators can be installed onto nearly any existing freeze dryer using the appropriate modifications and control interfaces. For large installations these modifications typically include, but are not limited to, the following:

- Installation of $1^1/_2$-inch warmed vapor inlet lines with redundant filters (also warmed) for independent delivery of vapor to both the chamber and the condenser.

- Connection of a vacuum line to both the chamber and condenser drain lines.

- Installation of an appropriately sized vacuum pump that is not affected by moisture. This pump is needed to dry the freeze dryer prior to sterilization.

- Modification of the piping on the freeze dryer to conform to GMP standards and to include validation access ports.

- Installation of appropriate vacuum and temperature sensors for monitoring the sterilization process.

- Interfacing the sterilant generator to the appropriate valves, pumps, and sensors either directly or indirectly through the freeze dryer control system.

A Typical VHP®DV 1000 Installation

Figure 15.15 is a piping schematic of a typical VHP®DV 1000 H_2O_2 vapor generator installation. The generator vapor output is connected into the filtered air inlet (vacuum break) piping as shown. Automatic

Figure 15.15. Piping schematic of a typical VHP®DV 1000 installation on a large-scale freeze dryer with a dry vacuum pump system.

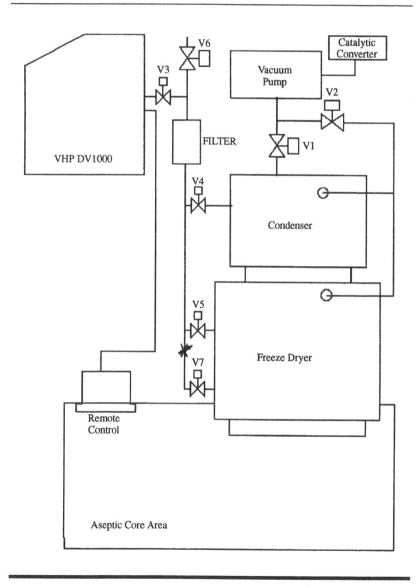

isolation valve V3 is used to prevent pressurized air from entering the generator from the air inlet piping.

Any of automatic valves (V4, V5, and V6) that were not present in the original vacuum break piping must be added. Controlled operation of these valves along with valve V3 will control the entry of air, sterilant vapors, or both, into the condenser and the chamber. Valves V4 and V5 may be alternately pulsed open to force the sterilant vapors to flow equally through more restrictive lines.

Redundant absolute (bacteria retentive) filters are typically utilized to filter the air that relieves the vacuum in the chamber and condenser. Pall Emflon® filters have been shown not to absorb or destroy the H_2O_2 vapor. The sterilant delivery lines and stainless steel filter housings should be warmed to 40–45°C to prevent sterilant condensation.

Either atmospheric air or a compressed gas (air, nitrogen, helium, etc.) may be used to relieve the vacuum in the freeze dryer. Whenever a compressed gas is utilized, valves V3 and V6 operate sequentially, but never simultaneously.

There is frequently a second filtered gas piping circuit into the freeze dryer chamber. This circuit is typically used to controllably bleed gas into the chamber during lyophilization to maintain the proper pressure differential between the condenser and chamber. Valve V7 and its accompanying piping is illustrative of this system.

The vacuum system will selectively withdraw air, sterilant vapors, and/or moisture from either the condenser vacuum outlet or from the drains in the condenser and the chamber. The drains can be expected to contain a significant amount of liquid, which will pass through valve V2 and directly into the vacuum pump whenever valve V1 is closed and valves V4, V5, and V6 are opened.

Vacuum systems that can be configured to accept this liquid include the following arrangements:

- A water ring vacuum pump in parallel with an oil seal vacuum system. The oil seal system would take over after the water ring pump has lowered the system pressure to approximately 25 torr.

- A water ring pump with an air ejector and blower with the capability to quickly evacuate the freeze dryer down to 1 torr.

- A "dry" vacuum system with a catalytic converter on its exhaust. This pump could also be used during lyophilization, eliminating the need for an oil seal pump.

The "dry" vacuum system and accompanying exhaust catalytic convert are included in the schematic contained in Figure 15.15. Figure 15.16 contains the combination liquid ring/oil seal pump installation.

Figure 15.16. Piping schematic of a typical VHP®DV 1000 installation on a large-scale freeze dryer with a combination liquid ring/oil seal pump system.

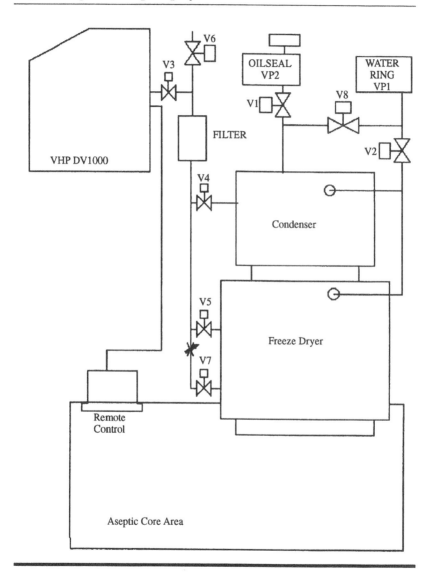

Isolation valve V8 has been added so that the bulk of the moisture (liquid and vapor) can be directed into the water ring vacuum pump system. All vapors (water and sterilant) that enter the water ring pump condense and are discharged with the cooling water. An activated carbon filter has been added to the outlet of the oil seal vacuum pump system to reduce the levels of any H_2O_2 residuals that might pass through and be exhausted.

A remote display/control can be located in the aseptic core area for the convenience of the operator. The remote control is recess mounted into the wall and is serviced from the back side without compromising the aseptic core area. It is sealed so that it can be wiped/washed down on the aseptic core side.

Detailed Operational Description

At the completion of a lyophilization cycle, the vials containing freeze-dried product are typically capped by lowering the shelves and pressing the loose fitting stoppers in place. The chamber door is then opened and the vials removed. At this time the freeze dryer chamber is cleaned and the ice on the condenser is melted. These tasks are best accomplished using warm ($45°C < T < 55°C$) water instead of steam. This defrost and CIP operation can be performed either manually or automatically. The chamber and condenser are typically warm and wet at this time as a result of the defrost and CIP procedures.

The four-phase sterilization cycle can now be initiated by an operator with an appropriate access code. The operator first activates the desired cycle, verifies that the cycle parameters are correct and initiates the cycle. The cycle proceeds (based on the installation configuration shown in Figure 15.15) according to the cycle charts shown in Figures 15.1 and 15.2.

Dry Phase

The DV 1000 control energizes the vacuum system and will continue to operate until the entire sterilization cycle is complete. The first dry pulse begins as valves V2, V5, and V6 open so that any moisture in the drain lines will be drawn out. After a short time interval, valve V1 opens, valves V5 and V6 close, and the freeze dryer is evacuated to the first vacuum set point (P1 = 10 torr). At this time valve V1 closes and valves V5 and V6 open, admitting filtered air until the pressure within the freeze dryer has reached vacuum set point P2 (650 torr), thus completing the first dry pulse.

The second, third, fourth and so on dry pulses, which are repetitions of the first dry pulse, follow. When the preset number of first stage dry pulses have been completed, the second stage of drying commences.

All valves are closed except V1 and V2. The vacuum pump withdraws air from within the chamber through valves V1 and V2 until the pressure within the freeze dryer reaches a third vacuum set point (P3 = 1 torr). At this time valves V1 and V2 close and valves V5 and V6 open, admitting filtered air until the pressure within the freeze dryer has reached pressure set point P2, thus completing the dry pulse.

The second, third, fourth, and so on dry pulses, which are repetitions of the previous second stage dry pulse, follow. The dry phase will continue until either the preset number of pulses have been completed or until a moisture sensor indicates that the system is dry. Then valves V1 and V2 will close, valves V5 and V6 will open, admitting filtered air until the pressure within the freeze dryer has reached pressure set point P4, then the cycle will advance to the leak test.

During the dry phase the shelves will preferably be warmed toward the high temperature set point (T_h = 45°C). This warming will facilitate the drying of the chamber and condenser. The vapor supply lines, in line filters, and so on will also be warmed up at this time. The chamber can also be warmed if there are provisions to do so in its design and construction.

Leak Test

All valves are closed except V1, V2, V4, V5, and V7. The vacuum pump withdraws air from within the chamber through valves V1 and V2 until the pressure within the freeze dryer reaches a fifth vacuum set point (P5 = 1 torr). Then valves V1 and V2 close and, after a short time delay, the pressure rise within the freeze dryer is monitored using a pressure sensor (MKS Baritron or equivalent) with a span of approximately 0 to 10 torr. If the rate of pressure rise is within the acceptable, preprogrammed limits (P6 Rise = 1 torr in 10 minutes), the system is deemed to be leak tight and dry and the cycle will advance to the sterilize phase. Set point P7 is typically set equal to P8 except during validation runs when P7 may be set to atmospheric pressure to provide access to the interior of the chamber at the completion of the leak test.

During the leak test the shelves will be cooled toward the low temperature set point (T_l = 25°C) and the chamber heating (if present) will be turned "off." The vapor supply lines, in-line filters, and so on will continue to be warmed.

Sterilization Phase

The DV 1000 control monitors the temperature of the shelves and vapor delivery lines, including line filters to verify that they are within the preprogrammed acceptable ranges before it will initiate this phase. Once they have been verified to be within their acceptable range, valves V1 and V2 open and the vacuum pump will evacuate the freeze dryer to the preinjection pressure set point (P8 = 1 torr). Valves V1 and

V2 close, valve V3 opens, and valves V4, V5, and V7 are sequenced to allow H_2O_2 sterilant vapors to flow from the DV 1000 generator into the freeze dryer chamber and condenser, exposing all of the delivery lines, air filters, and the interior of the chamber and condenser.

The amount of sterilant vapor that is introduced is monitored and controlled on a weight basis, as measured by an electronic balance integral to the VHP®DV 1000. Pressure sensors verify that the pressures within the vaporization system and the freeze dryer are within preprogrammed acceptable ranges. After the sterilant injection is complete, the pressure in the freeze dryer will rise to P9 (around 10 torr) and the first sterilant exposure countdown timer will be started.

When the first sterilant exposure countdown timer has reached zero, filtered air will be admitted at a controlled rate until the pressure within the freeze dryer has reached pressure set point P10 ($\approx$ 165 torr). As air enters the freeze dryer, it will push the sterilant that is trapped within any dead legs further into the dead leg and compress it, thereby increasing the sterilant concentration within the dead leg. After this vapor compression, a second sterilant exposure countdown timer will be started.

When the second sterilant exposure countdown timer has reached zero, valves V1 and V2 open and the vacuum pump will evacuate the freeze dryer to the preinjection pressure set point P8, thus completing the sterilant exposure pulse.

The second, third, fourth, and so on sterilant exposure pulses, which are repetitions of the previous pulse, follow until the preset number of sterilant exposure pulses has been completed. Then valves V1 and V2 will close and valve V5 and V6 will open, admitting filtered air until the pressure within the freeze dryer has reached pressure set point P12 ($\approx$ 500 torr), and the cycle will advance to the aerate phase.

Aerate Phase

The DV 1000 control begins the first aeration pulse as it opens valves V2, V5, and V6; any moisture in the drain lines will be drawn out. After a short time interval, valve V1 opens, valves V5 and V6 close, and the freeze dryer is evacuated to the first vacuum setpoint (P11 = 10 torr). At this time valve V1 closes and valves V5 and V6 open, admitting filtered air until the pressure within the freeze dryer has reached vacuum set point P12 (500 torr) completing the first dry pulse, followed by the second, third, fourth, and so on aeration pulses. A pause will occur after the preprogrammed number of pulses has been completed; the chamber will be vented to slightly below atmospheric pressure P14 ($\approx$ 720 torr). The operator may then enter an access code that signals the control to automatically vent the chamber to atmospheric pressure if an automatic residual monitoring device is not being utilized.

The operator may withdraw a sample of air from the freeze dryer chamber through a small access port in the chamber door to check the level of sterilant residuals. If the residual H_2O_2 vapor concentration is acceptable, the operator must so indicate using the control touch pad. The cycle will then advance to completion, will print out the batch end report, and de-energize the vacuum system. If the residual concentration is not acceptable, additional aeration pulses must be performed.

The extensions of aeration will continue until the residual concentration within the chamber is deemed acceptable by the operator or until 99 aeration pulses have been completed.

Notes on Validation of Lyophilizer Sterilization

The hydrogen peroxide vapor sterilization cycle consists of the dry phase, the leak test phase, the sterilize phase, and the aeration phase. Not only must each of these four phases have been validated for the sterilization cycle to be considered validated, but also certain preparatory steps (such as the defrosting of the condenser and the cleaning of the chamber) must be validated.

Defrost Step

The purpose of the defrost step is to remove ice that has accumulated in the condenser during the previous lyophilization run. The step must be able to remove both the maximum and minimum ice loads from the condenser and yet leave the condenser at approximately the same nominal temperature, since a H_2O_2 sterilization cycle will follow. All condensate must have been drained out at the completion of this phase.

Defrost is best accomplished using hot water in either an overflow or a flow-through mode. Alternately, the condenser can be flooded with water and steam introduced to maintain the temperature of the water. The direct use of steam is not recommended because it unnecessarily increases thermal stresses and can result in greater temperature variances throughout the condenser.

Cleaning Step

The purpose of the cleaning step is to remove contaminants from the internal environment of the freeze dryer so that they cannot find their way into the product during the lyophilization cycle. Potential contaminants would include particulate matter, broken vials, and/or residue from previously lyophilized products; heat transfer fluids; refrigerants; hydraulic fluids; dirt and grime from maintenance procedures; cracked thermocouple insulation; and residues from the cleaning agent itself.

A properly designed and validated automatic CIP system is preferable because it can be monitored and controlled by a computer, thereby eliminating the potential for operator error. Also, the effectiveness of the automated process would not be affected by operator variables. Manual or semiautomated cleaning procedures can also be effective if a defined sequence of steps is well documented and followed.

It is recommended that the operator inspect the inside of the chamber after it has been unloaded at the end of each lyophilization run. Manual removal is recommended for vials that may have fallen off the shelves during loading or unloading, broken vials, stoppers, and other large debris.

It is necessary to not only identify the contaminants that must be removed by the cleaning procedure but also to determine the expected amount and location of each contaminant. The maximum level of each contaminant that can be tolerated must also be determined, along with an appropriate, reliable sampling procedure. Then a cleaning procedure can be developed that will specify the following parameters:

- Type (aqueous or nonaqueous), composition, and concentration of the cleaning agent that is to be used

- Temperature of the solution

- The method of application, which should include flow rates, application pressures, and duration (time or total amount of solution utilized)

- Rinse solution specifications (i.e., WFI)

- Temperature of the rinse solution

- Method of application of rinse, which should include flow rates, application pressures, and duration (time or total amount of solution utilized)

The cleaning method challenge must verify that the desired cleaning effect has been achieved in a chamber and condenser contaminated in the appropriate locations with levels that equal or exceed those expected during normal operation.

Standard cleaning challenges have been developed and are often utilized to determine the effectiveness of cleaning procedures. The challenge typically consists of a colored solution that is "painted" onto the surfaces within the freeze dryer and allowed to dry. The standard cleaning procedure is then performed and the interior of the freeze dryer is inspected for evidence of visible residue. Some of the challenges will glow when exposed to ultraviolet light, so trace residue is easier to detect.

It is desirable for the chamber and condenser to be at the same nominal temperature at the end of the wash step whenever a H_2O_2 vapor sterilization cycle is to follow.

Sterilization Cycle

The purpose of the sterilization cycle is to kill any microorganisms (bacteria, bacterial spores, viruses, fungi, molds, etc.) that may be present within the freeze dryer. The effectiveness of a sterilization cycle is normally expressed as a probability of survival of a viable organism (or SAL). At the present time a SAL of 10^{-6} is widely accepted as the minimum standard for sterilization.

The recognized and acceptable approach to the development and validation of a sterilization cycle requires that a 6-log reduction be achieved using a half-cycle exposure to the H_2O_2 vapor sterilization agent. A full sterilization cycle would then produce a 12-log reduction, or an SAL of 10^{-6}.

The selection of an appropriate BI is critical to the sterilization validation process. *Bacillus stearothermophilus* spores are the organism of choice for the biological challenge. They have the highest resistance to H_2O_2 vapor of all of the organisms tested thus far, are easy to prepare and culture, and do not present a hazard to personnel.

Since most of the components within a freeze dryer are fabricated from stainless steel, it is recommended that the *Bacillus stearothermophilus* inoculum be placed on a stainless steel substrate and dried for use during the initial validation. Commercially available packaged BIs should also be used in conjunction with the stainless steel carriers since these indicators may be used to verify sterilization efficacy during normal operation and for the cursory annual validation verification.

The stainless steel substrate can be in any form but typically is made of 1-cm square coupons or thin, long, narrow strips. The inoculated coupons can be hung using "paper clip" hangers or placed on the shelves and floor. Long inoculated strips can be placed completely into dead legs. The long strips can be cut using a sterile scissors and the pieces dropped into test tubes containing sterile growth media for incubation and outgrowth evaluation.

Commercial BIs that are selected should not use paper, or any other absorptive material, as a substrate for the inoculum or for a component part of the packaging. Hydrogen peroxide residuals in the substrate can inhibit growth during culturing, and thus produce false negatives. Catalase, which breaks down H_2O_2 into water vapor and oxygen, can be added to the growth media if there is a concern about low level residuals producing false negatives.

Triplicate runs should be made during the validation of each of the individual phases, except for the sterilization phase where a triplicate half-sterilize phase and a triplicate full-sterilize phase should both be run.

The quality, consistency in microorganism count (population level), consistency in relative resistance, and stability of a packaged BI are important factors to consider when selecting a vendor. A certificate of quality should be kept on file for each lot of indicators. Typical stability data for the recommended storage conditions should be obtained and placed in the validation file.

The quality, composition, and stability of the liquid H_2O_2 must be consistent if the vapor generated from it is to be consistent. A certificate of quality that states the concentration of the solution (when filled) for each lot of containers should be kept on file. Stability data for typical containers of the solution at the normal storage temperature and for a temperature 10°C (or more) above this normal storage temperature should be obtained and put in the validation file.

All sensors (pressure, temperature, electronic balance, and moisture level) that are used to monitor the sterilization process must be calibrated at regular intervals (six months minimum) using standards that are traceable to NIST (or other similar agency) standards. A precalibration check must be conducted prior to the actual calibration to insure that the sensor was within acceptable operational tolerances. A postcalibration check will provide a record of the calibrated values that are to be compared to the precalibration check values, when the unit is recalibrated at the next regular interval.

Figures 15.3(a) and 15.3(b) contain typical pressure and temperature data for the freeze dryers during each of the four phases. The pressure parameters P1 through P14, T_l and T_h are all programmable parameters that can be set and saved, creating a custom cycle profile.

Dry Phase.   The purpose of the dry phase is to remove substantially all moisture from the interior of the freeze dryer and its piping. This moisture includes both standing water and a substantial portion of the vapor that is adsorbed (or absorbed) onto (into) the components of the freeze dryer. Standing water not only shields the surface below from contact with H_2O_2 vapor, but also absorbs (condenses) much of the vapor that comes into contact with it. Desorption and outgassing of excessive amounts of water vapor act like a virtual leak to many pressure (vacuum) sensors and could cause the freeze dryer to fail the leak test that precedes sterilization.

The dry phase begins with one or more evacuations through the drain lines to remove most of the water trapped therein. The moisture that remains after this step is removed by drawing, and maintaining, a deep vacuum on the entire freeze dryer so that rapid evaporation, desorption, and outgassing can occur. However, the cooling that accompanies evaporation may present a problem. The controlled warming of the chamber floor, condenser floor, the shelves, and low spots in the drain piping is effective at countering this problem.

If it is not possible to warm certain areas or components and unacceptable cooling occurs (i.e., standing water freezes or cold spots occur within the freeze dryer), it may be necessary to alternate air breaks and/or air flow-throughs with evacuations of a controlled duration and a controlled depth. The heat that is brought in with the air will help offset the evaporative cooling effects.

The use of a trace moisture sensor such as the Endress & Hauser HygroTwin 2850 to determine the end of the dry phase is ideal. This same sensor may also be of use during the lyophilization run, but it must be shielded to prevent it from being wetted during CIP. The simultaneous use of dual vacuum sensors, such as a Piranhi Gage and a Capacitance Manometer, can also be used to determine the end of the dry phase since the former has a different sensitivity to water vapor than the latter.

It is also possible to determine the duration of the evacuation time or the number of evacuation/air break pulses that will normally dry the chamber and then add additional time or pulses when programming an automatic dry phase. The leak test should automatically catch any unsuccessful dry phases and return to and extend the dry phase.

It is necessary to obtain a temperature map (profile) of the interior of the freeze dryer during the dry phase to verify that the temperature distribution will be within an acceptable tolerance ($\pm 5°C$) of the nominal presterilization temperature at the end of the leak test. The placement of the thermocouples for this profile can remain constant through all four phases of the sterilization cycle.

The dry phase is also useful for particulate control. Any particulate matter in the freeze dryer can become airborne when the vacuum is broken at the end of a lyophilization run. The air rushing in creates turbulence, which can pick up and deposit particles in unstoppered or open product.

The series of evacuations alternating with air breaks/air flow-throughs carried out during the dry phase will tend to purge the chamber of particulate matter by getting such matter airborne and sweeping it out through the vacuum system before any product is introduced into the freeze dryer. The particulate matter after ten evacuations/air breaks in a freeze dryer should be well below the level in a Class 100 clean room.

Leak Test (optional).   The purpose of the leak test is to verify that the interior of the freeze dryer is dry and that the system is vacuum tight. A vacuum sensor, such as an MKS Baritron Capacitance Manometer, which can detect pressure rises due to the presence of water vapor or air, should be used to verify that the vessel is dry and vacuum tight.

The entire system is evacuated to approximately 1 torr (1 mm Hg absolute), all valves are closed, and after a 30-second wait the pressure rise is monitored for an appropriate, preset time interval. If the pressure rise is below the preset value and if all temperature sensors are within the acceptable ranges, the microprocessor will advance the cycle to the sterilize phase. If the pressure rise exceeds the preset value, the microprocessor will return to an extended dry phase. A second leak test will follow the extended dry phase. If the cycle cannot advance to the sterilize phase after a preset number of attempts to pass the leak test, the sterilization cycle will abort and print out the appropriate messages.

The leak test may be considered optional if a moisture sensor is used to determine when the system is dry and if a leak test is performed as an integral part of the lyophilization run.

The sensors that monitor the chamber temperature, shelf temperatures, condenser temperature, and vapor inlet piping (including vent filters) temperatures must be within the acceptable range of their nominal values for the sterilization cycle to advance to the sterilize phase. If they are out of range, the microprocessor will delay advancing the cycle until the values are within range.

The nominal temperature for the chamber, shelves, and condenser will typically be between 25°C and 35°C. The acceptable variation from the nominal value is normally within ±5°C). The nominal temperature for the vapor inlet piping will depend upon the length of the piping, the thermal characteristics of the piping, and the internal volume of the freeze dryer. Short, plastic piping leading to a small pilot-size freeze dryer may not require heating. Long stainless steel sterilant inlet piping will have to be heated to prevent condensation.

To prevent condensation, the piping temperature must be at or above the instantaneous dew point conditions of the vapor. When the vapor passes very quickly through a short length plastic piping, there is a minimum amount of heat transfer and a minimum cooling of the vapor. When the vapor passes through long, restrictive stainless steel piping, there can be a substantial amount of heat transferred from the vapor to the piping if the piping is not heated. Large, production-size freeze dryers will typically require heating of the vapor inlet lines to a minimum of 45°C. The maximum temperature should not exceed 55°C, to minimize degradation of the sterilant before it reaches the interior of the freeze dryer.

Sterilize Phase.   The purpose of the sterilize phase is to effect sterilization by exposing the interior of the freeze dryer to H_2O_2 vapor sterilant while controlling all parameters that can affect the sterilization process. The parameters that must be controlled during this phase include the following:

- Sterilant concentration

- Freeze dryer temperature

- Water vapor concentration (level of saturation)

- Preinjection vacuum level

- Exposure time

The sterilant concentration is controlled by introducing the same amount of sterilant into the same freeze dryer each time. The amount of sterilant is controlled either volumetrically (VHP®DV 30) or by weight (VHP®DV 1000). Near Infrared sensors, such as those offered by UOP and Rosemount Analytical, or a Draeger polytron may also be able to provide a real time indication of the actual sterilant vapor concentration.

Validation of the sterilization phase should be performed while injecting an amount equal to the nominal amount to be injected minus an amount equal to the resolution of the sterilant delivery system (see Table 15.5).

The temperature of the freeze dryer is monitored and controlled during each of the steps preceding sterilization to assure that it is at the correct temperature at the beginning of the sterilize phase. Temperature sensors (RTDs) are permanently located in the chamber, on the shelves, and in the condenser to verify that the temperature is within the specified range. Sterilization at or slightly above ambient temperature will allow freeze dryers without temperature control capability (jacketed chamber and condenser, or electric blanket heater) to be sterilized with H_2O_2 vapor.

The humidity (amount of water vapor) in the freeze dryer is the same every time. It is essentially zero prior to the controlled introduction of the sterilant vapor, which is nearly two-thirds water vapor. According to Schumb et al. (1955), the low temperature water vapor tends to act as a stabilizer to H_2O_2 vapor, thereby increasing its half-life and its sterilization effectiveness.

The preinjection vacuum level is one of the control set points that is typically set at 1 torr. The sterilization capability of H_2O_2 vapor is unaffected by the presence, or absence, of air or other inert gases; however, the penetration capabilities (delivery) of the vapor into sensor dead legs, drain line piping, crevices, and so on is impeded by the presence of other molecules. Evacuation to 1 torr prior to each introduction of sterilant has been shown to provide repeatable penetration of dead legs.

The exposure time is controlled using timers and counters in the microprocessor software. The sterilize phase is comprised of a preset number of exposure pulses, each of which have an identical sterilant exposure time. When the countdown timer indicates that the last sterilize pulse has been successfully completed, the cycle will advance to the aeration phase.

Aeration Phase.   The purpose of the aeration phase is to not only remove the sterilant vapors that are present in the freeze dryer at the completion of the sterilization phase, but also to reduce the residuals that are generated afterward from outgassing and desorption. These residuals must be reduced to below acceptable levels before any product is introduced into the chamber. An acceptable level is one that will not harm the quality, composition, or stability of the product to be lyophilized under the conditions present during a normal lyophilization run.

Aeration typically consists of a series of evacuations, alternated with air breaks and/or air flow-throughs. The temperature of the chamber shelves and any other components that can be warmed should be elevated to above 45°C to facilitate the aeration process.

Validation of the aeration phase should be performed while injecting an amount equal to the nominal amount to be injected plus an amount equal to the resolution of the sterilant delivery system (see Table 15.5).

The level of H_2O_2 residuals can be measured using commercially available devices, such as Draeger detector tubes, a Draeger polytron, or an MDA Scientific Chemcassette™ monitor. The Chemcassete™ monitor and the Draeger polytron can be piped into the lyophilization system to provide automatic monitoring of the residuals in the freeze dryer chamber, in the aseptic core, or in the exhaust discharge from the lyophilizer evacuation system.

Samples of products to be lyophilized should be exposed to the maximum expected (threshold) residual level of H_2O_2 vapor for

Table 15.5. Injection Amount, Resolution, and Quantities Used in Validation

Generator Model	VHP®DV 30	VHP®DV 1000
Injection amount	1	70
Resolution	0.02	2
Quantity used for sterilization validation (worst case)	0.96–0.98	66–68
Quantity used for aeration validation (worst case)	1.02–1.04	72–74

several hours before, or during, lyophilization. The quality, composition, and stability of these samples should be monitored over time to verify that the product was not adversely affected.

It is important to note that an acceptable reduction in product concentration may occur almost immediately (or within the first 30 days) with no subsequent loss in concentration following exposure to a threshold level of residuals.

The aeration phase should be performed in triplicate after an increased injection sterilize phase to verify that the residuals within the chamber will be below the threshold level at the end of the aeration phase. It is recommended that duplicate sensors be used to verify the residual levels during these validation runs.

Minimal particulate matter should be in the lyophilizer at the beginning of the aeration phase because of the purging during the dry phase and the sterilize phase. The filtered air purges, combined with the fact that the H_2O_2 vapor is generated from a very pure liquid, should result in few chamber particulates at the end of the sterilize phase.

The series of evacuations alternating with air breaks/air flow-throughs that are carried out during the aeration phase will continue to purge the chamber of particulate matter before any product is introduced into the freeze dryer. The particulate matter after ten evacuations/air breaks during aeration in the freeze dryer should be approaching that in a Class 10 clean room.

FUTURE PROSPECTS FOR HYDROGEN PEROXIDE VAPOR TECHNOLOGY

In the future both barrier technology and H_2O_2 vapor sterilization will be widely introduced to new areas of aseptic manufacturing. The major advantage of barrier technology is the physical elimination of the chief source of product contamination, the human. Hydrogen peroxide vapor is a sterilant with excellent material compatibility, efficacy, safety, and environmental aspects. New applications for the barrier/H_2O_2 vapor sterilization technologies will be forthcoming as the industrial/pharmaceutical industries become educated about their combined advantages. Also, automation will play a major role in the future of advanced aseptic processing, from simple interfaces and control of sterile environments to complex product handling and transfer systems.

More specifically, in the areas of aseptic manufacturing, refinements to the H_2O_2 vapor sterilization process will allow for shorter exposure times and more rapid aeration of systems, either through enhanced temperature control, improved air handling, or other means yet to be developed. Automatic electronic environmental monitoring, some of which is now available, will be refined and more widely used to assure worker safety. Also, more accurate means for measuring H_2O_2 concentration within the enclosure during sterilization will simplify and improve the validation process, continual verification, and even feedback and control systems for the sterilization equipment itself.

Many previously nonsterilizable lyophilizers will be retrofitted or new lyophilizers will be designed with H_2O_2 vapor sterilization at both a lower capital and operational cost than their steam-sterilizable predecessors. Without the requirements of a pressure vessel and high temperature sterilization, the cost of building, as well as the physical stress of operating (with extreme temperature variations) will be reduced.

New technologies, such as H_2O_2 vapor sterilization system, will continue to provide crucial improvements to sterile process engineering. These advanced systems are an important step forward for the processing and delivery of safe and effective pharmaceutical products to patients.

REFERENCES

Akers, J. E., and J. P. Agalloco. 1993. Aseptic processing. A current perspective. In *Sterilization technology*, edited by R. F. Morrissey, and G. B. Phillips. New York: Von Nostrand Reinhold.

Bayliss, C. E., and W. M. Waites. 1979. The synergistic killing of spores of *Bacillus subtilis* by hydrogen peroxide and ultra-violet light irradiation. *FEMS Microb. Lett.* 5:331–333.

Bradley, A., S. P. Probert, C. S. Sinclair, and A. Tallentire. 1990. Airborne microbial challenges of blow/fill/seal equipment: A case study. *Paren. Sci. Tech.* July/August.

Carlsson, J., and V. S. Carpenter. 1980. The rec A$^+$ gene product is more important than catalase and superoxide dismutase in protecting *Escherichia coli* against hydrogen peroxide toxicity. *Bacter.* 14:319–321.

Edwards, L. M. 1993. Hydrogen peroxide gas sterilization of an enclosed vial filling system. *Pharma. Eng.* March/April.

Gold, P. 1993. Installation of barrier technology on conventional aseptic processing equipment. PharmTech Proceedings '93, Advanstar Communications, September, Atlantic City, NJ.

Heinmets, F., W. W. Taylor, and J. J. Lehman. 1954. The use of metabolites in the restoration of the viability of heat and chemically inactivaed *Eschericia coli. Bacter.* 67:5–12.

Ingraham, A. S. 1992. The chemistry of disinfectants and sterilants. In *Contemporary Topics* (American Association for Laboratory Animal Science) 31(2).

Kawasaki, C., H. Nagano, T. Ito, and M. Kondo. 1970. Mechanism of bactericidal action of hydrogen peroxide. *J. Food Hydg. Soc. Jpn.* 11:155–160.

Leo, F. 1993. Personal interview. Vice President, Research and Development, Automatic Liquid Packaging, Inc., Woodstock, IL.

Miller, T. E. 1954. Killing and lysis of gram–negative bacteria through the synergistic effect of hydrogen peroxide, ascorbic acid, and lysozyme. *Bacter.* 98:949–955.

Pollard, E. C., and P. K. Weber. 1967. Chain scission of ribonucleic acid and in deoxyribonucleic acid by ionizing radiation and hydrogen peroxide *in vitro* and in *Escherichia coli* cells. *Rad. Res.* 32:417–440.

Polyakov, A. A., I. B. Wauryzbaiz, F. K. Valeeva, and A. V. Kulilovshii. 1973. Morphological changes in *Escherichia coli* after the action of hydrogen peroxi de and a mixture of hydrogen peroxide with a surfactant. *Chem. Abs.* 82:642 (640b).

Rickloff, J. R., J. P. Dalmasso, and L. W. Lyhte. 1992. Hydrogen peroxide gas sterilization: A review of validation test methods. Presented at the PDA Annual Meeting, Philadelphia, PA.

Schumb, W. C., C. N. Satterfield, and R. L. Wentworth. 1955. *Hydrogen peroxide.* New York: Reinhold Publishing Corporation.

Sintim–Damoa, K. 1993. Other gaseous sterilization methods. In *Sterilization technology,* edited by R. F. Morrissey, and G. B. Phillips. New York: Van Nostrand Reinhold.

Appendix

<1116> Microbiological Evaluation of Clean Rooms and Other Controlled Environments

United States Pharmacopeial Convention, Inc.

<1116> **Microbiological Evaluation of Clean Rooms and Clean Zones**, page 4042 of *PF* 18(5) [Sept.–Oct. 1992]. The previous proposal is canceled and is replaced by a new proposal that was developed with the collaboration of the Microbiological Control Advisory Panel.

4M01350   (MCB)   RTS—15242-01

Manufacturers of pharmaceuticals or medical devices purporting to be sterile have two options to ensure the sterility of their final products. One is to sterilize the product-container-closure system terminally, that is, to subject the final filled and closed container to a validated sterilization process resulting in a very high sterilization process assurance level. Another option that has been used is aseptic processing, which involves the separate sterilization of the product and of the package (container and closure). Although a full discussion of sterility assurance is included under *Sterilization and Sterility Assurance of Compendial Articles* <1211>, a brief review of the subject is important in the context of this chapter. Sterility assurance in the pharmaceutical, diagnostic, and medical device industries is generally presented as a probability. However, from a regulatory, compendial, and legal perspective, the expectation is that every unit ready for commercial distribution is sterile. Regardless of the manufacturing technology employed or the acceptance criteria described in the literature, the objective in the manufacturing of sterile products is to ensure that every released unit is free of microbial contamination. This goal can be approached in aseptic manufacturing only if manufacturers maintain a high level of attention to detail, constant discipline, and continuous supervision, and if the product is properly formulated.

A large proportion of sterile products are manufactured by aseptic processing. Because aseptic processing relies on the removal and then the exclusion of microorganisms from the process stream and the prevention of microorganisms from entering open containers and closures during filling, inherent product bioburden as well as microbial bioburden of the manufacturing environment are important factors relating to the level of sterility assurance of these products. The current trend toward proteinaceous products, including biotechnology-derived products and other products of natural origin that cannot be subjected to terminal sterilization, places an even greater emphasis on suitable control of the aseptic environment, particularly when these products do not contain preservative systems.

The heat stability of the product itself is only one factor to be considered in the evaluation of manufacturing technology. The container-closure system may also be affected by exposure to terminal sterilization conditions. In addition, certain dosage forms, such as combination products, may not be amenable to heat treatments. These products may have clear advantages in terms of convenience of patient use or in terms of safety in administration. From the perspective of public health, the sterility of health care products must be considered at all steps leading to the administration of the product; therefore, the possibility of product contamination at the point of administration or in preparing a product for administration cannot be ignored. In some cases an aseptically filled product could afford a higher level of assurance of sterility than a terminally sterilized product that required manipulation prior to patient delivery. The decision path concerning manufacturing technology can be a complex one; the emergence of newer "advanced" aseptic processing technologies has further complicated the selection of sterile product manufacturing methodology. Fortunately, these technologies promise to raise to higher levels of product safety and quality processes that depend upon the exclusion of microbial contamination.

A long-standing principle and the current regulatory thinking is that if a product can be terminally sterilized, it should be terminally sterilized. The onus of proving that a product cannot be terminally sterilized (for example, because of stability problems), and the determination of when and if a product is a candidate for aseptic processing is on the manufacturers of sterile products. The purpose of this informational chapter is to review the various issues that relate to aseptic processing in the establishment, maintenance, and control of the microbiological quality of clean rooms and other controlled environments.

This informational chapter includes discussions on (1) the background of clean room classifications based on "particulate" target number; (2) a microbiological evaluation program associated with controlled environments; (3) training of personnel; (4) critical factors involved in the design and implementation of an environmental control program; (5) development of a sampling plan, including sampling sites; (6) the establishment of microbiological Alert and Action levels; (7) the various methodologies and instrumentation used for airborne and surface equipment sampling; (8) types of media and diluents used; (9) the identifcation of environmental microorganisms; (10) operational evaluation of the microbiological status via media fills; and (11) a glossary of terms used in this chapter. Excluded from this chapter is a discussion of controlled environments for use by licensed pharmacies in the preparation of sterile products for home use, which is covered under *Sterile Drug Products for Home Use* <1206>.

There are alternative methods to assess and control the microbiological status of controlled environments that are not contained in this chapter. The contents of this chapter are informational, and other methods that are equivalent may be employed. The numerical values for quantifying a controlled environment's contamination are not intended to represent absolute values. Given the wide variety of sampling equipment and methods in use, one cannot reasonably suggest that the attainment of these values guarantees the needed level of microbial control, or that excursions beyond these values indicates a loss of control. Microbiological sampling and analysis by its very nature has a significant associated variability and a potential for inadvertent contamination of sampling media and devices. The ways, processes, and methods indicated in this chapter are not normative or requirements, but only informational.

Establishment of Clean Room Classifications

The design, construction, and mode of operation of clean rooms and clean zones are covered in detail in Federal Standard 209E. This standard of air cleanliness is based on specific concentrations of airborne particles. Methods for verification of air cleanliness class and monitoring the airborne particles also are included. This document does not apply to particles deposited on equipment or supplies within the clean room and, among others, is not intended to characterize the viable nature of the particles.

The application of Federal Standard 209E to clean rooms and clean zones in the pharmaceutical industry has been used by manufacturers of clean rooms to provide a specification for building, commissioning, and maintaining these clean rooms. However, even with a large body of data available in the pharmaceutical industry, there is no agreement on a relationship between the number of airborne particles in a clean room and the concentration of airborne viable microorganisms.

The criticality of the number of airborne particles in the electronic industry makes the application of Federal Standard 209E a necessity, while the criticality of the presence of airborne microorganisms in clean rooms in the pharmaceutical industry makes that industry less inclined to apply all the provisions of that standard.

The reasoning that the fewer particulates present in a clean room, the less likely it is that airborne microorganisms will be present can be accepted broadly and can provide the aseptic processing manufacturers and the builders of clean rooms and other controlled environments with building criteria as a point of departure in the establishment of an aseptic processing line.

Federal Standard 209E, as used in the pharmaceutical industry, deals with classification of clean rooms based on limits of airborne particles with sizes equal to or larger than 0.5 μm. For comparison purposes, Table 1 on Airborne Particulate Cleanliness Classes in Federal Standard 209E has been adapted to the pharmaceutical industry needs. It is generally accepted that if fewer particulates are present in an operational clean room or other controlled environments, the microbial count under operational conditions will be less.

Importance of a Microbiological Evaluation Program for Controlled Environments

It is widely recognized that the monitoring of total particulate count in controlled environments, even with the use of electronic instrumentation on a continuous basis, does not provide appropriate information on the microbiological status of the environment in these processing systems. The basic liminitation of particulate counts is that they measure particles of 0.5 μm (or less); while microorganisms are not free-floating or single cells, they are frequently associated with particles of 10 to 20 μm. Particulate counts as well as microbial counts within a controlled environment may vary with the sampling loction and the activities being conducted during sampling. Monitoring of the environment for particulates and for microorganisms is an important control function because they both relate to the compendial requirements for *Particulate Matter* and *Sterility* under *Injections* <1>.

Table 1. Airborne Particulate Cleanliness Classes.*

| Class Name | Size of particles equal to and larger than 0.5 μm | | |
S1	U.S. Customary	(m³)	(ft³)
M1	—	10.0	0.283
M1.5	1	35.3	1.00
M2	—	35.3	1.00
M2.5	10	353	10.0
M3	—	1,000	28.3
M3.5	100	3,530	100
M4	—	10,000	283
M4.5	1,000	35,300	2,000
M5	—	100,000	2,830
M5.5	10,000	353,000	20,000
M6	—	1,000,000	28,300
M6.5	100,000	3,530,000	100,000
M7	—	10,000,000	283,000

*Adapted from U.S. Federal Standard 209E, September 11, 1992—"Airborne Particulate Cleanlinesss Classes in Clean Rooms and Clean Zones."

An environmental microbial monitoring program for controlled environments should assess the effectiveness of cleaning and sanitization and monitor operator practices that could have an impact on the microbial status of the controlled environment. Microbial monitoring, regardless of how sophisticated the system may be, will not and need not identify and quantitate all microbial contaminants present in these controlled environments. However, routine microbial monitoring should provide sufficient information so that decisions can be made with respect to the adequacy of programs designed to control the microbial status of controlled environments within the established parameters.

Environmental microbial monitoring and analysis of the results by defined individuals will permit decisions related to activities being performed in these environments. To allow for the collection of meaningful data, the environment should be sampled during normal operations. Environmental microbial sampling should occur when materials are in the area, processing activities are ongoing, and a full complement of operating personnel is on site.

Microbial monitoring of controlled environments should include quantitation of the microbial content of air, surfaces, equipment, sanitization containers, floors, walls, and personnel garments (e.g., gowns, boots, gloves, and masks). The objective of the microbial monitoring program is to obtain representative estimates of microbial status of the environment at a moment in time. When data are compiled and analyzed, data trends may become evident and should provide appropriate information against which decisions can be made by trained and responsible individuals. While it is important to review environmental results on a daily basis, it is also critical to review results over extended periods to determine whether trends can be ascertained. The microbial status of controlled environments can be assessed and evaluated on the basis of these trend data. Periodic reports or summaries should be issued to alert the managers responsible.

When the specific microbial level of a controlled environment is exceeded, a documentation review and investigation should occur. These should include a review of area maintenance documentation, including sanitization documentation; a review of the inherent physical or operational parameters or both; a review of the training status of personnel involved; reinforcement of training of personnel to emphasize the microbial control of the environment; additional sampling at increased frequency; additional sanitization; additional product testing; and an evaluation of the need for an additional *media fill*.

Based on the review of the investigation and testing results, a decision can be made on the significance of the "out-of-specification" event and the acceptability of the operations or products processed under that condition. Any investigation and the rationale for the course of action should be documented and included as part of the overall quality management system of the controlled environment.

A controlled environment such as a clean zone or clean room is characterized by certification according to a relevant clean room operational standard. Parameters that are tested in a clean room include filter integrity, air velocity, air directions, air changes, and pressure differentials. These parameters potentially affect the microbiological status of the clean room operation. The design, construction, and operation of clean rooms varies greatly, making it difficult to generalize requirements for these parameters. Adjunct to the above physical test parameters is a risk assessment analysis based on air-flow patterns that can provide information useful in the design and development of clean rooms and implementation of microbiological environmental control programs. This approach to a hazard- and critical-point analysis may be more feasible than the use of a viable microorganism challenge. A method for conducting a particulate challenge test to the system by increasing the ambient particle concentration in the vicinity of critical work areas and equipment has been developed by Ljungquist and Reinmuller.[1] It consists of a two-step procedure. The first step is to use smoke generation equipment to visualize the air movements throughout a clean room or a controlled environment. The presence of vortices or turbulent zones can be identified, and the air-flow pattern may be fine-tuned to eliminate or minimize these effects. The second step in the procedure is to generate particulate matter close to the critical zone and *sterile field* by means of air-current test tubes. This evaluation is done under simulated production conditions, but with equipment and employees in place.

Proper testing and optimization of the physical characteristics of the clean zone or clean room is essential prior to completion of the validation of the microbiological environment program. Assurance that the clean room is operating adquately and according to its physical specifications will give a higher probability that the microbiological status of the clean room will be appropriate for aseptic processing. These tests should be repeated during routine certification of the clean room or controlled environment and whenever changes are made to the operation, including, but not limited to, the room's personnel flow, processing, operation, material flow, or equipment layout.

Training of Personnel

Training of all individuals working within aseptic processing areas, sterility testing sites, or in any other areas in controlled environments, is critical. This training is also critical for personnel responsible for the microbial monitoring program, where contamination of the clean working area could inadvertently occur during microbial sampling by personnel who do not use appropriate aseptic precautions. In highly automated operations, the monitoring personnel may be the employees who have the most direct contact with the critical zones within the processing area.

---

[1]Interaction Between Air Movements and the Dispersion of Contaminants: Clean Zones with Undirectional Air Flow, *Journal of Parenteral Science and Technology*, 47(2), 1993.

Because microbiological sampling often requires complex manipulations, the potential for microbial contamination due to inappropriate sampling techniques should be of concern. To avoid sampling-induced bias or the potential for extrinsic contamination of products, a formal personnel training program is required. This formal training must be documented for all personnel assigned to controlled environments.

Management of the facility must assure that all individuals responsible for the administration and general operation of the environmental control program, including the samplers and laboratory technicians, are well versed in basic microbiological principles. This training should include basic bacteriology and mycology. Relevant subdisciplines include microbial physiology, disinfection and sanitation, media selection and preparation, growth and metabolic considerations, taxonomy, and sterilization principles. Personnel involved in microbial identification will require specialized training on all required laboratory methods and on any automated instruments that are used. Additional training on the management of the environmental data collected must be provided to personnel. The staff must have a clear understanding of the alert and action levels, of trend analysis, and of report writing. Knowledge and understanding of applicable standard operating procedures is critical, especially those standard operating procedures relating to specific corrective measures that must be taken when environmental conditions so dictate. Understanding of regulatory compliance policies and each individual's responsibilities with respect to good manufacturing practices (GMPs) should be an integral part of the training program.

The maintenance of aseptic environmental conditions by both production operators and environmental sampling personnel during the manufacturing of sterile products requires that appropriate training be given to both groups.

The major recognized source of microbial contamination of aseptically processed products is the personnel involved in processing. This is the premise that justifies the development and use of automation and barrier technology in aseptic processing, where personnel are essentially removed from contact with the environment of the process. Contamination occurs from the spreading of microorganisms by individuals, particularly those with active infections as well as from individuals who are asymptomatic or convalescent. Individuals with obvious respiratory or pulmonary infections as well as with obvious skin infections should not be allowed in controlled environments.

These facts underscore the importance of good personal hygiene and a careful attention to detail in the gowning procedure used by employees entering the controlled environment. Once these employees are properly gowned, they must be careful to maintain the integrity of their gloves and suits at all times. Since the major threat of contamination of product being aseptically processed comes from the operating personnel, the control of microbial contamination associated with these personnel is one of the most important elements of the environmental control program.

The importance of thorough and well-documented training of personnel involved in controlled environments as well as good employee aseptic technique cannot be overemphasized. However, the environmental monitoring program by itself, no matter how well conceived and implemented, will not be able to detect all transgressions in aseptic processing that could compromise the microbiological quality of the product. These above factors, including aseptic gowning, a sound environmental monitoring program, and a properly classified clean room, cannot by themselves give total assurance of sterility in aseptic manufacturing. The aseptic process requires periodic media-fill studies to revalidate the process and to ensure that the appropriate operating controls and training are effectively maintained.

Critical Factors Involved in the Design and Implementation of a Microbiological Environmental Control Program

An environmental control program should be capable of detecting adverse processing conditions in a timely fashion that would allow for meaningful and effective corrective actions. It is the responsibility of the manufacturer to develop, initiate, implement, and document such a microbial environmental monitoring program. It is essential that the environmental control program be designed and implemented in a manner that precludes the possibility of the sampling activities themselves contributing to the contamination of the product.

Although general factors involved in an environmental control program will be discussed, it it imperative that such a program be tailored to specific facilities and conditions. For instance, the selection of media used to detect and quantitate microorganisms should be made with the microflora of the specified facility in mind. A general microbiological growth medium such as Soybean Casein Digest Medium should be suitable in most cases. However, this medium may be supplemented with additives to overcome or to minimize the effects of sanitizing agents used in the controlled environment, or of antibiotics if they are aseptically processed in these environments. The selected media should be capable of detecting and quantifying microorganisms of human, water, or soil origins. The detection and quantitation of yeasts and molds is also required on a periodic basis. Thus, the development of a monitoring program shold include initial screening with a variety of media to determine the characteristic microflora present in the facility. It is, of course, generally accepted that one cannot design a control program that could detect all species of bacteria, yeasts, and molds that might be present in the controled environment. In addition to Soybean Casein Digest Medium, the use of a general mycological medium such as Sabouraud's, Modified Sabouraud's, or Inhibitory Mold Agar is also recommended. In general, testing for obligatory anaerobes is not performed frequently. However, should conditions that are suitable for the presence and growth of anaerobes exist, the ability of the selected media to detect and quantitate these anaerobes or microaerophilic microorganisms should be validated.

Another factor involved in the design of a control program is the selection of time and incubation temperatures once the appropriate media have been selected. Typically, incubation temperatures in the $22.5 \pm 2.5°$ and $32.5 \pm 2.5°$ have been used with an incubation time of 48 to 72 hours. These times and temperatures may vary depending on the predominant microflora or the types of media selected.

Another critical area relates to the sterilization methods used to prepare growth media for the environmental program. Sterilization cycles must be validated and, in addition, media must be tested for sterility and for growth promotion prior to or concurrent with use.

The sterility of the media can be assessed by incubation of a representative number of prepared plates or tubes at the appropriate time and temperature. The growth promotion of media should be determined with microorganisms listed in the section *Growth Promotion* under *Sterility Tests* <71>. In addition, representative microflora isolated from the controlled environment should also be used to test media for growth promotion. Media must be able to support growth when inoculated with less than 100 cfu of the challenge organisms.

An appropriate environmental control program should include validation of methods for microbiological sampling of the environment and of sampling sites. For instance, the environmental air sampler that will be used must be capable of being operated in an aseptic manner. As indicated under training and because air samplers generally require extensive manipulations, prior training of technicians before they actually before they actually become part of the control program is a requirement. It is also expected that the sampling devices operate in a manner that is not disruptive to the manufacturing environment. The use of devices that create air wakes or turbulence disruptive to unidirectional air flow should be avoided, especially within the critical zone or near open product containers or components in the aseptic processing environment. Sampling devices that are large or bulky also can disrupt unidirectional air flow and can increase the potential for sloughing or contaminants from personnel involved in setting up and operating these devices. Wherever possible, air samplers that use remote tubing probes should be employed. Finally, sampling devices and sampling sites should not be invasive to the manufacturing process. For instance, surface sampling (contact plates or swabs) should not ordinarily be done during manufacturing. Samplers as well as sampling devices must not violate the *sterile field* in any way during aseptic product manufacturing. The use of swab techniques and surface sampling in sterile fields or within sterile isolation systems is not desireable. The boundaries of the sterile field must be defined as being sufficiently distant from the product so that turbulence created by hands and arms of sampling technicians or by sampling devices does not affect the unidirectional air flow as it approaches the production line or area.

The methods used for identification of isolates shold be validated with indicator microorganisms (see *Microbial Limits* <61>) as well as with the most common environmental isolates within the controlled environment being monitored. Where rapid identification procedures are employed, data should be available demonstrating equivalence with compendial methods.

Establishment of Sampling Plan and Sites

During initial start-up or commissioning of a clean room or other controlled environment, the locations for air and surface sampling should be determined. Consideration should be given to the proximity to the product and whether air and surfaces that may contain microorganisms might be in contact with a product. Such areas should be considered product contact areas or critical areas requiring more monitoring than non-product-contact areas. Locations where intensive or high-volume movement of personnel occur should be considered as critical areas for monitoring. In a parenteral vial filling operation, areas would typically include loading of the stopper bowl, placement of vials on the filling line, and the sanitization of bottles or vials that personnel routinely handle.

The frequency of sampling will depend on the criticality of specified sites and the subsequent treatment received by the product after it has been aseptically processed. Table 2 shows suggested frequencies of sampling in decreasing order of frequency of sampling and in relation to the criticality of the area of the controlled environment being sampled.

The rationale for the relative importance of routine microbiological monitoring is based on the nature of the product being manufactured and on the nature of the processes involved. As manual interventions during the manufacturing operations increase, and as the potential for presonnel contact with the product increases, the

Table 2. Frequency of Sampling on the Basis of Criticality of Controlled Environment.

Sampling Area	Frequency of Sampling
Class 100 or better room designations	Each shift
Supporting areas to Class 100 (e.g. Class 10,000)	Daily
Other support areas (Class 100,000)	Twice/week
Potential product/container contact areas	
Other support areas to aseptic processing areas but non-product contact (Class 100,000 or unclassified)	Once/week
Terminally Sterilized Drug Products[1]	
Product contact areas[2]	Once/1 week
Non-product contact area[2]	Once/2 weeks
Terminally Sterilized Devices[1]	
Product contact areas[2]	Once/2 weeks
Non-product contact area	Once/3 weeks
Sterile Enteral Product	
Product contact areas[2]	Once/2 weeks
Non-product contact areas	Once/3 weeks

[1]When the product has a product bioburden testing and analysis program.

[2]If historical data that cover over a 12-month period reflect consistent results with no major shifts or changes in bioburden count, the sampling frequency can be reduced to once every 3 weeks.

relative importance of an environmental monitoring program increases. Environmental monitoring is more critical for products that are aseptically processed than for products that are processed in an almost aseptic manner then terminally sterilized. For those products that are terminally sterilized, the determination and quantitation of microorganisms resistant to the subsequent sterilization treatment is more critical than the microbiological environmental monitoring of the surrounding manufacturing environments. This in no way indicates that that routine monitoring of controlled environments for products that are terminally sterilized is not important, but that its relative value to a bioburden program is lower. However, if the terminal sterilization cycle is not based on the overkill cycle concept but on the bioburden prior to sterilization, the value of the bioburden program is critical.

Establishment of Microbiological Alert and Action Levels in Controlled Environments

An *Alert level* in microbiological environmental monitoring is that level of microorganisms that gives an early warning of potential drift from normal operating conditions. Exceeding the Alert level is not necessarily grounds for definitive corrective action, but it should at least prompt a follow-up investigation.

An *Action level* in microbiological environmental monitoring is that level of microorganisms that when exceeded requires immediate follow-up and corrective action.

During initial start-up of a facility, at least several weeks of data on microbial environmental levels should be reviewed to establish a baseline.

Although Alert levels are specific for the capabilities of a given facility, the target for Action level should be uniform across facilities and across manufacturers' controlled environments. This will ensure a consistent controlled environment from the microbial environmental perspective, similar to the consistent controlled environment from the total particulate perspective obtained through the application of Federal Standard 209E or similar material or international guidance documents.

Microbial Considerations and Action Levels for Controlled Environments

Classification of clean rooms and other controlled environments is based on Federal Standard 209E based on total particulate counts for these environments. The pharmaceutical and medical devices industries have generally adopted the classification of Class 100, Class 10,000, and Class 100,000, especially in terms of construction specifications for the facilities.

Although there is no direct relationship established between the 209E classes of controlled environments and microbiological levels, the pharmaceutical industry has been using microbial levels corresponding to these classes for a number of years, and these levels have been those used for regulatory compliance.[2] These levels have been shown to be readily achievable with the current technology for controlled environments. There have been reports and concerns about differences in

[2]NASA, 1967—Microbiology of Clean Rooms.

these values obtained using different sampling systems, media variability, and incubation temperatures. Overall, however, these microbial levels (see Tables 3, 4, and 5) have been flexible enough to accommodate and minimize these variables. The values shown in these tables represent individual test results and, as such, are not intended to be averaged or manipulated since this will provide a false sense of the appropriateness of the microbial environmental status of the controlled environment. If documented sampling or testing errors

Table 3. Air Cleanliness Levels in Colony-Forming Units (cfu) in Controlled Environments (Using a Slit-to-Agar Sampler or Equivalent).

Class*		cfu per cubic meter of air**	cfu per cubic feet of air
SI	U.S. Customary		
M3.5	100	Less than 3	Less than 0.1
M5.5	10,000	Less than 20	Less than 0.5
M6.5	100,000	Less than 100	Less than 2.5

*As defined in Federal Standard 209E, September 1992.

**A sufficient volume of air should be sampled to yield finite results.

Table 4. Surface Cleanliness Levels of Equipment and Facilities in cfu in Controlled Environments.

Class*		cfu per Contact Plate*
SI	U.S. Customary	
M3.5	100	3
M5.5	10,000	5
		10 (floor)

*Contact plates areas vary from 24 to 30 cm². When swabbing is used in sampling, the area covered should be greater than or equal to 24 cm² but no larger than 30 cm².

Table 5. Surface Cleanliness of Operating Personnel Gear in cfu.

Class			cfu per Contact Plate*	
SI	U.S. Customary	Gloves	Mask/Boots/Gowns	
M3.5	100	3	5	
M5.5	10,000	20	10	

*See in Table 4 under (*).

occur, the results obtained should be discarded, and adequate retraining and precautions to eliminate these errors should be implemented.

Methodology and Instrumentation for Quantitation of Viable Airborne Microorganisms

It is generally accepted by regulatory agencies as well as by industry that airborne microorganisms in controlled environments can influence the microbiological quality of the intermediate or final products manufactured in these areas. Also, it generally is accepted that, as with other microbiological assays, that estimation of the airborne microorganisms can be affected by instruments and procedures used to perform these assays. The importance of defining examples of instrumentation and methodology used in environmental microbial monitoring is generally accepted. Where alternative methods or equipment is used, the general equivalence of the results obtained should be ascertained. The technology for airborne microbial quantitation has shown few changes since the development of impingement samplers over fifty years ago. Advances in technology in the future are, however, expected to bring innovations that would offer greater precision and sensitivity than the current available methodology.

In 1959, the United States Communicable Disease Center, now the Centers for Disease Control (CDC), published a document on "Sampling Microbiological Aerosols." It is a useful source of the historical review of these technologies but does not cover some of the more recent units designed for airborne sampling that are commonly used in the industry. Several basic types of air samples are described in the CDC document and include impaction samplers, liquid impingement samplers, filtration samplers, centrifugal samplers,

482 **Aseptic Pharmaceutical Manufacturing II**

sedimentation samplers, and electrostatic precipitation samplers. Although units of all of the above types have some merit in the quantitation of airborne viable microorganisms bioburden, not all of them are currently commercially available.

The most commonly used samplers in the U.S. pharmaceutical and medical device industry today are the impaction and centrifugal samplers. A number of commercially available samplers are described below.

Slit-to-Agar Air Sampler (STA)—This sampler is the instrument upon which the microbial levels given for the various controlled environments are based. The unit is activated by an attached source of vacuum that is controllable. The air intake is obtained through a standardized slit below which is a slowly revolving Petri dish containing a nutrient agar. The dish slowly revolves at set speeds. Particles in the air containing variable microorganisms impact on the agar surface. To accommodate sampling in small or restrictive areas, a remote air intake is usually available. Its use is preferred to minimize disturbance of the laminar flow pattern.

Sieve Impactor—The unit consists of a container designed to accommodate a Petri dish containing a nutrient agar. The cover of the unit is perforated, with the perforations of a predetermined size. A vacuum pump draws air through the cover and the particles in the air containing microorganisms impact on the agar medium in the Petri dish. Some samplers are available with a cascaded series of containers with each of the covers of the consecutive units containing perforations of decreasing size. These units allow for the determination of the distribution of the size ranges of particulates containing variable microorganisms.

Centrifugal Sampler—The unit consists of a propeller or turbine that pulls air into the unit and then propels the air outward to impact on a tangentially placed nutrient agar strip set on a flexible plastic base.

Sterilizable Microbiological Atrium—The unit is a variant of the single-stage sieve impactor. The unit's cover contains uniformly spaced orifices approximately 0.25 inch in size. The base of the unit accommodates one Petri dish containing a nutrient agar. A vacuum pump controls the movement of air through the unit and multiple-unit control center as well as a remote sampling probe are available.

Surface Air System Sampler—This integrated unit consists of an entry section that accommodates an agar contact plate. Immediately behind the contact plate is a motor and turbine that pulls air through the unit's perforated cover over the agar contact plate and beyond the motor, where it is exhausted. Multiple mounted assemblies are also available.

Gelatin Filter Sampler—The unit consists of a vacuum pump with an extension hose terminating in a filter holder that can be remotely lcoated in the critical space. The filter consists of random fibers of gelatin capable of retaining airborne microorganisms. After a specified exposure time, the filter is aseptically removed and dissolved in an appropriate diluent and then plated on an appropriate agar medium to estimate its microbial content.

One of the major limitations of mechanical air samplers is the limitation in sample size of air being tested. Where the microbial level in the air of a controlled environment is expected to contain not more than 3 cfu/M^3, it is necessary to ensure that several M^3 of air be tested if results are to be assigned a reasonable level of precision and accuracy. Typically, slit-to-agar samplers have an 80-liter-per-minute sampling capacity (the capacity of the surface air system is somewhat higher) that would require an exposure time of 15 minutes to achieve one M^3 of air sampling. It would be preferable to use at least a 1-hour sampling period to obtain a representative sample of the environment. Although one manufacturer of the slit-to-agar sampler has indicated a capability of sampling of 700 liters of air per minute that would then be able to sample one M^3 in less than 2 minutes, the potential for disruption of the laminar flow in the critical area and the creation of a turbulence would increase the probability of contamination due to the sampling methodology.

For centrifugal air samplers, a number of studies have shown that the sampler demonstrates a selectivity for larger particles. Large particles are probably more likely to include viable microorganisms. For instance, human skin fragments at least 10 μm in size will usually contain microorganisms. The use of this type of sampler may result in higher airborne counts than the other types of air samplers because of the inherent selectivity for larger particles. This phenomenon of selectivity for larger particles is not evident when controlled microbial aerosols of single microbial cells are studied in recovery studies comparing different types of samplers.

The centrifugal sampler has been shown also to disrupt the linearity of the airflow in the controlled zone where it is placed for sampling. Newer models of the centrigual sampler have been developed that have reduced, but not eliminated, this propensity. Regardless of the type of sampler used, the use of a remote probe requires that the extra tubing does not have an adverse effect on the viable airborne count. This effect should either be eliminated or, if this is not possible, a correction factor should be introduced in the reporting of results.

The use of liquid impingers is not recommended because they are not readily available commercially. In addition, spurious results may be obtained due to either proliferation of sampled microorganisms in the aqueous medium or fragmentation of the clumps of microorganisms due to the turbulence created and the sudden impact on a liquid medium.

Methodology and Equipment for Sampling of Viable Surfaces for Quantitation of Microbial Containmants in Controlled Environments

Another component of the microbial environmental control program in controlled environments is surface sampling of equipment, facilities, and personnel gear used in these environments. The standardization of surface sampling methods and procedures has not been as widely addressed as the standardization of air sampling procedures. To minimize disruptions to critical operations, surface sampling should be done at the conclusion of operations. Surface sampling may be accomplished by the use of *contact plates* or by the *swabbing method*. Contact plates filled with nutrient agar are used when sampling regular or flat surfaces and are directly incubated at the appropriate time for a given incubation temperature for quantitation of viable counts. Specialized agar can be used for specific quantitation of fungi, spores, etc.

The swabbing method is used for sampling of irregular surfaces, especially for equipment. Swabbing can also be used to supplement contact plates for regular surfaces. The swab is then placed in an appropriate diluent and the estimate of microbial count is done by plating of an appropriate aliquot on or in specified nutrient agar. The area to be swabbed is defined using a sterile template of appropriate size. In general, it is in the range of 24 to 30 cm^2. The microbial estimates are reported per contact plate or per swab.

Culture Media and Diluents Used for Sampling or Quantitation of Microorganisms

The type of medium, liquid or solid, that is used for sampling or quantitation of microorganisms in controlled environments will depend on the procedure and equipment used. An all-purpose medium is Soybean-Casein Digest Agar when a solid medium is needed. Other media, liquid or solid, are listed below.

Liquid Media*	
Media	Per liter of water
Tryptose saline	
Tryptose	1 g
Sodium Chloride	5 g
Adjust pH to 7.0 ± 0.2	

Continued on next page.

Continued from previous page.

Media	Per liter of water
Peptone Water	
Peptone	20 g
Sodium Chloride	5 g
Adjust pH to 7.0 ± 0.2	
Buffered Saline	
Sodium Chloride	8.5 g
Disodium Phosphate (anhydrous)	5.8 g
Potassium Dihydrogen Phosphate (anhydrous)	3.5 g
Adjust pH to 7.0 ± 0.2	
Buffered Gelatin	
Gelatin	2 g
Disodium Phosphate (anhydrous)	4 g
Adjust pH to 7.0 ± 0.2	
Enriched Buffered Gelatin	
Gelatin	2 g
Disodium Phosphate (anhydrous)	4 g
Brain-Heart Infusion	37 g
Adjust pH to 7.0 ± 0.2	
Buffered Water (Stock)	
Potassium Dihydrogen Phosphate	34 g
Adjust pH to 7.2. Add 1 to 2 mL of stock to 1 liter of distilled water.	
Brain Heart Infusion	
Infusion from Calf Brain	200.0 g
Infusion from Beef Heart	250.0 g
Peptone	10.0 g
Sodium Chloride	5.0 g
Disodium Phosphate	2.5 g
Dextrose	2.0 g
Final pH 7.4	
Soybean-Casein Digest Medium	
Pancreatic Digest of Casein	17.0 g
Papaic Digest of Soybean Meal	3.0 g

Continued on next page.

Continued from previous page.

Media	Per liter of water
Sodium Chloride	5.0 g
Dibasic Potassium Phosphate	2.5 g
Dextrose	2.5 g
pH after sterilization 7.3	
Soybean-Casein Digest Agar Medium	
Pancreatic Digest of Casein	15.0 g
Papaic Digest of Soybean Meal	5.0 g
Sodium Chloride	5.0 g
Agar	15.0 g
pH after sterilization 7.3	
Nutrient Agar	
Peptone	5.0 g
Beef Extract	3.0 g
Agar	15.0 g
Final pH 6.8	
Tryptone Glucose Extract Agar	
Tryptone	5.0 g
Beef Extract	3.0 g
Glucose	1.0 g
Agar	15.0 g
Final pH 7.0	
Lecithin Agar	
Peptone	5.0 g
Dextrose	1.0 g
Beef Extract	3.0 g
Lecithin	1.0 g
Polysorbate 80	7.0 g
Agar	150.0 g
Final pH 7.0	
Brain Heart Infusion Agar	
Infusion from Calf Brain	200.0 g
Infusion from Beef Heart	250.0 g
Peptone	10.0 g
Sodium Chloride	5.0 g

Continued on next page.

Continued from previous page.

Media	Per liter of water
Disodium Phosphate	2.5 g
Dextrose	2.0 g
Agar	15.0 g
Final pH 7.4	
Contact Plate Agar	
Beef Extract	3 g
Tryptone	5 g
Dextrose	1 g
Agar	15 g
Sorbitan Monooleate	7 g
Lecithin	1 g

*Liquid and solid media are sterilized in an autoclave using a validated cycle.

Alternative media to those listed can be used provided that they are validated for the purpose intended. Commercially available prepared media can also be used, but they also have to fulfill the requirements for the media that are listed here and must be validated for the purpose intended (see *Critical Factors Involved in the Design and Implementation of Environmental Control Program*).

Identification of Microbial Isolates from the Environmental Control Program

The control program includes an appropriate level of identification of the flora obtained by sampling. A knowledge of the normal flora in controlled environments aids in determining the usual microbial flora anticipated for the specific facility and the evaluation of the effectiveness of cleaning and sanitizaiton procedures, methods, .and agents. The information gathered by an identification program can also be useful in the investigation of the source of contamination, especially when the Action levels are exceeded, or in making documented decisions on the disposition of products produced under these conditions.

Identification of isolates from critical areas should take precedence over identification of microorganisms from noncritical areas. Regardless of the use of manual, semimanual, or automated systems of microorganisms, identification methods should be validated (see

Critical Factors Involved in the Design and Implementation of Environmental Control Program).

Operational Evaluation of the Microbiological Status of Clean Rooms and Other Controlled Environments

The microbiological status of a controlled environment is monitored through an appropriate environmental monitoring program. Additional information on the evaluation of the microbiological status of the controlled environment can be obtained by the use of *media fills*. An acceptable media-fill shows that a successful simulated product run can be conducted on the manufacturing line at that point in time. However, other supporting programs must exist, such as appropriate construction of facilities, environmental monitoring, training of personnel, and processing steps prior to aseptic processing.

When an aseptic process is developed and installed it is generally expected to qualify the microbiological status of the process by running three consecutive media fills. A media fill utilizes growth medium in lieu of specific products to detect the growth of microorganisms. There are several issues in the concept of media fill that need to be addressed: the media-fill procedures, media selection, fill volume, incubation, storage and inspection of filled units, interpretation of results, and possible corrective actions required.

Since a media fill is designed to stimulate an actual aseptic processing of a specified product, it is important that conditions that will occur during a normal run also be in effect during the media fill. This includes the full complement of personnel and all the processing steps and materials that constitute a normal processing. Further, during the conduct of media-fill runs various predocumented interventions that are known to occur during actual product runs should be planned.

Alternatively, in order to add a safety margin, a "worst-case" condition can be used. Examples may include frequent stop-and-go sequences, unexpected repair of processing system, replacement of filters, etc. If a media-fill run is conducted under idealized conditions (i.e., normal conditions), it might provide a biased view of the true microbiological control status of the process. It would be inappropriate to allow under normal production conditions a lengthy and highly manipulative repair sequence unless the capability of the process to withstand such a challenge has been evaluated during a media-fill run. The qualification of an aseptic process need not be done for every product, but must be done for each processing line used. since the geometry of the container (size as well as opening of the container) and the speed of the line are factors that are variable in the use of an aseptic processing line, appropriate combination of these factors, preferably at the extremes, should be used in the qualification of the line.

The 1987 FDA Guideline on Sterile Drug Products Produced by Aseptic Processing indicates that media-fill runs be done to cover all production shifts for line/product/container combinations. This requirement must be considered not only for qualification media-fill runs, but also for periodic reevaluation or revalidation. It is an appropriate practice to vary the time of day during which the runs are conducted. It is inappropriate to conduct them only at the beginning of a work day, and they must also be done in a manner that simulates the end of a given manufacturing lot. For example, if production occurs for 24 consecutive hours without line clearance or sanitization of the environment, a media-fill run should be done immediately after the end of this cycle. The assurance of sterility provided by the process must be consistent throughout the batch (i.e., for every unit of product filled), and process-capability data should be collected to demonstrate that this is the case.

In general, an all-purpose, rich medium such as Soybean Casein Broth that has been checked for growth promotion with a battery of indicator organisms (see *Sterility Tests* <71>) at a level of below 100 cfu/unit, can be used. Additional organisms that could be added are predominant isolates from the controlled environment where aseptic processing is to be conducted. Following the aseptic processing of the medium, the filled containers are incubated at 20° to 25° and at 30° to 35°. Following incubation for 7 days at each temperature, the medium-filled containers are usually inspected for growth; the growth is confirmed microscopically when the contaminants are identified to determine the possible sources of contamination.

Critical issues in performing media fills are the number of fills to qualify an aseptic process, the number of units per media fill, the intepretation of results, and implementation of corrective actions. Historically, many firms employ three media-fill runs during initial qualification or start-up of a facility to demonstrate repetitive consistency that the aseptic processing line is in control. As previously stated, a number of other supporting programs must also demonstrate satisfactory control. The minimum number of units per media fill is considered to be 3,000, with a contamination rate of not more than 0.1% being the criterion for acceptance of a successful media-fill run. It should be recognized that many firms in the United States and other countries are filling more than 3,000 units in a single media-fill run.[3]

Although commonly referred to in the industry since the late 1970's, the sterility assurance level of 10^{-3} ascribed to an aseptic processing and verified by three media-fill runs is not a sterility assurance level in the sense that it is understood in terminal sterilization

---

[3]A Parenteral Drug Association Survey (Technical Monograph 17) showed that out of 27 respondents, 50% were filling more than 3,000 units per run.

where a 10^{-6} sterility assurance level can be determined experimentally. It is basically a contamination rate limit. The contamination rate that has been widely accepted in the industry and by the regulatory agencies is 1 contaminated unit per 1000 (0.1%). There has been some controversy surrounding the application of a statistical measure of confidence that should be applied to this 0.1% contamination rate. However, it is recognized that repeated media runs are required in order to confirm the statistical validity of the observed contamination rate for the process.

PDA Technical Monograph Number 17,[3] "A Survey of Current Sterile Manufacturing Practices," indicated that many manufacturers believe that their aseptic processes are capable of contamination rates below 0.1%. Over the last several years there have been several reports indicating that lower contamination rates are routinely attained in performing media fills.

The occurrence of contamination in production or during a media-fill run is not a random event. Therefore, contamination does not follow a Poisson distribution. The most critical source of contamination in the clean room is the personnel. It would be then expected that contamination in media-fill and production runs would result from human proximity to, or interaction with, the production process. Thus, the appearance of contamination would not be random at all but would be strongly correlated with line interventions or aseptic handling steps. It further follows that the rate of contamination would depend upon how well these activities are done by the aseptic processing area operators. The widespread use of isolator systems for sterility testing has demonstrated that elimination of personnel does reduce contamination in aseptic handling. Reports of high levels of sterility assurance in restricted-access barrier production systems and highly automated production systems indicate that random airborne contamination is rarely a cause of contaminated units in media-fill runs.

An Overview of the Emerging Technologies for Advanced Aseptic Processing

Because of the strong correlation between human involvement and intervention and the potential for product contamination in aseptic processing, production systems in which personnel are removed from critical zones have been designed and implemented. Methods developed to reduce the likelihood of human-borne contamination reaching product include equipment automation, various types of barriers, and isolator systems. Facilities that employ these advanced aseptic

---

[3]A Parenteral Drug Association Survey (Technical Monograph 17) showed that out of 27 respondents, 50% were filling more than 3,000 units per run.

processing strategies are already in opeation. The operation of these facilities may in some cases differ only in degree from those utilized in conventional clean rooms and other controlled environments. However, in facilities where personnel have been completely excluded from the critical zone, room classification based on particulate and environmental microbiological monitoring requirements may be significantly different from conventional practice.

The following are definitions of some of the systems currently in place to reduce the contamination rate in aseptic processing:

Barriers—In the context of aseptic processing systems, a barrier is a device that restricts contact between operators and the aseptic field enclosed within the barrier. These systems are used in hospital pharmacies, laboratories, and animal care facilities, as well as in aseptic filling. Barriers may not necessarily be sterilized and do not always have sterile transfer systems that allow passage of materials into or out of the system without exposure to the surrounding environment. Barriers range from plastic curtains around the critical production zones to the hinged plexiglass enclosures found on modern aseptic-filling equipment. Barriers may also incorporate such elements as glove ports, half-suits, and rapid-transfer ports. The enclosure characteristically found on form, fill, and seal equipment could be termed a barrier.

Blow/Fill/Seal—This type of system essentially combines the blow-molding of container and seal with the filling of product in one piece of equipment. Generally, it is used for the manufacture of products in unit-dose containers. From a microbiological point of view, the sequence of forming the container, filling with sterile product, and formation and application of the seal are achieved aseptically in an uninterrupted operation with minimal exposure to the environment. These systems have been in existence for about 30 years and have demonstrated the capability of achieving contamination rates below 0.1%. Contamination rates of 0.001% have been cited for blow/fill/seal systems when combined media-fill data are summarized and analyzed.

Isolator—This technology is used for a dual purpose. One is to protect the product from contamination from the environment, including personnel, during filling and closing, and the other is to protect personnel from deleterious or toxic products that are being manufactured.

Isolator technology is based on the principle of placing previously sterilized components (containers/products/closures) into a sterile environment. These items will remain sterile during the whole processing operation, since no personnel or nonsterile components are brought into the isolator. The isolator barrier is an absolute barrier that does not allow for interchanges between the protected and unprotected environments. Isolators either may be physically sealed against the entry of external contamination or may be effectively sealed by the application of continuous overpressure. Manipulations of materials by personnel are done via use of gloves, half-suits, or total suits. All air entering the isolator must pass through either an HEPA or UPLA filter, and exhaust air typically exits through a filter of at least HEPA grade. Peracetic acid, hydrogen peroxide, hydrogen peroxide vapor, formaldehyde vapors, or other appropriate chemical sterilants may be used for sterilization of the isolator unit's internal environment. The sterilization of the interior of isolators and all contents are usually validated to a sterility assurance level of 10^{-6}.

Equipment, components, and materials are introduced into the isolator chamber through a number of procedures: use of a double-door autoclave with an airtight seal to the enclosure wall when the contents, materials, and components are not heat-sensitive; through continuous introduction of components via a conveyor belt passing through a sterilizing tunnel, also for non–heat sensitive materials; and with the use of a transfer container system through a docking system in the isolator enclosure. Information on the use of isolator systems for sterility testing is found under *Sterility Tests <71>*.

The requirements for the classification of environments surrounding these newer technologies for aseptic processing depend on the type of technology used.

Blow/Fill/Seal equipment that restricts employee contact with product may be placed in a classified clean room, especially if some form of employee intervention is possible during production.

Barrier systems will require some form of classified clean room. Because of the numerous barrier system types and applications, the requirements for the environment surrounding the barrier system will vary. The design and operating strategies for the environment around these systems will have to be developed by the manufacturers in a logical and rational fashion. Regardless of these strategies, the capability of the system to produce sterile products must be validated to operate in accordance with pre-established criteria.

Isolator systems could be operated in unclassified environments when used for sterility testing but not for manufacturing. Because air enters the isolator through integral filters of HEPA quality or better, and their interiors are sterilized typically to a sterility assurance level of 10^{-6}, isolators must not exchange air with surrounding environment unless that environment contains sterile air and is free of human operators. However, it has been suggested that when the isolator is in a controlled environment, the potential for contaminated product is reduced in the event of a pinhole leak in the suit or glove.

The extent and scope of an environmental microbiological monitoring of these advanced systems for aseptic processing depends on the type of system used. Manufacturers have to balance the frequency of environmental sampling systems that require human intervention with the benefit accrued by the results of that monitoring. Since barrier systems are designed to reduce human intevention to a minimum, remote sampling systems should be used in lieu of personnel intervention. In general, once the validation establishes the effectiveness of the barrier system, the frequency of sampling to monitor the microbiological status of the aseptic processing area could be reduced, as compared to the frequency of sampling of classical aseptic processing systems.

Isolator systems require relatively infrequent microbiological monitoring. Continuous total particulate monitoring can provide assurance that the air filtration system within the isolator is working properly. The methods for quantitative microbiological air sampling described in this chapter may not have sufficient sensitivity to test the environment inside an isolator. Experience with isolators indicates that under normal operations pinhole leaks or tears in gloves represent the major potential for microbiological contaminaton; therefore, frequent testing of the gloves for integrity and surface monitoring of the gloves is essential. Surface monitoring withi the isolator may also be beneficial.

GLOSSARY

Airborne Particulate Count (also referred to as *Total particulate count*)—Particles detected are 0.3 μm, 0.5 μm, and larger. When a number of particles is specified, it is the maximum allowable number of particles per cubic meter of air (or per cubic foot of air).

Airborne Viable Particulate Count (also referred to as *Total airborne aerobic microbial count*)—When a number of microorganisms is specified, it is the maximum number of colony-forming units (cfu) per cubic meter of air (or per cubic foot of air) that is associated with a Cleanliness Class of controlled environment based on the *Airborne particulate count*.

Aseptic Processing—A mode of processing pharmaceutical and medical products that involves the separate sterilization of the product and of the package (containers/closures or packaging material for medical devices) and the transfer of the product into the container and its closure under microbiologic critically controlled conditions.

Air Sampler—Devices or equipment used to sample a measured amount of air in a specified time to determine the particulate or microbiological status of air in the controlled environment.

Air Changes—The frequency per unit of time (minutes, hours, etc.) that the air within a controlled environment is replaced. The air can be recirculated partially or totally replaced.

Action Levels—Microbiological levels in the controlled environment, specified in the standard operating procedures, which when exceeded should trigger an investigation and a corrective action based on the investigation.

Alert Levels—Microbial levels, specified in the standard operating procedures, which when exceeded should result in an investigation to ensure that the process is still within control. Alert levels are specific for a given facility and are established on the basis of a baseline developed under an environmental monitoring program. These Alert levels can be modified depending on the trend analysis done in the monitoring program. Alert levels are always lower than Action levels.

Bioburden—Total number of microorganisms detected in or on an article prior to a sterilization treatment.

Clean Room—A room in which the concentration of airborne particles is controlled to meet a specified airborne particulate Cleanliness Class. In addition, the concentration of microorganisms in the environment is monitored; each Cleanliness Class defined is also assigned a microbiological level of air, surface, and personnel gear.

Clean Zone—A defined space in which the concentration of airborne particles and microorganisms is controlled to meet specific Cleanliness Class levels.

Controlled Environment—Any area in an aseptic process system for which airborne particulate and microorganism levels are controlled to specific levels, appropriate to the activities conducted within that environment.

Commissioning of a Controlled Environment—Certification by engineering and quality control that the environment has been built according to the specifications of the desired cleanliness class and that, under conditions likely to be encountered under normal operating conditions (or worst-case conditions), it is capable of delivering an aseptic process. Commissioning includes media-fill runs and results of the environmental monitoring program.

Corrective Action—Actions to be performed that are in standard operating procedures and that are triggered by exceeding Action levels.

Environmental Isolates—Microorganisms that have been isolated from samples from the environmental monitoring program and that represent the microflora of an aseptic processing system.

Environmental Monitoring Program—Documented program, implemented through standard operating procedures, that describes in detail the procedures and methods used for monitoring particulates as well as microorganisms in controlled environments (air, surface, personnel gear). The program includes sampling sites, frequency of sampling, and investigative and corrective actions that must be followed if Alert or Action levels are exceeded. The methodology used for trend analysis is also described.

Equipment Layout—Graphical representation of an aseptic processing system that denotes the relationship between and among equipment and personnel. This layout is used in the *Risk Assessment Analysis* to determine sampling site and frequency of sampling based on potential for microbiological contamination of the product/container/ closure system. Changes must be assessed by responsible managers, since unauthorized changes in the layout for equipment or personnel stations could result in increase in the potential for contamination of the product/container/closure system.

Federal Standard 209E—"Airborne Particulate Cleanliness Classes in Clean Rooms and Clean Zones" is a standard approved by the Commissioner, Federal Supply Services, General Service Administration, for the use of "All Federal Agencies." The Standard establishes classes of air cleanliness based on specified concentration of airborne particulates. These classes of air cleanliness have been developed, in general, for the electronic industry "super-clean" controlled environments. In the pharmaceutical industry, the Federal Standard 209E is used to specify the construction of controlled environment. Class 100, Class 10,000, and Class 100,000 are generally represented in an aseptic processing system. If the classification system is applied on the basis of particles equal to or greater than 0.5 μm, these Classes are now represented in the SI system by Class M3.5, M5.5, and M6.5, respectively.

Filter Integrity—Characteristics of a filter that ensure the functional performance of a filter used for liquid or gas in an aseptic processing system.

Material Flow—The flow of material and personnel entering controlled environments should follow a specified and documented pathway that has been chosen to reduce or minimize the potential for microbial contamination of the product/closure/container systems. Deviation from the prescribed flow could result in increase in potential for microbial contamination. Material/personnel flow can be changed, but the consequences of the changes from a microbiological point of view should be assessed by responsible managers and must be authorized and documented.

Media Growth Promotion—Procedure that references *Growth Promotion under Sterility Tests* <71> to demonstrate that media used in the microbiological environmental monitoring program, or in *media-fill* runs, are capable of supporting growth of indicator microorganisms and of environmental isolates from samples obtained through the monitoring program.

Media Fill—Microbiological evaluation of an aseptic process by the use of growth media processed in a manner similar to the processing of the product and with the same container/closure system being used.

Out-of-Specification Event—Temporary or continuous event when one or more of the requirements included in standard operating procedures for controlled environments are not fulfilled.

Product Contact Areas—Areas and surfaces in a controlled environment that are in direct contact with either products, containers, or closures and the microbiological status of which can result in potential microbial contamination of the product/container/closure system. Once identified, these areas should be tested more frequently than non–product-contact areas or surfaces.

Risk Assessment Analysis—Analysis of the identification of contamination potentials in controlled environments that establish priorities in terms of severity and frequency and that will develop methods and procedures that will eliminate, reduce, minimize, or mitigate their potential for microbial contamination of the product/container/closure system.

Sampling Plan—A documented plan that describes the procedures and methods for sampling of a controlled environment, identifies the sampling sites, the frequency and number of samples, the analysis of data and the interpretation of results.

Sampling Sites—Documented geographical location, within a controlled environment, where sampling for microbiological evaluation is taken. In general, sampling sites are selected because of their potential for product/container/closure contacts.

Standard Operating Procedures—Written procedures describing operations, testing, sampling, interpretation of results, and corrective actions that relate to the operations that are taking place in a controlled environment and auxiliary environments. Deviations from standard operating procedures should be noted and approved by responsible managers.

Sterile Field—In aseptic processing or in other controlled environments, it is the space at the level with or above open product containers, closures, or product itself, where the potential for microbial contamination is highest.

Sterility—An acceptably high level of probability that a product processed in an aseptic system does not contain viable microorganisms.

Swabs—Devices provided that are used to sample irregular as well as regular surfaces for determination of microbial status. The swab, generally composed of a stick with an absorbent extremity, is moistened before sampling and used to sample a specified unit area of a surface. The swab is then rinsed in sterile saline or other suitable menstruum and the contents plated on nutrient agar plates to obtain an estimate of the viable microbial load on that surface.

Trend Analysis—Data from a routine microbial environmental monitoring program that can be related to time, shift, facility, etc. This information is periodically evaluated to establish the status or pattern of that program to ascertain whether it is under adequate control. A trend analysis is used to facilitate decision-making for requalification of a controlled environment or for maintenance and sanitization schedules.

Index

mass spectrometer, 120, 128, 147–148
mass spectrometry, 147–148, 283
mass transfer, 126, 257
Master Cell Bank. *See* MCB
MCB, 246, 247, 254
McReynold's constants, 132
m/e. *See* mass/charge ratio
MECC, 134
media fill, 14, 210, 472, 488–490, 496
  challenges in, 94–95
  failure of, 21
  monitoring, 88
  SCL and, 371
  specifications for, 20–21, 91–92
  sterilization of, 435, 446
  validation of, 26, 36, 39
  voluntary standards for, 27
medicinal therapy, 245. *See also* parenteral therapy/solutions
melting point, 120, 314, 315, 363
membrane filter, 3, 7, 126
membrane filtration, 163, 170, 171, 177, 179, 203
Memorandum of Understanding, 61
mercury, 350
mercury lamp, 144
mercury sulfate, 118
metabolism, 256, 257, 416
metabolites, 254, 256, 261
metastasis, 156
methane, 133
methanol, 119, 120, 129, 257
methionine, 282, 416
methotrexate sodium, 294
methylene chloride, 130
metrology, 104
micellar chromatography, 129
micellar electrokinetic capillary chromatography. *See* MECC
microbial death process, 3
microbial survivor probability, 6. *See also* SAL
microbicide, 415, 432
microbore column, 128
microcarrier, 263
Micrococcus sp., 211
microcolony, 186, 187
microporous polyethylene, 139
microsphere, 124
milling, 94
mineral oil, 138, 363
minerals, 258
Mini Aseptic Filling System. *See* MAFS™

minibore column, 128
MIR. *See* multiple internal reflectance
misbranding, 338, 339
miscibility, 124
moist heat sterilization, 86, 102, 264
moisture determination, 120
molar absorptivity, 136
mold, 21, 25, 88, 197, 198, 416, 457
molybdenum, 154, 170
monobasic sodium phosphate, 296
monochromator, 144
monoclonal antibody, 156, 163, 248–249, 250, 251
Monod equation, 256
morphine, 1
MoU. *See* Memorandum of Understanding
MS. *See* mass spectrometry
multidimensional NMR, 147
multiple internal reflectance, 357
murine hybridoma, 251. *See also* hybridoma
murine myeloma, 251
MWCB, 246, 247, 254, 259, 260, 262
Mycobacterium smegmatis, 417
mycoplasma, 32, 258, 283
myocardium, 156

NaCl. *See* sodium chloride
native protein, 246
NDA, 15, 19, 89, 96, 99, 346, 356, 358
near-infrared spectrometry, 135, 140–141
neat liquid, 139, 141
nebulization, 226, 233
nebulizer, 225, 230
negative pressure, 159, 160, 165, 319, 399
nephelometry, 144
New Drug Application. *See* NDA
$(NH_4)_2SO_4$. *See* ammonium sulfate
NIH syndrome, 57
NIR. *See* near-infrared spectrometry
nitrate, 125
nitrocellulose, 134
nitrogen, 132, 133, 155, 170, 253, 257, 305, 328, 333, 430, 450
nitroglycerine, 355
nitroprusside sodium, 295
nitrous oxide, 133, 142
NMR, 135, 145–147
nominal filter, 268
nondispersive infrared spectrometry, 123
nonprotonated substance, 137

Norcardia species, 417
normal phase chromatography, 127, 130
Northern Blot, 144
nuclear magnetic resonance, 155
nuclear magnetic resonance spectrometry. *See* NMR
nuclear medicine, 155, 172
    department, 156, 169
    procedures, 170
nuclear reactor, 154, 155
nuclear relaxation, 147
nucleation curve, 314, 315
nucleic acids, 134, 283, 427
nucleotide, 254
Nujol mulls, 138, 139
nylon, 134, 355, 390, 437, 439

open junction, 118, 119
open-manual production, 369, 370
operational qualification. *See* OQ
operator quality control, 48
ophthalmic ointment/cream, 125, 339,
    343, 347, 348, 353, 355, 359
ophthalmic solution, 121
opium, 1
optical isomerization, 349
optical rotation, 122
optical sensor, 121
OQ, 103–104, 111, 113, 304, 305
organic SEC, 130, 131
organosilane, 127
orifice sensor, 122
Orthoclone OKT®3, 251
osmolality, 121–122
osmolarity, 343
osmometry, 121–122
OTC product, 362
otics, 347
outgassing, 304, 446, 459
out-of-specification, 472, 496
overkill, 7, 106–107
overtone band, 140
oxidation, 282, 284
oxidized cellulose, 119
oxyacetylene, 142
oxygen, 155, 170, 255, 256, 257, 353, 416,
    426, 427, 430
oxygen transfer, 263, 267
ozone sterilization, 399

packaging. *See* aseptic packaging/
    labeling
packing capacity, 126
PAGE, 134

paraformaldehyde, 416
parenteral therapy/solutions
    contamination of, 188–191, 196–
        203, 211
    excipients in, 117
    labeling for, 339, 340
    lyophilization of, 291, 297
    origins of, 1–2
    packaging of, 344, 347
    particles in, 122
    radiopharmaceuticals in, 155, 163, 179
Parison molding station, 242
particle counting/sizing, 122, 259
particulate radiation, 153
partition coefficient, 124
pass-through tunnel, 392, 394
path length, 136, 138, 139, 140, 141
pathogen, 197, 311
PD pump. *See* positive displacement
    pump
peanut meal, 258
PEG. *See* polyethylene glycol
penicillin, 2, 295, 363
Penicillin sp., 197, 198
Penicillium chrysogenum, 256, 417
pentamidine isethionate, 296
pentane, 133
pentyl, 130
peptides, 245
    FT–Raman, 144
    HIC of, 130
    IXC of, 131–132
    labeling of, 163
    mass spectrometry of, 148
    radiopharmaceuticals with, 156, 163
    SEC of, 130–131
    stabilization of, 285
peptide mapping, 283
peptone, 258, 485, 486
peptone water, 485
peracetic acid, 276, 334, 416, 435
perchloric acid, 119, 125
perfluorokerosene, 138
performance qualification. *See* PQ
perfusion, 348
peristaltic pump, 278
permeation, 343, 354, 355, 357
permeation test, 343, 344
peroxide plasma, 426–427
peroxydisulfate, 123
peroxysulfate, 123
personnel
    access to premises by, 159–160, 165
    dos and don'ts for, 220–221